Disorder and Fracture

NATO ASI Series

Advanced Science Institutes Series

A series presenting the results of activities sponsored by the NATO Science Committee, which aims at the dissemination of advanced scientific and technological knowledge, with a view to strengthening links between scientific communities.

The series is published by an international board of publishers in conjunction with the NATO Scientific Affairs Division

A	**Life Sciences**	Plenum Publishing Corporation
B	**Physics**	New York and London
C	**Mathematical**	Kluwer Academic Publishers
	and Physical Sciences	Dordrecht, Boston, and London
D	**Behavioral and Social Sciences**	
E	**Applied Sciences**	
F	**Computer and Systems Sciences**	Springer-Verlag
G	**Ecological Sciences**	Berlin, Heidelberg, New York, London,
H	**Cell Biology**	Paris, and Tokyo

Recent Volumes in this Series

Series B: Physics

Disorder and Fracture

Edited by

J. C. Charmet and
S. Roux

Ecole Supérieure de Physique et Chimie Industrielles de Paris
Paris, France

and

E. Guyon

Université de Paris-Sud
Orsay, France

Plenum Press
New York and London
Published in cooperation with NATO Scientific Affairs Division

Proceedings of a NATO Advanced Study Institute
on Disorder and Fracture,
held May 29-June 9, 1989,
in Cargèse, France

Library of Congress Cataloging-in-Publication Data

NATO Advanced Study Institute on Disorder and Fracture (1989 :
  Cargèse, France)
    Disorder and fracture / edited by J.C. Charmet and S. Roux and E.
  Guyon.
        p.   cm. -- (NATO ASI series. Series B, Physics ; v. 235)
    "Proceedings of a NATO Advanced Study Institute on Disorder and
  Fracture, held May 29-June 9, 1989, in Cargèse, France"--CIP verso
  t.p.
    Includes bibliographical references and index.
    ISBN 0-306-43688-4. -- ISBN 0-306-43576-4 (pbk.)
    1. Fracture mechanics--Congresses. 2. Order-disorder models-
  -Congresses.   I. Charmet, J. C. II. Roux, Stéphane. III. Guyon,
  Etienne. IV. Title. V. Series.
  TA409.N42   1989
  620.1'126--dc20   90-45733
                                                              CIP

© 1990 Plenum Press, New York
A Division of Plenum Publishing Corporation
233 Spring Street, New York, N.Y. 10013

Printed in the United States of America

SPECIAL PROGRAM ON CHAOS, ORDER, AND PATTERNS

This book contains the proceedings of a NATO Advanced Research Workshop held within the program of activities of the NATO Special Program on Chaos, Order, and Patterns.

Foreword

Fracture, and particularly brittle fracture, is a good example of an instability. For a homogeneous solid, subjected to a uniform stress field, a crack may appear anywhere in the structure once the threshold stress is reached. However, once a crack has been nucleated in some place, further damage in the solid will in most cases propagate from the initial crack, and not somewhere else in the solid. In this sense fracture is an unstable process. This property makes the process extremely sensitive to any heterogeneity present in the medium, which selects the location of the first crack nucleated. In particular, fracture appears to be very sensitive to disorder, which can favor or impede local cracks. Therefore, in most realistic cases, a good description of fracture mechanics should include the effect of disorder. Recently this need has motivated work in this direction starting from the usual description of fracture mechanics.

Parallel with this first trend, statistical physics underwent a very important development in the description of disordered systems. In particular, let us mention the emergence of some "new" concepts (such as fractals, scaling laws, finite size effects, and so on) in this field. However, many models considered were rather simple and well adapted to theoretical or numerical introduction into a complex body of problems. An example of this can be found in percolation theory. This area is now rather well understood and accurately described. Nevertheless, examples of the exact application of this model to the real world are extremely scarce. Indeed, in most situations the geometric disorder present in a medium is not totally random, but rather generated through complex physical processes which introduce naturally correlations which are difficult to predict. Conversely, when heterogeneities are not correlated with each other, their concentration has very little likelihood of reaching a percolation threshold. But the very large effort devoted to the subject has permitted the elaboration of many concepts and tools which can certainly be used in much more complex cases. This is the reason why the physics of disorder is mature enough to be extended to more realistic cases, such as fracture of disordered solids, and indeed a few attempts have been made in this direction quite recently.

The conjunction of these two facts has motivated a natural coming together of both the mechanics and statistical physics communities on the problem of fracture and disorder, in order to share their concerns and their knowledge, their problems and their tools. This was the genesis of the meeting held in Cargèse in June 1989, and this book gathers together the texts of the lectures given during this Institute.

Obviously this is only a first step along the path that we have to follow in order to begin to bring together the know-how and effort to attack the problems which we have to face in material science. Other meetings will have to be organized, common projects prepared, well-defined questions in a language understandable by both communities formulated, and so on. This renders both important and timely the need to prepare a good basis for common exchanges. This we have tried to achieve in this volume. It is naturally too early to try to guess what will result from such an approach, but it seems likely that many applications which combine instabilities and disorder will benefit from extensions of progress accomplished in fracture mechanics, and thus will be useful both in mechanics and in statistical mechanics.

These papers are grouped in sections covering "Tools", "Diffusion Limited Aggregation Model", "Statistical Models of Fracture", "Rheology and Fracture", "Materials and Applications", ranging from theoretical concepts to practical applications. Each section has a brief introduction which summarizes the links between different chapters which can be read independently.

We wish to thank the NATO Scientific Panel on nonlinear phenomena for their support of the Institute and this volume. We acknowledge also support from the CNRS, DRET and NSF.

The Editors

Contents

IV. Rheology and Fracture

V. Materials and Applications

Index   301

I Tools

Fracture, in crude terms, is nothing but the generation of an interface in a medium where eventually some disorder is present. This interface is however not random. It is therefore important to know the basic exact and conjectured results that have been obtained concerning the conformation of interfaces with various constraints in a random environment, or subject to a thermal noise. The first lecture, by Kardar, presents the most common problems and the main results. The contexts in which these models were first considered are rather far away from the field of fracture. Nevertheless these results may apply for different approaches already used for fracture. For instance, it is natural to compare the fracture surface with the one that minimizes the breaking energy, once a random distribution of local breaking energy is given. This is exactly a "Random Bond" problem, as treated in Kardar's lecture.

In addition, some tools that may be useful to characterize the roughness of the rupture surface are presented. In particular, the evolution of the mean roughness with the system size is certainly a very precious tool, which can be checked against theoretical expectations.

In the second lecture, by Roux and Hansen, the multifractality analysis is presented. This tool has appeared in the past few years as a very simple way to characterize fully the scaling properties of a local quantity in a system at criticality. It has been used in various contexts, including turbulence in hydrodynamics, growth probabilities in various aggregation phenomena, current distribution in random resistor networks at the percolation threshold, etc. Starting from this last example — which by the way could be considered as an extreme case of fracture (in a *very disordered* material) — it is shown how to apply this notion to off–threshold problems. As a practical application of these notions, the histogram of the elastic modulus or compliance jumps during a fracture process on a simple model is presented. This histogram presents a characteristic shape characterized by power–law behaviors.

The final lecture, by Redner, of this first part is more directly related to fracture since it deals with fragmentation. The main results presented are on the scaling properties of the fragment size distribution, assuming a Smoluchowski master equation. This formalism can be thought of as a mean-field limit of a real fragmentation process. It allows the comparison between the fragmentation process and the reverse case of gelation processes where this type of approach has proved useful in the past. It is very general, and according to the precise process used in a real case, some terms of the Smoluchowski equation can be specified. Therefore, this lecture presents a very general framework in which any kind of fragmentation process can fit as long as spatial and temporal correlations are not important. This lecture closes the section devoted to the tools.

Chapter 1

Fluctuations of Interfaces and Fronts

Mehran Kardar

Department of Physics, Massachusetts Institute of Technology
Cambridge, Massachusetts 02139, USA

Many surfaces occurring in nature are characterized by self-affine fluctuations. The width w of the interface grows with the observation length L as $w \sim L^\zeta$. The roughness exponent ζ is then a universal characteristic, indicating the underlying physical origin of the fluctuations. The examples considered include interfaces pinned in random media, growth fronts in deposition, and surface of an evolving sandpile. In these cases the origin of the dynamically organized criticality can be traced to symmetries or conservation laws.

1. Introduction

One of the important recent advances in statistical physics involves the concepts of scaling and universality.[1] Simply put, scaling refers to power law relations observed between various quantities, while universality implies that the exponents of these power laws are fundamental constants, robust to the details of the underlying interactions. They depend only on gross features such as symmetries and provide an important clue to the physical origin of scaling. While the theoretical basis for universality is the renormalization group,[1] the geometrical visualization of scaling is provided by the concept of fractals.[2] Fractals, as discussed by H. Herrmann and S. Roux in these proceedings,[3] are invariant under changes of scale. Yet another form of self-similarity is provided by self-affine structures in which different longitudinal and transverse scaling factors are necessary. If the coastline is the prototype of fractal scaling, then a mountain landscape is a good example of self-affinity.[4] To reproduce the original landscape from a portion of the picture, different scale factors have to be employed in the horizontal and vertical directions. The logarithmic ratio of transverse to longitudinal scale factors provides the (universal?) self-affine (roughness) exponent ζ.

In these notes we shall see how the exponent ζ can be calculated in a number of physical circumstances. We start by looking at the static problem of an interface pinned by the impurities in a random medium. We then examine the dynamic processes of fluctuations in a growing interface, or in an evolving sandpile. In the process, we shall gain some insight into the origin of self-similar behavior in physical systems.

Disorder and Fracture, Edited by J. C. Charmet *et al.*
Plenum Press, New York, 1990

2. Thermal Roughening

Interfaces arise in many contexts such as phase boundaries between liquids, gases, and/or solids, or the domain wall in a magnet. Two dimensional examples include steps on a crystal surface, as well as boundaries between different phases in adsorbed layers. All these problems can be addressed in the same framework by considering interfaces embedded in d-dimensional space.[5] Usually the equilibrium (or lowest energy configuration) is a flat surface. Small deformations about this configuration are described by a function $h(\mathbf{x})$, denoting the deviation in height h , as a function of the remaining $(d-1)$ coordinates $\mathbf{x}$. A basic property of the interface is its surface tension (denoted by ν), which is the free energy cost per unit area for creating a surface. Deformations lead to an increased energy, estimated by

$$\mathcal{H}[h(\mathbf{x})] = \nu \int d^{d-1}\mathbf{x} \left[\sqrt{1+(\nabla h)^2} - 1\right] \approx \frac{\nu}{2} \int d^{d-1}\mathbf{x}(\nabla h)^2. \tag{1}$$

Checking that the local area is $\sqrt{1+(\nabla h)^2}$ is a simple exercise in differential geometry, and the result is expanded for slow variations such that $(\nabla h)^2 \ll 1$. This assumption will be checked for self-consistency later on. For interfaces in magnets more care is necessary and the interested reader is referred to the review article by M.E. Fisher.[6]

At finite temperatures T, deformations appear spontaneously due to thermal fluctuations. The extent of these fluctuations is estimated using the equipartition theorem. The normal modes of the surface are capillary waves, indexed by a wave number $\mathbf{q}$, $\left(h(\mathbf{q}) = \int d^{d-1}\mathbf{x}\, e^{i\mathbf{q}\cdot\mathbf{x}}h(\mathbf{x})\right)$, and

$$\mathcal{H} = \int d^{d-1}\mathbf{x}\, \frac{\nu}{2}(\nabla h)^2 = \int \frac{d^{d-1}\mathbf{q}}{(2\pi)^{d-1}} \cdot \frac{\nu\mathbf{q}^2}{2}|h(\mathbf{q})|^2. \tag{2}$$

Due to thermal fluctuations each mode acquires $kT/2$ in energy, and hence an amplitude $\langle|h(\mathbf{q})|^2\rangle = kT/(\nu\mathbf{q}^2)$. The corresponding extent of fluctuations in real space is given by

$$\left\langle|h(\mathbf{x})|^2\right\rangle = \int \frac{d^{d-1}\mathbf{q}}{(2\pi)^{d-1}} \cdot \frac{kT}{\nu\mathbf{q}^2}. \tag{3}$$

The integral has lower and upper cuttoffs proportional to $1/L$ and $1/a$, where a is a lattice spacing, and L is the system size. For $d \geq d_u = 3$, the integral is finite and L independent as $L \to \infty$. *The surface is flat.*

For $1 < d < 3$, the integral is dominated by its lower cutoff and $|h(\mathbf{x})|^2 \sim L^{(3-d)/2}$. The surface is now rough with typical fluctuations H scaling as $H \sim L^{\zeta_0}$, with the thermal roughening exponent

$$\zeta_0 = (3 - d)/2. \tag{4}$$

We see that the power law exponent is universal, only depending on dimensionality, while the amplitude depends on physical parameters such as temperature and surface tension. In $d = 3$, $\zeta_0 = 0$ but $\langle h^2 \rangle \sim \ln(L)$ and the interface is still rough. For $d \leq d_l = 1$, H/L is no longer small and the gradient expansion in Eq.(1) is no longer valid. Thus self-consistency requires an exponent $\zeta < 1$. $d_u = 3$ is the upper, and $d_l = 1$ is the lower critical dimension for thermal roughening.

The above roughening exponent can also be obtained by simple dimensional analysis. To deform a surface of size L by a height H requires roughly an elastic energy $\nu(H/L)^2 L^{d-1} \sim \nu H^2 L^{d-3}$. The *disordering energy* E_d is the amount of energy available from fluctuations to cause such a distortion. In the above example it comes from thermal fluctuations, and $E_d \sim kT$. Equating the two energies leads to $kT \sim \nu H^2 L^{d-3}$ and the roughness exponent $\zeta_0 = (3 - d)/2$ is recovered. Only in the critical dimension of $d_u = 3$ the argument fails to give the logarithmic roughness.

3. Impurity Roughening

In the presence of quenched impurities the interface fluctuates to take advantage of defects with low energy, and to avoid regions of high energy. The resulting fluctuations, and the roughness exponent, are expected to be larger than thermal roughening. The "Imry-Ma"[7] estimate of roughening is obtained by balancing the surface tension cost of distortion, with a typical disordering energy. The later will depend on the action of the impurities.

The impurities most commonly studied are preferentially attracted to one phase. In the magnetic analogy they are called "random fields."[7] Suppose that on a scale L the interface deforms by an amount H to take advantage of a local region with excess impurity concentration. The volume swept by the interface has volume HL^{d-1}. If the impurities are uncorrelated, the central-limit theorem predicts a typical energy gain from the impurities scaling as $E_d \sim (HL^{d-1})^{1/2}$. Balancing this gain with the surface tension cost of $H^2 L^{d-3}$, leads to $H \sim L^{\zeta_{RF}}$ with

$$\zeta_{RF} = \frac{(5 - d)}{3}. \tag{5}$$

The upper critical dimension is now shifted to $d_u = 5$, and the lower critical dimension to $d_l = 2$. Below two dimensions the ordered phases are destroyed by interface fluctuations.

Another class of impurities has no preference for either bulk phase, but locally attracts or repels the interface itself. In the magnetic terminology, these are "random bonds."[9] In this case deforming the interface at scale L brings it into contact with L^{d-1} new bonds, and the typical energy gain can be argued[9] to be $E_d \sim L^{(d-1)/2}$. The scaling argument now leads to $H \sim L^{\zeta_{RB}}$ with

$$\zeta_{RB} = \frac{(5 - d)}{4} \tag{6}$$

with upper and lower critical dimensions of 5 and 1 respectively.

Is there a way to test these scaling predictions? Numerical and exact results in $d = 2$ provide a good testing ground. Although finding the interface configuration is a complex optimization problem, a transfer matrix approach allows calculations to be performed in polynomial time in L. Fig.(1) demonstrates a collection of optimal interface configurations generated in the presence of random bond impurities.[10] These configurations can then be analyzed for the roughening exponent ζ. In the case of random field impurities the result $\zeta_{RB} \approx 1$ is verified,[11] while for random bonds impurities[12,13] the result $\zeta_{RB} \approx 2/3$ disagrees with the value of $3/4$ from Eq.(6). Indeed in two dimensions, exact results obtained by a number of methods[14,15] confirm the values $\zeta_{RF} = 1$ and $\zeta_{RB} = 2/3$.

In view of the results in $d = 2$, the scaling results in Eqs.(5) and (6) are suspect. This led to a more careful theoretical examination of the roughness exponent.[5,16,17] A number of theoretical predictions, as well as the result of a numerical study based on extending the transfer matrix method[18], for $d = 3$ are summarized in Table 1. The random field exponent $\zeta_{RB} = 2/3$ from Eq.(5) appears to be correct, while for random bonds $\zeta_{RB} \approx 0.5 \pm 0.1$. More details can be found in Ref.[18]. It is interesting to speculate a connection between the rough random bond interfaces and fluctuations obtained on a crack surface growing into a material with impurities. Indeed there are visual similarities between the paths in Fig.(1), and configurations of maximum stress, observed in packings of glass beads by Bideau et al.[3] However, in the absence of concrete analytic mappings, or of even quantitative experimental identification of ζ, this remains only a speculation.

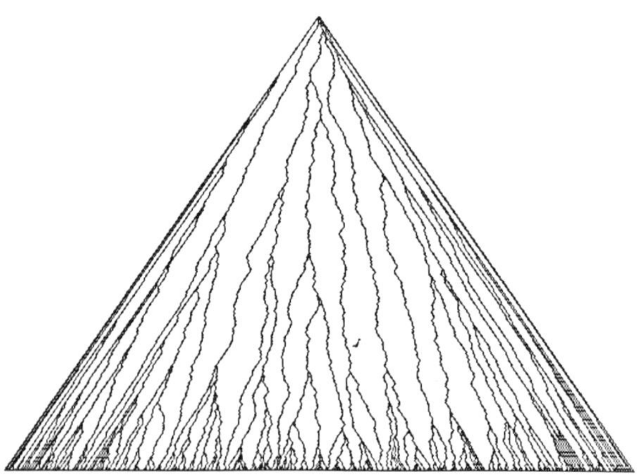

Figure 1 Configurations of optimal interfaces (of length 500) in a random bond problem, connecting the apex of the triangle to various points on its base.

Table 1 Theoretical suggestions, and numerical results for the roughness exponents of an interface in $d = 3$, subject to random field (ζ_{RF}), or random bond (ζ_{RB}) quenched impurities.

Approach	ζ_{RF}	ζ_{RB}
Imry-Ma[8,9]	2/3	1/2
Supersymmetry[17]	1	$--$
Flory[5]	2/3	0.40
$\varepsilon = 5 - d$ expansion[16(a)]	2/3	0.42
$\varepsilon = 5 - d$ expansion[16(b)]	2/3	0.44
Numerical results[18]	0.59 ± 0.07	0.50 ± 0.08

4. Dynamics of Surface Evolution

Up to now the focus was on the static (equilibrium) configurations of interfaces in random media. Here we shall explore the dynamic (non-equilibrium) evolution of such surfaces, in the absence of quenched impurities. Again we shall be concerned with the degree of surface roughness, as well as characteristic relaxation times. The object of interest is the surface height $h(\mathbf{x}, t)$, at position $\mathbf{x}$, and time t. How do we construct an equation of motion for h ? In circumstances when near equilibrium configurations relax to equilibrium ones described by a Hamiltonian $\mathcal{H}[h]$, the usual Langevin approach can be applied.[19] The "force" $F = \delta\mathcal{H}/\delta h$ is equated to a dissipative term (proportional to $\Gamma^{-1}\partial h/\partial t$) to get an equation

$$\frac{\partial h}{\partial t} = -\Gamma F[h] + \eta(\mathbf{x}, t). \tag{7}$$

The noise $\eta(\mathbf{x}, t)$ is related to thermal fluctuations as usual, and makes Eq.(7) stochastic. By contrast $\eta = 0$ for a *deterministic* evolution.

However, in many non-equilibrium (and non-reversible) situations, one is concerned with open systems for which no apparent $\mathcal{H}[h]$ can be identified. In such cases, finding the force $F[h]$ is non-trivial. Here I propose that $F[h]$ can be constructed by appealing to symmetries and conservation laws. In the same spirit as Landau theory of magnetism[1,19], all terms consistent with symmetries of the problem should be included in $F[h]$. Examples of such symmetries are translations ($h(x) \to h(x) + h_0$) and reflections ($h(x) \to -h(x)$) of the height profile. The former excludes any function of h from appearing in F; and hence $F \equiv F[\nabla h]$. The latter disallows terms such as $(\nabla h)^2$ from appearing in Eq.(7). Isotropy in the $\mathbf{x}$ space removes terms such as ∇h . If all three symmetries are present, the lowest order term in a gradient expansion becomes $\nabla^2 h$, and

$$\frac{\partial h}{\partial t} = \nu\nabla^2 h + \eta(\mathbf{x}, t) + \text{higher order terms}. \tag{8}$$

The higher order terms are usually irrelevant and will be ignored for now. The diffusion Eq.(8), is of course none other than the Langevin equation for the Hamiltonian $\mathcal{H} = \nu/2 \int d^{d-1}\mathbf{x}(\nabla h)^2$ in Eq.(1). Since $F \sim -\delta\mathcal{H}/\delta h$ in this case, the usual Fokker-Planck formalism[19] then implies that dynamic evolution according to Eq.(8) will lead to a stationary probability distribution given by the Boltzmann weight of this Hamiltonian, $i.e.$

$$P[h] \sim \exp\left[-\frac{\nu}{2}\int d^{d-1}\mathbf{x}(\nabla h)^2\right]. \tag{9}$$

As the discussion leading to Eq.(4) indicates, the resulting surfaces are rough, characterized by the thermal roughening exponent of $\zeta_0 = (3-d)/2$. Furthermore, the linear diffusion Eq.(8) can be solved exactly to give the time dependence of various modes. In particular, a mode of wavelength λ has a characteristic relaxation time $\tau \sim \nu^{-1}\lambda^2$. In fact a general two point correlation function will have the form

$$\langle |h(\mathbf{x},t) - h(\mathbf{x}',t')|\rangle \sim |\mathbf{x} - \mathbf{x}'|^\zeta f(|t - t'|/|\mathbf{x} - \mathbf{x}'|^z), \tag{10}$$

where $\zeta = \zeta_0 = (3-d)/2$, and $z = z_0 = 2$ is the so-called *dynamic scaling exponent.*[19]

5. Pinned Interfaces

The diffusion equation (8) describes an interface subject to many symmetry operations. From this point on, we explore what happens when some of these symmetries are absent. In many cases ($e.g.$ surface of fluid subject to gravity, or a magnetic domain in a spacially varying field[20], the interface is pinned to a particular height, say $h = 0$). The invariance under translations ($h \to h + c$) is now absent, and any function of h can be added to Eq.(8). The lowest order term will be linear, $i.e.$ $-h/\tau$, where the sign has been chosen to insure stability. The new equation of motion

$$\frac{\partial h}{\partial t} = -\frac{h}{\tau} + \nu\nabla^2 h + \eta(\mathbf{x},t) + \text{higher order terms}, \tag{11}$$

has a characteristic time scale τ , and a corresponding length scale $\lambda \sim \sqrt{\nu\tau}$. The capillary modes are now quenched for wavelengths larger than λ. Typical correlations will decay as $\exp(-x/\lambda)$ in space, and as $\exp(-t/\tau)$ in time.

The linear term in Eq.(11) thus destroys the scale invariant and self-similar nature of fluctuations by introducing characteristic time and length scales into the problem. Thus for the interface to evolve into a self-similar stationary state, it is imperative that a linear term be absent from the equation of motion. Translational symmetry is one way of achieving this — as we shall see later, a conservation law is another.

6. Growing Interfaces

The rapid growth of crystals by deposition, or molecular beam epitaxy, is an important technological process.[21] It also provides the simplest example for a non equilibrium evolution process. There have in fact been several detailed computer simulation studies of such growth processes.[22–27] We would like to understand the dynamic scaling of fluctuations inherent to this type of growth. The presence of a particular growth direction removes the $h \to -h$ symmetry, and allows us to introduce even terms in h in the equation of motion. To lowest order, we get[28,29]

$$\frac{\partial h}{\partial t} = v + \nu \nabla^2 h + \frac{\lambda}{2}(\nabla h)^2 + \eta(\mathbf{x},t) + \text{h.o.t.} \ . \tag{12}$$

Clearly v is related to the average growth velocity. In fact, the coefficients of all even terms must be proportional to v , as they all vanish in the symmetric case, with no prefered growth direction. The constant v is easily removed by transforming to a moving frame $(h \to h - vt)$, and the first non-trivial term is the nonlinear contribution $(\nabla h)^2$. Geometrically this term can be justified by noting that growth by addition of particles proceeds by a parallel transport of the surface gradient in the normal direction.[20,30]

Further evidence of relevance of Eq.(12) is provided by examining deterministic growth, i.e. with $h = 0$. This corresponds to a slow and uniform snowfall, on an initial profile at $t = 0$ described by $h_0(\mathbf{x})$. The nonlinear Eq.(12) can be solved with the aid of a "Cole-Hopf" transformation,

$$W(\mathbf{x},t) = \exp\left[\frac{\lambda}{2\nu}h(\mathbf{x},t)\right]. \tag{13}$$

W now satisfies the diffusion equation $\partial W/\partial t = \nu \nabla^2 W$, which is solved subject to the initial condition $W(\mathbf{x}, t = 0) = \exp(\lambda h_0(\mathbf{x})/2\nu)$. Thus the growing profile can be obtained *exactly* at all times from

$$h(\mathbf{x},t) = \frac{2\nu}{\lambda} \ln\left\{\int d^{d-1}\mathbf{x}' \ \exp\left[-\frac{|\mathbf{x}-\mathbf{x}'|^2}{2\nu t} + \frac{\lambda}{2\nu}h_0(\mathbf{x}')\right]\right\}. \tag{14}$$

It is instructive to examine the $\nu \to 0$ limit, in which the integral in Eq.(14) can be performed by the saddle point method. For each $\mathbf{x}$ we have to identify an $\mathbf{x}'$ which maximizes the exponent, leading to the collection of paraboloids described by

$$\lim_{\nu \to 0} h(\mathbf{x},t) = \max\left[h_0(\mathbf{x}') - \frac{|\mathbf{x}-\mathbf{x}'|^2}{2\nu t}\right]_{\mathbf{x}'}. \tag{15}$$

Such parabolic sequences (see Fig.(2)) are quite common in many layer by layer growth processes in nature, from biological to geological formations. The nonlinearity in Eq.(12) thus mathematically accounts for their origin. As seen from Fig.(2), growth leads to a reduction in roughness and smaller parabolas eventually give way to larger ones. What is the typical size of the parabolas at times t? In maximizing the exponent in Eq.(14), we have to balance a reduction $|\mathbf{x} - \mathbf{x}'|^2/2\lambda t$, by a possible gain from $h_0(\mathbf{x}')$ in selecting a point away from $\mathbf{x}$. The scaling is thus set by the roughness of the initial profile. For a self-affine initial landscape with $\langle |h_0(\mathbf{x}) - h_0(\mathbf{x}')| \rangle \sim |\mathbf{x} - \mathbf{x}'|^\zeta$, we balance $(\delta\mathbf{x})^2/t$ with $(\delta\mathbf{x})^\zeta$, obtaining a typical size

$$(\delta\mathbf{x}) \sim t^{1/(2-\zeta)}. \tag{16}$$

In $d = 2$, if the initial profile looks like a random walk with $\zeta = 1/2$, we then obtain $\delta\mathbf{x} \sim t^{2/3}$ for typical extent of parabolas at time t.

7. Stochastic Growth

Eqs.(14) and (15) indicate that the surface relaxes to a flat profile under deterministic evolution. By contrast the stochastic equation (i.e. noisy with $\eta(\mathbf{x}, t) \neq 0$) leads to rough structures. The average correlation functions will behave as in Eq.(10) with exponents ζ and z to be determined. This task is simplified by the mappings of Eq.(12) to two other problems. The first connection is based on the

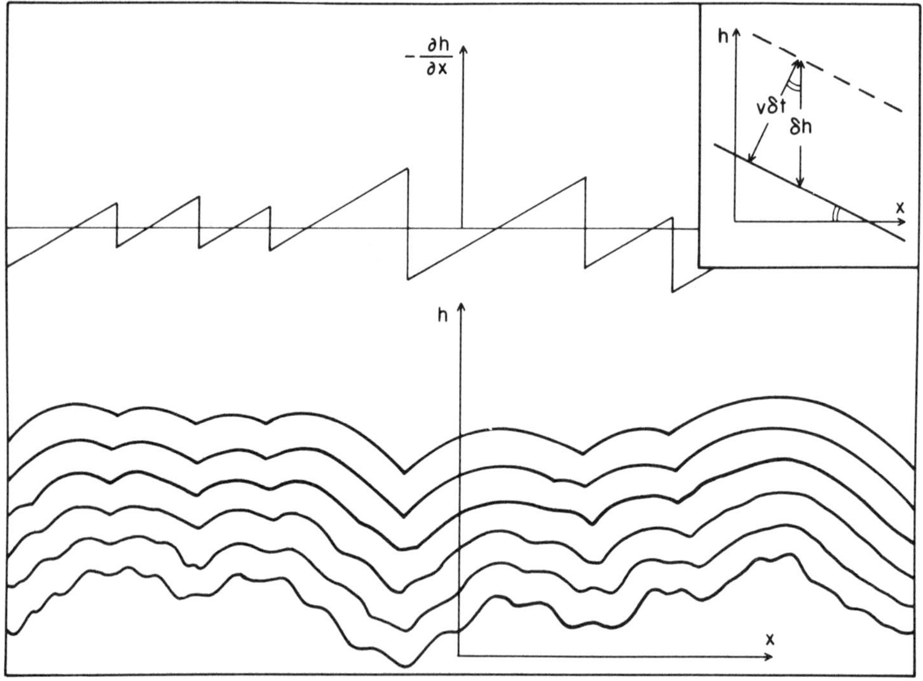

Figure 2 Successive layers of an interface relaxing to a flat configuration by a deterministic growth mechanism similar to Eqs.(14), (15). The gradient of the surface profile develops "shock waves."

"Cole-Hopf" transformation of Eq.(13), which results in

$$\frac{\partial W}{\partial t} = \nu \nabla^2 W + \left(\frac{\lambda}{2\nu}\eta(\mathbf{x},t)\right) W. \tag{17}$$

Note that the noise η which was *additive* in the original equation, becomes *multiplicative* in Eq.(17). It can be interpreted as a diffusion equation with random sources and sinks. The solution to Eq.(17) can be set in path integral form as

$$W(\mathbf{x},t) = \int_{(0,0)}^{(\mathbf{x},t)} d\mathbf{x}(t') \, \exp\left\{-\int_0^t dt' \left[\frac{\nu}{2}\left(\frac{d\mathbf{x}}{dt'}\right)^2 + \frac{\lambda}{2\nu}\eta(\mathbf{x}(t'),t')\right]\right\}, \tag{18}$$

which can be interpreted[10] as the net Boltzmann weight for polymers directed along the t direction, extending from $(\mathbf{0},0)$ to $(\mathbf{x},t)$. Each polymer is subject to a random potential $\lambda\eta(\mathbf{x},t)/2\nu$ from quenched impurities along its path. In $d = 2$ this is equivalent to the interface of a random bond Ising model, with typical configurations seen in Fig.(1).[12,15]

The second mapping defines a vector $\mathbf{v} = \nabla h$ (the slope), which satisfies

$$\frac{\partial \mathbf{v}}{\partial t} + \lambda \mathbf{v} \cdot \nabla \mathbf{v} = \nu \nabla^2 \mathbf{v} - \nabla \eta(\mathbf{x},t). \tag{19}$$

Eq.(19) is the *Burgers' equation*[30] for a vorticity free fluid. It has been extensively studied, as the deterministic version leads to "*shock waves*,"[30] clearly corresponding to the cusps and parabolas in Fig.(2). The stochastic version was treated by Forster, Nelson, and Stephen[31] using dynamical renormalization group methods.[19] The procedure is conceptually related to examining the equation at successively larger scales (in space and time). The results can be summarized by a single "*flow equation*" for the effective parameter $\bar{\lambda} = \lambda\sqrt{D/\nu^3}$

$$\frac{d\bar{\lambda}}{dl} = \frac{(3-d)}{2}\bar{\lambda} + \frac{(2d-5)}{4(d-1)}\bar{\lambda}^3. \tag{20}$$

In $d = 2$, the above equation has only one stable fixed point at finite $\bar{\lambda}$. The corresponding exponents are $\zeta = 1/2$ and $z = 3/2$. From the mapping to Eq.(18), we see that the exponent $1/z = 2/3$ describes the scaling of transverse fluctuations in the directed polymer of Fig.(1); a result given earlier. In $d = 3$, which is the critical dimension of the theory, $\bar{\lambda}$ is *marginally relevant*, and flows to $\bar{\lambda} \to \infty$. In $d > 3$, there are two regimes. Small nonlinearity maps onto $\bar{\lambda} = 0$, *i.e.* equivalent to Eqs.(8) and (9) with $\zeta = \zeta_0 = (3-d)/2$ and $z = z_0 = 2$. Large nonlinearities $(\bar{\lambda} > \bar{\lambda}_c)$ flow to infinity as in $d = 3$.

Unfortunately, the renormalization group procedure provides no information about ζ and z when $\bar{\lambda} \to \infty$ in $d \geq 3$. The only exact result is the identity $\zeta + z = 2$ which can be shown to follow from a hidden Galilean symmetry in Eq.(12)[29,31]. To obtain information on the strong-coupling $(\bar{\lambda} \to \infty)$ exponents, we resort to numerical simulations. Simulations have mainly been performed on two growth models. In *ballistic deposition*[22] particles are dropped randomly, fall down vertically, and stick upon their first contact with the cluster. In the *Eden model*[23] a surface site is selected at random, and one of its unoccupied neighbors is occupied. After a large number of particles are deposited, the width w is measured, and fitted to the dynamic scaling form $w(L,t) \sim L^\zeta f(t/L^z)$. The list of simulations is quite extensive[22–27], and will not be reproduced here. Some common features emerging from the simulations are:

(i) In $d = 2$, the predictions $\zeta = 1/2$ and $z = 3/2$ are verified.

(ii) All simulations are consistent with the identity $\zeta + z = 2$.

(iii) In $d = 3$ there is anomalous roughening with $\zeta \approx 0.2$ to 0.4.

(iv) Based on results for $d = 4, 5$, various conjectures have been proposed for $\zeta(d)$ in the strong-coupling limit.[32,33]

There are also recent simulations on the directed polymer problem in Eq.(18) which are consistent with the above conclusions based on growth models.[34]

8. "Self-Organized Criticality" and Sand Profiles

Self-Organized Criticality (SOC) is a term introduced by Bak, Tang, and Wiesenfeld (BTW)[35], to describe certain self-similar systems. Their motivation is as follows: spacial organization is very common in nature , as examplified by the many systems that show fractal behavior.[2] Temporal organization, on the other hand, is required to account for the prevalent "$1/f$" noise.[36] There may be a relation between spacial and temporal correlations, as in the case of approach to a continuous or critical phase transition.[1,19] However, unlike in phase transitions, there is in nature no parameter to be tuned to achieve criticality. BTW therefore called[35] such phenomena SOC. They also noted that a common feature of systems in nature exhibiting "$1/f$" noise (e.g. flow of sand in an hour glass, or of the river Nile), is that they represent *dissipative transport* in *open* and *extended* systems. As a simple possible extended dynamical system showing SOC, they introduced a cellular automaton model to describe the organization of *sandpiles*.[35]

Sand (discretized) is added from outside to a box with an open side. If the height difference between two adjacent sand columns exceeds a critical value, some sand drops to the lower column thereby dissipating energy. Once a steady state is achieved, extra added sand causes avalanches. An avalanche spreads over sizes s , and lasts a time τ. Computer simulations[35] indicate that the distributions $D(s)$ and $D(\tau)$ are self-similar in that there are no characteristic time or length scales for avalanches. Can we understand the self-similarity of this stationary state, and its scaling laws? The problem can be treated[37] by considering the evolution of the sand profile $f(\mathbf{x},t)$. It is more profitable to examine fluctuations $h(\mathbf{x},t)$ around some equilibrium configuration. There appear to be few symmetries to help us construct the equation of motion for h . There is no translational or reflection symmetry for h , and the presence of a transport direction even breaks the isotropy in $\mathbf{x}$ space. How can we therefore avoid adding a term $-h/\tau$ as in Eq.(11), thereby destroying self-similarity?

The clue is the existence of a *conservation law*.[19] When the surface relaxes after addition of extra sand particles, sand only moves from one place to another, thereby conserving the total number of particles. The conservation of $\int d^{d-1}\mathbf{x}\,h(\mathbf{x},t)$ implies the existence of a local current $\mathbf{J}$, and an equation of motion of the form $\partial h/\partial t + \nabla \cdot \mathbf{J} = 0$. The local current can be a nonlinear function of h, and because of symmetries of the problem the simplest allowed nonlinearity is $h^2\hat{T}$, where $\hat{T}$ is the transport direction. The full stochastic evolution equation is[37]

$$\frac{\partial h}{\partial t} = \nu_\perp \nabla_\perp^2 h + \nu_\parallel \partial_\parallel^2 h - \frac{\lambda}{2}\partial_\parallel h^2 + \eta(\mathbf{x},t). \tag{21}$$

Due to anisotropy two surface tensions $\nu_\parallel$ and $\nu_\perp$ are introduced. The noise

$$\langle \eta(\mathbf{x},t)\eta(\mathbf{x}',t')\rangle = 2D\delta^{d-1}(\mathbf{x}-\mathbf{x}')\delta(t-t') \tag{22}$$

is non-conservative, which is a new feature, setting Eq.(21) apart from those classified by Halperin and Hohenberg.[19] The nonlinear Eq.(21) can now be studied by the dynamic renormalization group method mentioned earlier.[19] In dimensions $d > 5$, the nonlinearity is irrelevant and behavior is similar to the diffusion equation. This is also true at the upper critical dimension of five. For $d < 5$, the nonlinearity is relevant. Surprisingly, however, there are sufficient number of constraints to allow an *exact* calculation of exponents in all dimensions.[37] Because directions parallel and perpendicular to transport ($x_\parallel$ and $\mathbf{x}_\perp$ respectively) are not equivalent, all correlation functions become anisotropic. Eq.(10) is now generalized to

$$\left\langle \left| h(x_\parallel,\mathbf{x}_\perp,t) - h(x'_\parallel,\mathbf{x}'_\perp,t')\right| \right\rangle \sim \left|x_\parallel - x'_\parallel\right|^\zeta f\left(\frac{t-t'}{\left|x_\parallel - x'_\parallel\right|^z}, \frac{|\mathbf{x}_\perp - \mathbf{x}'_\perp|}{\left|x_\parallel - x'_\parallel\right|^u}\right). \tag{23}$$

The new exponent u describes scaling of $\mathbf{x}_\perp$ with $x_\parallel$. The exact values of these exponents are

$$z = \frac{6}{8-d}, \ \zeta = \frac{2-d}{8-d}, \ u = \frac{3}{8-d}. \tag{24}$$

Starting from these exponents, one can calculate the fluctuations in the sand current escaping from the edge of the box. They are found to have $1/f^\phi$ fluctuations with $\phi = 1/z$. We are currently conducting simulations to test these results.

9. Conclusions

We have seen that in many physical circumstances interfaces exhibits self-similar fluctuations. The scaling of these fluctuations can be calculated numerically, or analytically in a number of cases. Examples discussed in these notes included interfaces in random systems, growing interfaces, and evolving sand piles. Various lectures at the Cargese workshop on Fracture have emphasized and exhibited the prevalence of scaling and self-similarity.[3] Examples include crack patterns, the scaling laws in earthquakes, and critical strengths. It would be interesting to see if some of the methods of continuum field theory (gradient expansions, symmetry considerations, renormalization group, etc.) can be applied to addresss such problems.

The recent experience with statistical field theories indicates that scaling forms and power laws have special significance. The exponent of the power law is frequently a universal number[1], characteristic of the underlying interactions of the problem. Critical exponents at a critical phase transition often serve as a guide to the microscopic ordering taking place.[19] However, as the examples of interface growth, and sandpile evolution indicate, scaling forms and self-similar fluctuations also appear as a consequence of symmetries and conservation laws in dynamical systems. The latter examples are more likely to occur than a critical point, and may indeed be the origin of SOC phenomena noted by BTW.[35] Once the underlying mechanism for self-similarity is identified it may be possible to construct field theoretic models for the phenomena, and to obtain the universal critical exponents by renormalization group techniques. How far such an approach can be extended, of course, remains to be seen.

Acknowledgements

This work is the outcome of collaborations with T. Hwa, E. Medina, D.R. Nelson, G. Parisi, and Y.-C. Zhang. The author acknowledges support from the NSF through grant No. DMR-86-20386, and a fellowship from the Sloan Foundation. Many thanks are due to J.C. Charmet, E. Guyon, H. Herrmann, S. Roux, and other organizers of the 1989 Cargese Meeting on "Disorder and Fracture."

References

[1] K.G. Wilson, Rev. Mod. Phys. **55**, 583 (1983); K.G. Wilson, Sci. Am. **241**, 158 (August 1979).

[2] B. Mandelbrot, *The Fractal Geometry of Nature* (Freeman, San Francisco, 1982).

[3] These Proceedings.

[4] *The Science of Fractal Images*, edited by H. Peitgen and D. Saupe (Springer-Verlag, New York, 1988).

[5] M. Kardar, J. Appl. Phys. **61**, 3601 (1987).

[6] M.E. Fisher, J. Chem. Soc., Faraday Trans. **II 82**, 1569 (1986).

[7] Y. Imry and S.-K. Ma, Phys. Rev. Lett. **35**, 1399 (1975).

[8] J. Villain, Phys. Rev. Lett. **52**, 1543 (1984);
R. Bruinsma and G. Aeppli, Phys. Rev. Lett. **52**, 1547 (1984);
G. Grinstein and J. Fernandez, Phys. Rev. B **29**, 6389 (1984).

[9] T. Nattermann, J. Phys. **C 18**, 6661 (1985).

[10] M. Kardar and Y.-C. Zhang, Phys. Rev. Lett. **58**, 2087 (1987).

[11] J.F. Fernandez, G. Grinstein, Y. Imry, and S. Kirkpatrick, Phys. Rev. Lett. **51**, 203 (1983).

[12] D.A. Huse and C.L. Henley, Phys. Rev. Lett. **54**, 2708 (1985).

[13] M. Kardar, Phys. Rev. Lett. **55**, 2923 (1985).

[14] M. Kardar and D.R. Nelson, Phys. Rev. Lett. **55**, 1157 (1985);
M. Kardar, Nucl. Phys. **B 290** [FS20], 582 (1987).

[15] D.A. Huse, C.L. Henley, and D.S. Fisher, Phys. Rev. Lett. **55**, 2924 (1985).

[16] (a) D.S. Fisher, Phys. Rev. Lett. **56**, 1964 (1986);
(b) T. Halpin-Healy, Phys. Rev. Lett. **62**, 442 (1989).

[17] A. Niemi, Phys. Rev. Lett. **49**, 1808 (1982).

[18] M. Kardar and Y.-C. Zhang, Europhys. Lett. **8**, 233 (1989).

[19] S.-K. Ma, *Modern Theory of Critical Phenomena* (Benjamin-Cummings, Reading, 1976);
P.C. Hohenberg and B.I. Halperin, Rev. Mod. Phys. **49**, 435 (1977); and references therein.

[20] J.D. Weeks, Phys. Rev. Lett. **52**, 2160 (1984).

[21] H. Leamy, G. Gilmer and A.G. Dirks, in *Current Topics in Materials Science*, edited by E. Kaldis (North-Holland, Amsterdam, 1980), Vol. 6.

[22] F. Family and T. Vicsek, J. Phys. **A 18**, L 75 (1985).

[23] M. Plischke and Z. Racz, Phys. Rev. **A 32**, 3825 (1985).

[24] R. Jullien and R. Botet, Phys. Rev. Lett. **54**, 2055 (1985).

[25] J.G. Zabolitzky and D. Stauffer, Phys. Rev. **A 34**, 1523 (1986).

[26] P. Meakin, P. Ramanlal, L.M. Sander, and R.C. Ball, Phys. Rev. **A 34**, 5091 (1986).

[27] M. Plischke, Z. Racz, and D. Lui, Phys. Rev. **B 35**, 3485 (1987).

[28] M. Kardar, G. Parisi, and Y.-C. Zhang, Phys. Rev. Lett. **56**, 889 (1986).

[29] E. Medina, T. Hwa, M. Kardar, and Y.-C. Zhang, Phys. Rev. **A 39**, 3053 (1989).

[30] J.M. Burgers, *The Nonlinear Diffusion Equation* (Riedel, Boston, 1974).

[31] D. Forster, D.R. Nelson, and M.J. Stephen, Phys. Rev. **A 16**, 732 (1977).

[32] D.E. Wolf and J. Kertész, J. Phys. **A 20**, L257 (1987).

[33] J.M. Kim and J.M. Kosterlitz, Phys. Rev. Lett. **62**, 2289 (1989).

[34] W. Renz, Unpublished results presented at Statphys 17, Brazil (1989).

[35] P. Bak, C. Tang, and K. Wiesenfeld, Phys. Rev. Lett. **59**, 381 (1987).

[36] P. Dutta and P.M. Horn, Rev. Mod. Phys. **53**, 497 (1981).

[37] T. Hwa and M. Kardar, Phys. Rev. Lett. **62**, 1813 (1989).

Chapter 2

Introduction to Multifractality

Stéphane Roux[†] and Alex Hansen[‡]

Laboratoire de Physique et Mécanique de la Matière Hétérogène
Ecole Supérieure de Physique et Chimie Industrielles de Paris
URA CNRS 857, 10 rue Vauquelin, F-75231 Paris Cédex 05, France

We introduce the basic concepts of multifractality by considering the example of the current distribution inside a random resistor network right at the percolation threshold. An infinite set of critical exponents governs the scaling of the moments of the distribution of currents with the system size. The "Legendre transform" of this series is shown to be connected to the histogram of the current distribution, through the formalism of multifractality.

The off-threshold distribution of currents for finite size lattices can also be described in this framework. We show on a simple "toy-problem" that the histogram of a multifractal quantity integrated away from criticality to the critical point exhibits very generally two power-law behaviors, for large and small values. Applications of this property to various models of rupture are discussed.

1. Why is multifractality useful?

There exists a large collection of models in statistical physics where the exhaustive characterization of the *distribution* of a local physical quantity requires the introduction of an *infinite* series of independent critical exponents. Examples range from turbulence (in the velocity correlation functions) to growth models (in the local probability of growth).[1−3] In those cases the use of the multifractal framework is not only useful, but almost unavoidable.

The description of the infinite set of critical exponents can be handled in different ways. Either by giving the scaling exponents as a function of the order of the moment considered, or by its Legendre transform,[4] or by a set of generalized dimensions.[5] Obviously, all these forms are mathematically equivalent, and are easily related one to the other. However, we will rather concentrate on the Legendre transform since it provides a simple and physical description of the set of exponents.

[†] also at: CERAM, ENPC, La Courtine, BP 105, F-93194 Noisy-le-Grand Cédex, France

[‡] permanent address: Dept. of Physics, University of Oslo, P.O.box 1048 Blindern, 0316 Oslo 3, Norway

Disorder and Fracture, Edited by J. C. Charmet *et al.*
Plenum Press, New York, 1990

Despite the use of sometimes lengthy equations, the basic elements are very simple. In the following, we will concentrate on the example of the random resistor network at (and later away from) the percolation threshold.

2. Model and basic results

We introduce a regular lattice whose bonds are ohmic resistors. The lattice is randomly diluted (the bonds of the lattice are removed) up to the percolation threshold. At this point we call $p = p_c$ the concentration of present bonds. Two opposite borders are hooked up to a current source, which imposes a current I through the network.

At the percolation threshold, the number of bonds, N_∞, that belong to the cluster which connects the two opposite electrodes, is known to vary with the lattice size $L \times L$ as:

$$N_\infty \propto L^{D_f} \tag{1}$$

where D_f is the fractal dimension of the "infinite" cluster.* However, only part of this cluster (and in fact a vanishingly small fraction of it) is conducting the current. This subset is called the *backbone*. The remaining part of the connected cluster consists of dead-ends linked to the backbone by a single bond (or site). The fractal dimension of the backbone is $D_b \approx 1.62^{[6]}$ as compared to $D_f = 91/48 \approx 1.89^{[7]}$.

The conductivity of the lattice of the percolation threshold is known also to vary as a power-law of the lattice size

$$G \propto L^{-t/\nu} \tag{2}$$

where t is the conductivity critical exponent, and ν the correlation length exponent. Numerically, in two dimensions, $\nu = 4/3^{[7]}$, and $t/\nu = 0.9745.^{[8]}$

The backbone can be seen as the support of (i.e. the set of bonds which contributes to) the zeroth moment of the current distribution. The resistance of the network is equal to the energy dissipated under a unit current. It is thus equal to the second moment of the local current distribution.

Other moments of the current distribution are directly accessible, in particular, the fourth one[9] (and higher even order[10]) can be shown to be directly related to the *noise* of the lattice. More precisely, suppose each resistor in the lattice independently fluctuates in time, by a small amount δr. The global resistance will also fluctuate by δR. The ratio $(\delta R/\delta r)$ can be shown[9] to be proportional to the fourth moment of the current distribution:

$$\delta R/\delta r \propto \frac{\left(\sum i^4\right)}{I^4} \tag{3}$$

---

* For a finite-size lattice, the "infinite cluster" is the cluster which connects the two electrodes.

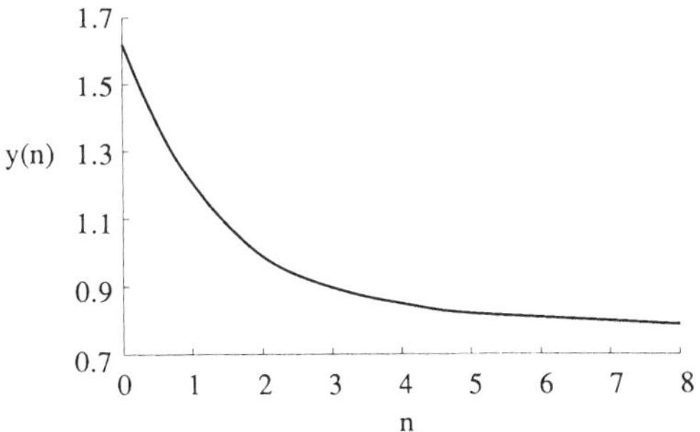

Figure 1 Scaling exponents $y(n)$ as a function of the moment order for the current distribution in a random resistor network at the percolation threshold.

This quantity scales with the lattice size with a new exponent[9] independent from the previous ones.

Finally, the rupture properties of the resistor network will be dependent on the largest currents flowing through the lattice. We know that at the percolation threshold, there exist some "singly connected bonds" which are bottlenecks where all the current is forced to flow. If those bonds are cut, then the lattice is no longer continuous. The number of the singly connected bonds varies as $L^{D_{scb}}$. Coniglio[11] has shown that $D_{scb} = 1/\nu$. Those bonds govern the infinite order moment of the current distribution.

To summarise the results we have quoted so far, we can state that, at the percolation threshold, and for an unit current flowing through the lattice, every moment, $\mathcal{M}_n$ of order n of the local current distribution can be characterized by a scaling exponent $y(n)$:

$$\mathcal{M}_n = \sum |i|^n \propto L^{y(n)} \tag{4}$$

where $y(0) = D_b$, $y(2) = t/\nu$ and $y(\infty) = 1/\nu$.

More generally, for any moment order n (not necessarily integer), we introduce the continuous function $y(n)$. Figure 1 shows this function obtained from numerical simulations. The aim of the next section is to present an alternative way to represent this function $y(n)$, which is obviously mathematically equivalent. However this formulation will allow us to get some insight in the meaning of those scaling exponents.

3. Multifractal formalism

We introduce the distribution $n(i, L, p_c)\frac{di}{i}$ of current i flowing through a lattice of size L at the percolation threshold p_c. The moment of order m of this distribution is given by

$$\mathcal{M}_m = \int i^m \, n(i, L, p_c) \, \frac{di}{i} \tag{5}$$

For the simplicity of notation, we introduce the new variable † α.

$$\alpha = \log(i)/\log(L) \tag{6}$$

and for the distribution

$$n(i, L, p_c) = L^{f(\alpha, L)} \tag{7}$$

Using those notations, $\mathcal{M}_m$ can be written as

$$\mathcal{M}_m = \int L^{m\alpha + f(\alpha, L)} \, \frac{di}{i} \tag{8}$$

We now assume that the distribution $n(i, L, p_c)$ is such that we can use the steepest descent method to estimate this integral. Accordingly, we seek the value of i, or α, for which the exponent of the integrand is most slowly varying. This happens when the exponent has an extremal value, hence its derivative is zero. This leads to the following condition

$$\frac{\partial f(\alpha, L)}{\partial \alpha} = -m \tag{9}$$

Let us call $\alpha^*(m, L)$ the value of α which fulfills this condition. The scaling of the mth moment is given by

$$\mathcal{M}_m \propto L^{m\alpha^*(m,L) + f(\alpha^*(m,L), L)} \tag{10}$$

or by identification with Eq.4,

$$y(m) = m\alpha^*(m, L) + f(\alpha^*(m, L), L) \tag{11}$$

The derivative of this last equation with respect to m gives

$$\begin{aligned}
\frac{\partial y(m)}{\partial m} &= \alpha^*(m, L) + \frac{\partial m\alpha + f(\alpha, L)}{\partial \alpha} \cdot \frac{\partial \alpha^*(m, L)}{\partial m} \\
&= \alpha^*(m, L)
\end{aligned} \tag{12}$$

because of the stationarity condition Eq.(9). This last equation shows that $\alpha^*(m, L)$ is independent of L. We thus call it $\alpha^*(m)$. Thus finally we have obtained the two important relations

$$\alpha^*(m) = \frac{\partial y(m)}{\partial m} \tag{13}$$

and

$$f(\alpha^*(m)) = y(m) - m\alpha^*(m) \tag{14}$$

---

† Sometimes the opposite convention of sign is chosen.

where we have dropped the L dependence of f, since the right hand side of the latter equation does not depend any longer on L. More precisely, the variation of the distribution $\log(n(i, L, p_c))/\log(L)$ with the lattice size L, is entirely taken into account by the use of the reduced variable α and f.

From now on we will drop the notation α^* to write α since confusion is no longer possible. As underlined before, this computation is nothing but a mere change of notations, and the function $f(\alpha)$, referred to as the multifractal spectrum, is strictly equivalent to the series of exponents $y(m)$. However, the simple mathematical derivation presented above shows a few interesting features,[4] which we now underline:

- The function $f(\alpha)$ can be seen through Eqs.(13 and 14) as the *"Legendre transform"* of the function $y(m)$.
- An alternative way to obtain this spectrum is to use Equations (6) and (7). There, one notes that this multifractal spectrum appears as the *log-log histogram* of the current distribution, up to a simple scale factor $1/\log(L)$ on both axes. This last scale factor captures the lattice size dependence. Thus all histograms fall onto a unique curve for all sizes, as shown on Figure 2.
- The $f(\alpha)$ spectrum is *universal* in the same way as the set of critical exponents, $y(m)$, is universal. Therefore, whatever is the lattice local structure, at the percolation threshold, the $f(\alpha)$ is identical. It only depends on the space dimension.
- Since the $f(\alpha)$ spectrum can be seen as the log-log histogram of the current, it characterizes *exhaustively* the distribution of the currents.
- The set of bonds that contributes to a given moment is characterized by a precise

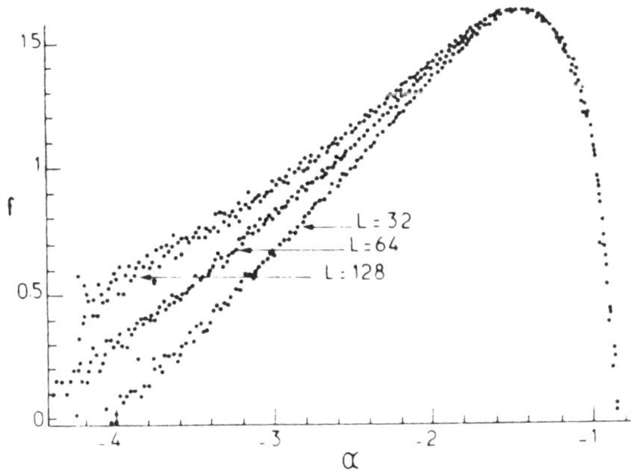

Figure 2 Rescaled log-log histogram of the current distribution in a random resistor network at percolation threshold, for three different lattice sizes. The large current part of the histogram is identical for all sizes. The collapse is bad for small currents, due to strong finite size effects.

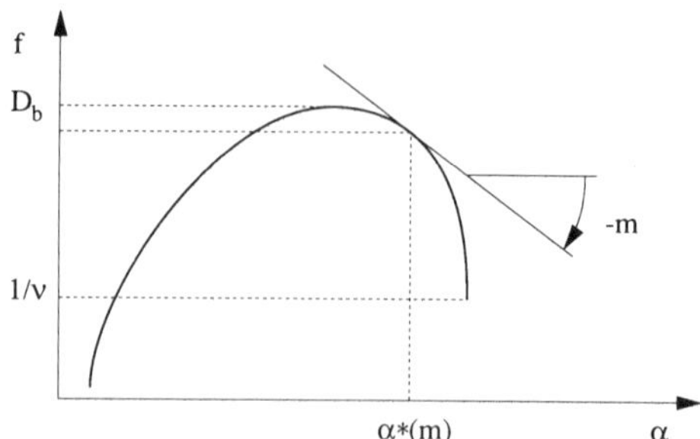

Figure 3 The generic properties of the representation of the multifractal spectrum are schematized.

value of α, $(\alpha^*(m))$, and thus of the current itself. Therefore, the scaling of these currents i with the lattice size is given by $i \propto L^{\alpha^*(m)}$. We note that this is strictly valid for a large enough lattice. In this case, the bonds contributing dominantly to the second and the fourth moments for instance are *independent*. This property gives a concrete physical meaning to the variable α.

• The *fractal dimension* of this latter set of bonds which contributes to the mth moment, is given through Eq.(7) as $f(\alpha^*(m))$. This property complements the previous one to give the fundamental advantage of this representation of the multifractal spectrum.

• The tangent to the $f(\alpha)$ curve which has a slope $-m$, is in contact with the curve exactly at the value of α which contributes to the mth moment (Eq.(9)). A simple consequence of this property is that the $f(\alpha)$ spectrum is *convex* (see Figure 3).

• At the point where the spectrum reaches its maximum, the value of α is $\alpha^*(0)$ due to the previous property. Since the moment of order 0 is governed by a critical exponent equal to the fractal dimension of the backbone, $y(0) = D_b$, we deduce from Eq.(14) that the value of $f(\alpha)$ at the apex of the curve is D_b.

• When the moment order m goes to infinity, we are only sensitive to the singly connected bonds, and thus $f(\alpha(m))$ tends to $1/\nu$.

Numerically, the Legendre transform of the scaling exponents $y(m)$ is generally a more precise way of obtaining the multifractal spectrum on a finite size lattice, rather than the log-log histogram of the current distribution. However, since α is related to the derivative of $y(m)$ with respect to m, it seems *a priori* difficult to obtain a precise estimate of it, knowing the difficulty to estimate already the scaling exponent itself. In fact this difficulty can easily be avoided by noting the following property[12]: The derivative of $\mathcal{M}_m$ with respect to m is given by the log-moments $\mathcal{L}_m$:

$$\mathcal{L}_m = \frac{\partial \mathcal{M}_m}{\partial m} = \sum |i|^m \log(|i|) \tag{15}$$

It scales with the lattice size as

$$\mathcal{L}_m \propto L^{y(m)} \frac{\partial y(m)}{\partial m} = L^{y(m)} \alpha^*(m) \tag{16}$$

Thus, $\alpha^*(m)$, can be obtained *for a single lattice* by taking the ratio between the log moment and the usual one:

$$\alpha^*(m) = \frac{\mathcal{L}_m}{\mathcal{M}_m} \tag{17}$$

The scaling of exponents $y(m)$ still requires at least two sizes to be computed, since they give the scaling of the moments with the lattice size.

Up to now we have considered the moments of the current distribution computed in the "constant current ensemble," since we imposed that the boundary conditions were given in terms of a unit current going through the lattices. However, this requirement is not necessary, and other ensembles can be introduced. In particular, we can consider the constant voltage ensemble, where the voltage drop across the lattice is kept constant. Due to the linearity of the problem it is straightforward to relate a local current in the constant current ensemble (subscript I) to the constant voltage ensemble (subscript V)

$$i_V = Gi_I \tag{18}$$

where G is the conductance of the lattice. For the moments, similarly

$$(\mathcal{M}_m)_V = G^m (\mathcal{M}_m)_I \tag{19}$$

using the scaling of the conductance with the lattice size (exponent $y_I(2)$) we can relate the sets of scaling exponent between the two ensembles by

$$y_V(m) = y_I(m) - my_I(2) \tag{20}$$

thus the multifractal spectrum reads

$$\alpha_V(m) = \alpha_I(m) - y_I(2) \tag{21}$$

and

$$f_V(\alpha_V(m)) = f_I(\alpha_I(m)) \tag{22}$$

Therefore, the multifractal spectrum in the constant voltage ensemble is deduced from the one in the constant current ensemble by a simple translation parallel to the α axis by an amount $y_I(2) = t/\nu$.

Other tools have been devised to deal with the infinite collection of critical exponents appearing in the current distribution (and other problems). In particular Hentschel and Procaccia[5] introduced the general fractal dimensions, D_q, by

$$D_q = \frac{y(q)}{(q-1)} \tag{23}$$

which is meaningful for a normalized distribution, say a probability, such that the first moment of the distribution is constant, (i.e. $y(1) = 0$). In this case, $D_1 = (\partial y(q)/\partial q)|_{q=1}$. For a non-normalized distribution, this definition has to be adjusted

$$D_q = \frac{y(q) - y(q_0)}{(q - q_0)} \tag{24}$$

for a given value of q_0.

The main difference between this latter representation and the Legendre transform is apparent here: D_q is a "secant" exponent, which can be compared to the "tangent" one α in the previous formalism. Although both formalisms contain the same mathematical information, the Legendre transform appears to be simpler to interpret physically and we will stick to this presentation in the following.

The same kind of analysis has been carried out for other transport properties in percolation networks such as elasticity.[13]

Up to this point, the multifractality concept introduced so far is a way to characterize the scaling behavior of the distribution of a local quantity (*e.g.* currents in our example). However, it might appear as being a theoretical concept since it is very difficult to have access to the complete distribution of this local quantity. Eventually, only a few moments of the distribution are directly measurable. However, we will show in the following that some properties resulting from the existence of a multifractal distribution at the percolation point may be observable under very general circumstances. For this, we introduce first another physical quantity, namely the conductance jump, that has also a multifractal behavior at the percolation threshold in the random resistor network problem.[14]

Consider a diluted lattice at the percolation point such that the lattice is still continuous from one border to the other. If one removes from the lattice one single bond, the conductance of the lattice will change by an amount δG. This change is recorded and the bond is put back in the lattice. Another bond is picked, and the conductance jump is recorded. The same procedure is repeated until all bonds have been tried. In fact only those which belong to the backbone will give a non-zero contribution. These jumps which arise from the removal of one single bond are distributed according to a multifractal distribution. It is intuitive that these conductance jumps occurring for a bond cutting will be correlated with the current flowing through it. However, no exact argument allows to relate one multifractal spectrum to the other. We will make use of this multifractal character as an application at the end of this text.

4. Away from criticality

We will now turn to a problem, which will prove to have a very general range of application. In particular, we will end this section with the application of these concepts to the case of some rupture models which display a similar kind of behavior as the toy-problem.

We have seen that the distribution of currents right at percolation threshold is multifractal. Away from the percolation threshold, the distribution of currents will appear to be identical to that observed at threshold, if the lattice size is smaller than the correlation length, ξ. In this case we will call $N(i, L, p_c) \frac{di}{i}$ the distribution of currents. Above this scale, we consider that the medium is homogeneous. In this latter case, we can easily relate the distribution of currents, $N(i, L, p)$ at a scale $L > \xi$, to that observed at a scale ξ, $N(i, \xi, p_c)$. More precisely, if $N(i, L, p)$ refers to the total number of bonds (*i.e.* , not the proportion) that carry a current i in a lattice of size L, for a concentration of present bonds of p, we have

$$N(i, L, p) = (L/\xi)^{f_0} . N(i(L/\xi)^{-\alpha_0}, \xi, p_c) \tag{25}$$

The first factor in the right hand side of this equation comes from the homogeneity of the lattice at a scale of ξ, and thus the exponent $f_0 = d$ coming into play is the space dimension. The change of the currents is given by the total current flowing at a scale of ξ, knowing the current imposed at a scale L, which gives simply $\alpha_0 = 1 - d$. This formula contains all the information needed to extract the distribution of current away from threshold, knowing what occurs at threshold (which we have seen to be given by the multifractal spectrum). Indeed, if we plot this distribution in a log-log scale, we will obtain a curve related to the $f(\alpha)$ spectrum by a simple dilation. This property arises naturally from the rewriting of the latter equation in the following form:

$$\log(i)/\log(L) = \alpha_0 - x\{\alpha_0 - \alpha\} \tag{26}$$

$$\log(N(i, L, p))/\log(L) = f_0 - x\{f_0 - f(\alpha)\} \tag{27}$$

where

$$\alpha = \log(i(L/\xi), \xi), p_c) \tag{28}$$

and

$$x = \log(\xi))/\log(L) \tag{29}$$

The log–log histogram will thus be obtained from the $f(\alpha)$ spectrum by a dilation of center (α_0, f_0) and of ratio x.[15]

The toy–problem now consists in starting from a full lattice of resistors ($p = 1$), and progressively removing bond after bond. During this dilution process, every time a resistor is removed from the lattice, one records the current flowing the bond for a unit current injected through the lattice. The distribution of current is of course reequilibrated every time a bond is cut. This process can be done up to the percolation point where the lattice finally is no longer connected. The question we ask is the following: What is the distribution of currents recorded during this process? Apparently, this question seems odd in the sense that part of the currents will come from the vicinity of the critical percolation point, but a large amount of data will be obtained far from this point, and thus, the mixture of all information disregarding the dilution of the lattice at the point where the current was measured may "scramble" all interesting aspect of this question. This intuitive first guess is not founded. On the contrary, one can show that the resulting distribution displays at its two ends, two universal behaviors related to the multifractal distribution of currents *at* the percolation threshold. The important point being that this distribution is naturally obtained "blindly," without knowing the exact location of the critical point beforehand, and nevertheless brings useful information specific to this critical point. This unusual feature will reveal itself as being the most powerful tool for applications of this concept to more complex cases.

Let us call $\mathcal{N}(i, L)$ the distribution of currents in bonds removed from the lattice from the intact state to the percolation point. This distribution can easily be written in terms of the above mentioned distribution $N(i, L, p)$:

$$\mathcal{N}(i, L) = \int_{p_c}^{1} N(i, L, p) \frac{dp}{(1 - p_c)} \propto \int_{1}^{\infty} N(i, L, p(\xi)) \xi^{-1 - 1/\nu} d\xi \qquad (30)$$

where the weighting in the last integral comes from the singular behavior of the correlation length at the percolation threshold.

$$\xi \propto (p - p_c)^{-\nu} \qquad (31)$$

Using now the relation between the off-threshold current distribution and the multifractal spectrum, we obtain the expression

$$\mathcal{N}(i, L) \propto L^{f_0} \int_{1}^{\infty} \exp\{\log(\xi) f(\alpha_0 + \beta) - f_0 - 1/\nu\} \frac{d\xi}{\xi} \qquad (32)$$

where $\beta = \log(iL^{-\alpha_0})/\log(\xi)$. This last integral can easily be estimated by using a steepest descent method analogous to the one we have used before. The derivative of the argument of the exponential with respect to $\log(\xi)$ (taking into account the fact that β depends on $\log(\xi)$) selects the value of ξ —and thus of p— that contributes dominantly to a given value of the current. The maximum of the argument of the exponent gives the following condition:

$$f(\alpha 0 + \beta) - \beta f'(\alpha_0 + \beta) - f_0 - 1/\nu = 0 \qquad (33)$$

This condition can be interpreted geometrically in a simple way (see Figure 4). The values of β that satisfy this equation are such that the multifractal spectrum at the point of abscissa $\alpha = \alpha_0 + \beta$, has a tangent which goes by the point of coordinates $(\alpha_0, f_0 + 1 + 1/\nu)$. There exist only two such values, one for $\beta < 0$ and one for $\beta > 0$ as shown on figure 4, and respectively denoted β_- and β_+. These values are universal since the multifractal spectrum is itself universal. They are characteristic of the conductivity problem. In the following, we will mainly use the value of the slope $\varphi_\pm = f'(\alpha_0 + \beta_\pm)$, at the two tangent points. These slopes are also universal in the same sense.

This fact is strikingly simple. Let us suppose that we consider a given current i. This value of the current will have been recorded in the dilution process *only* (in the thermodynamic limit) at a very precise value of p. If this current i is larger

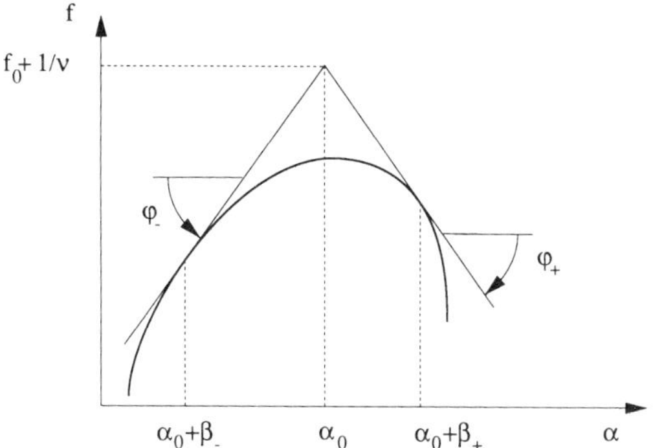

Figure 4 Geometrical construction of the condition expressed in Eq.(33). This provides an easy way to obtain $\varphi_\pm$ from the multifractal spectrum right at criticality.

than the current flowing in the bonds of the lattice for $p = 1$, then the correlation length at this point is such that $\beta = \beta_+$, or $\xi = (iL^{\alpha_0})^{1/\beta_+}$.[15]

We have now all the necessary information to conclude. The integral Eq.(32) can be written

$$\mathcal{N}(i, L) \propto L^{f_0} \exp\{\beta_\pm f'(\alpha_0 + \beta_\pm) \log(\xi)\}$$
$$\propto L^{f_0} \exp\{\varphi_\pm \log(iL^{-\alpha_0})\} \tag{34}$$
$$\propto i^{\varphi_\pm} L^{f_0 - \alpha_0 \varphi_\pm}$$

where the subscript $+$ or $-$ is chosen according to the fact that the current i is larger $(+)$ or smaller $(-)$ than the current flowing in the lattice for $p = 1$. Therefore, the distribution of current will consist of two power-laws. For small current, the number of bonds carrying a current i will increase with the current as a power-law with an exponent φ_-, whereas for large currents, it will also follow a power-law but with a negative exponent φ_+.

Numerical simulations of this toy problem work out quite nicely. In Fig. 5 we show the result of the random dilution of 2000 lattices of size $L = 24$. The slope on the right hand side of the wedge is about $\varphi_+ = -1.6$. Comparing this value to one that may be gotten from previously published data on the current histogram right at p_c [16], we find, using the geometrical construction of Fig. 4 and the discussion following Eq. (33), $\varphi_+ = -1.6 \pm 0.1$. This part of the histogram is due to the large currents appearing in the random dilution process. The part of the histogram in Fig. 5 to the left of the wedge stems from the small currents appearing. No apparent exponent is extractable from this part of the histogram. This is, however, not a surprise since the small currents appearing in the random resistor network at the percolation threshold most probably do not behave in a multifractal way! The reason for this is that the smallest currents that may appear in the network fall off exponentially rather than algebraically with the lattice size. This has as an effect that the spectrum in the thermodynamic limit has a horizontal asymptote[16] (on the small current side), and thus, the same should also hold for the integrated distribution. Due to the finite size of the lattices studied, only a small curvature can be seen on the left hand side of Fig. 5. However, the larger the lattice, the more pronounced the curvature, and the smaller the final slope.

This result is quite general. The only property needed is the development of a multifractal quantity as one approaches a critical point. If this is the case, then the record of this quantity from any initial state to the critical point will exhibit the behavior explicited above, i.e. for small and large values of the quantity, two power-laws will be observed with exponents characteristic from the critical point, whose geometrical construction is given above.

As a particular application of those concepts, we can quote one case which is more practical than the toy problem treated above. We have mentioned previously that the conductance jumps (and the resistance jumps) right at the percolation threshold are indeed multifractal. Therefore, if one records the series of conductance jumps which were encountered during a dilution from $p = 1$ to threshold, one expects to find such a distribution. The large conductance jumps and resistance jumps part of those distributions have indeed been observed both experimentally[17] during mercury injection of a porous medium, and numerically for two dimensional[18] and three dimensional[19] percolative systems. Unfortunately, the multifractal spectrum of resistance and conductance jumps right at the percolation threshold[14] is by far less well studied than the current distribution. (This is the reason why we considered the toy problem presented above).

However, as mentioned before, this result is not confined to the percolation case we quoted above. For instance, in some models of rupture of disordered media (fuse network[20] and elastic fragile networks[21]), it has been observed recently

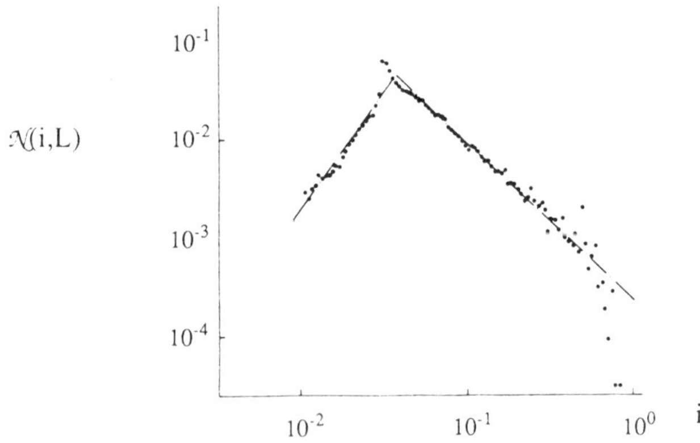

Figure 5 log-log plot of the distribution of current in the bonds picked during the random dilution process from $p = 1$ to p_c. The two straight lines are guides to the eyes.

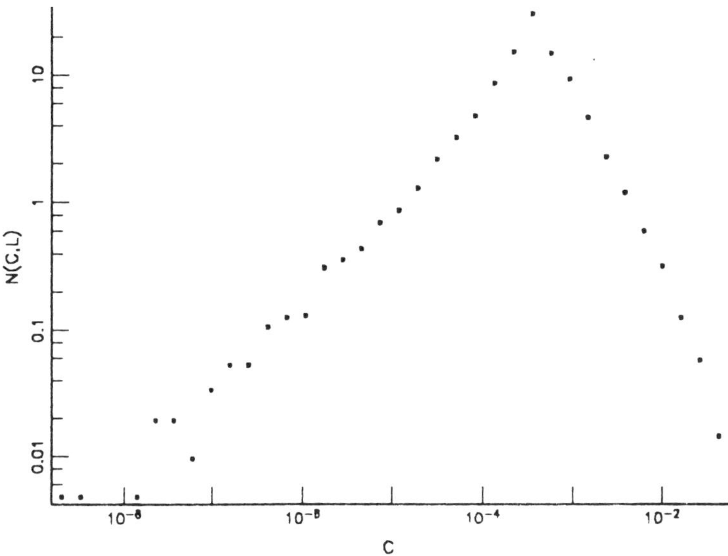

Figure 6 Log-log plot of the distribution of conductance jumps during the rupture process from the intact lattice to the final rupture.

that the distribution of current or forces right at the final breaking point is multifractal. Therefore, the conductance and resistance jumps (or elastic constant and compliance jumps) will also be multifractal at this point. As a direct consequence of our result, we expect that the distribution of conductance or resistance jumps integrated from the intact state to the final rupture point will display the two power-law behaviors for small and large currents.

Let us here report some numerical studies of just these quantities in a breakdown model. This model proceeds just as the random dilution one, but instead of removing bonds at random, they are chosen with a probability proportional to the current they carry to some power η.[22] In the limit $\eta = 0$, we recover the random dilution, or percolation, model. In Fig. 6 we show a (logarithmically binned) histogram of the conductance jumps throughout the entire breakdown process for 10 samples of size $L = 64$, and $\eta = 1$. Two distinct power-law regimes separated by a wedge are clearly visible. They reveal a multifractal structure of the jumps at the point where the lattice breaks apart.

This last example reveals a very powerful tool to analyse the development of a multifractal distribution, without having to analyse the local distribution of the quantity of interest right at the critical point.

Acknowledgements

We acknowledge useful discussions with E.Guyon and S.Redner. We also thank all the participants of the Institute for many exciting exchanges.

References

[1] J. Feder, *Fractals*, Plenum, (1988)

[2] T. Vicsek, *Fractal Growth Phenomena*, World Scientific Pub. (1988)

[3] see for the general case the review by G. Paladin and A. Vulpiani, Phys. Rep. **156**, 147, (1987)

[4] T.C. Halsey, M.H. Jensen, L.P. Kadanoff, I. Procaccia and B.I. Shraiman, Phys. Rev. **A 33**, 1141, (1986)

[5] H.G.E. Hentschel and I. Procaccia, Physica **D 8**, 440, (1983)

[6] L. de Arcangelis, S. Redner and A. Coniglio, Phys. Rev. **B 31**, 4725,(1985) and Phys. Rev. **B 34**, 4656, (1986)

[7] D. Stauffer, *Introduction to Percolation*, Taylor and Francis, (1986)

[8] J.M .Normand, H.J. Herrmann and M. Hajjar, J. Stat. Phys. **52**, 441, (1988)

[9] R. Rammal, C. Tannous, P. Breton and A.M.S.-Tremblay, Phys. Rev. Lett. **54**, 1718, (1985)

[10] R. Rammal, C. Tannous and A.M.S.-Tremblay, Phys. Rev. **A 31**, 241, (1985)

[11] A. Coniglio, J. Phys. **A 15**, 3829, (1982)

[12] A. Hansen, Ph. D. Thesis, Cornell University, (1986)

[13] A. Hansen and S. Roux, J. Stat. Phys. **53**, 759, (1988)

[14] S. Roux, A. Hansen an*d S. Redner, *preprint*

[15] S. Roux and A. Hansen, Europhys. Lett. **8**, 729, (1989). Note that an error is present in this reference, in an additional factor of one in the expression of the limit exponents φ_+ and φ_-. The expressions proposed in the present text are correct.

[16] G.G. Batrouni, A. Hansen and S. Roux, Phys. Rev. **A 38**, 3820, (1988)

[17] A.H. Thompson, A.J. Katz and R.A. Raschke, Phys. Rev. Lett. **58**, 29, (1987)

[18] G G. Batrouni, B. Khang and S. Redner, J. Phys. **A 21**, L23, (1988)

[19] J.N. Roux and D. Wilkinson, Phys. Rev. **A 37**, 3921, (1988)

[20] L. de Arcangelis and H.J. Herrmann, Phys. Rev. **B 39**, 2678, (1989)
L. de Arcangelis, A. Hansen, H.J. Herrmann and S. Roux, Phys. Rev. **B 40**, 877, (1989)

[21] H.J. Herrmann, A. Hansen and S. Roux, Phys. Rev. **B 39**, 637, (1989)
A. Hansen, S. Roux and H.J. Herrmann, J. Physique **50**, 733, (1989)

[22] E.L. Hinrichsen, A. Hansen and S. Roux, *preprint*

Chapter 3

Statistical Theory of Fragmentation

Sidney Redner

Center for Polymer Studies and Department of Physics
Boston University, Boston, MA 02215, USA

A general discussion of the kinetics of fragmentation processes is given. For a linear process, where fragmentation is driven by an external force, a scaling theory to describe the evolution of the cluster-size distribution is developed. General homogeneous systems, in which the break-up kernel is characterized by a homogeneity index λ, are treated. When $\lambda > 0$, corresponding to larger clusters more likely to break than small clusters, the scaled cluster size distribution, $\phi(x)$, decays with the scaled mass, x, as $x^{-2} \exp(-ax^{\lambda})$, as $x \to \infty$. For small x, $\phi(x)$ approaches the log normal form, $\exp(-a \ln^2 x)$, if the breakup kernel has a small-size cutoff, while $\phi(x)$ has a power law form in the absence of a cutoff. The conditions necessary for the existence of scaling are discussed. This reasoning also indicates that there is a "shattering" transition where mass is lost to a "dust" phase of zero-mass particles for $\lambda < 0$. For a collision induced, non-linear fragmentation process, the asymptotic behavior of "splitting" models, in which a two-particle collision results in one or both incident particles splitting in two, is analyzed. Scaling is found to hold for different ranges of the homogeneity index for these models. Comparisons and contrasts are made between the kinetics of linear and collision-induced fragmentation.

1. Introduction

A basic goal of this chapter is to provide a general theoretical account of the distribution of fragment sizes that results from fragmentation processes. Such processes are ubiquitous in nature, and they underlie phenomena such as polymer degradation, breakup of liquid droplets, crushing or grinding of rocks, and combustion.[1−8] These systems are often driven by a continuous external agent, such as repeated sledgehammer blows on a brittle piece of rock. The leads to a continuously evolving cluster size distribution in which the typical size decreases with time. As the typical size goes to zero, the lack of a characteristic size scale suggests that scaling and universality can be invoked to describe the nature of this distribution for sizes larger than the typical size.

I will discuss a purely kinetic, rate equation approach to describe the evolution of the cluster size distribution due to continuous fragmentation. This work has been performed during the course of an enjoyable collaboration with Z. Cheng.

Disorder and Fracture, Edited by J. C. Charmet *et al.*
Plenum Press, New York, 1990

The rate equation approach has been extensively investigated in the mathematical
and applied literature (see, *e.g.* , Refs. [9-12]), and there has been considerable
progress within a statistical mechanical viewpoint by the recent work of Ziff and
coworkers.[13−15] The basic viewpoint that I adopt is that scaling theory applied to
the rate equations provides a simple, yet powerful way obtaining universal information about the nature of the solutions to the rate equations.[16,17] This methodology
closely parallels recent developments made in understanding the kinetics of the inverse process of aggregation (see, *e.g.* , Refs. [18-21]). The rate equations are a
mean-field description, in which it is assumed that each fragment experiences the
average environment. In such an approach, the mass is the only relevant quantity
that characterizes a fragment, *i.e.* , the cluster size distribution is taken to be spatially uniform. The microscopic details of the breakup processes are embodied by
two kernels in the rate equations. One describes the *overall* rate at which a particle
breaks, and a second which gives the *relative* masses of the products emerging from
a breakup event.

Within the rate equation approach one may further classify fragmentation by
whether the breakup is driven only by a homogeneous external agent, which is
a purely *linear* process, or whether additional influences can also play a role in
particle breakup. One such possibility is fragmentation induced by collisions between fragments, which might arise in an explosive-type process, or perhaps in the
breakup of small eddies in a turbulent fluid flow.[22] As an idealization of such a
situation, I will consider a *non-linear* fragmentation model, in which collisions between fragments are responsible for particle breakup, with fragmentation products
continuously participating in repeated collisions and fragmentations.[23] While such
a collision-induced model may be too idealized to be of direct practical relevance,
the phenomenology of such processes is quite rich. Furthermore, there are parallels
with the rate equations of the inverse process of aggregation, since both models
are driven by the same type of bimolecular rate. Thus studies of collision-induced
fragmentation may provide a useful first step in understanding non-linear effects in
fragmentation kinetics.

2. Rate Equation Approach

Denote by $c(x, t)$ the concentration of particles of mass x at time t. The time
dependence of $c(x, t)$ is accounted for by writing the system of rate equations that
describes how $c(x, t)$ evolves as a result of the break-up process. As mentioned
previously, this approach represents an approximation of a mean-field character,
as fluctuations in the spatial positions of the fragments and in their shape are
ignored. While neglecting fluctuations may be quite drastic in some situations,
such an approximation appears to be valid for comminution processes such as ball
milling.[7] In such a system, there is thorough mixing of rock pieces by the turning
of an enclosing circular drum, with fragmentation being caused by the falling a a
steel ball on the rock. One should envision such a system when appealing to the
rate equation approach.

Below, I outline some basic properties of the rate equations for both linear and
non-linear fragmentation.

2.1 Linear fragmentation model

For a process in which break-up is driven by an external source, the evolution of the cluster size distribution is described by the linear integro-differential equation[9−17]

$$\frac{\partial}{\partial t} c(x,t) = -a(x)\, c(x,t) + \int_x^\infty c(y,t)\, a(y)\, f(x|y)\, dy. \tag{1}$$

Here, $a(x)$ is the *overall* rate at which x breaks, *i.e.* , $a(x)\,dt$ is proportional to the probability that an x-mer breaks in a time interval dt, and $f(x|y)$ is the *relative* breakup rate, *i.e.* , the conditional probability, or rate, at which x is produced from the breakup of y. The first term on the right hand side of Eq. (1) thus accounts for the "loss" of x-mers due to their breakup, while the second term accounts for the "gain" of x-mers due to the breakup of particles with mass $y > x$.

The cluster size distribution that results from Eq. (1) is determined by the details of the reaction kernels $a(x)$ and $f(x|y)$. I shall restrict ourselves to *homogeneous* kernels, as this case encompasses most physically attainable and relevant situations. Homogeneity implies that $a(x)$ has the power law form, $a(x) = x^\lambda$, thus defining the homogeneity index λ. Homogeneity also implies that $f(x|y)$ depends only on the ratio x/y, *i.e.* , $f(x,y)$ has the form $y^{-1} b(x/y)$. Mass conservation imposes the condition $\int_0^1 x\, b(x)\, dx = 1$, and we choose $\int_0^1 b(x)\, dx$, which is the average number of fragments produced in a breakup event, to be finite.

2.2 Collision-induced fragmentation model

For breakup driven by collisions between fragments, the interparticle interaction is now specified by the collision kernel $K(x,y)$, which gives the rate at which an x-mer and a y-mer meet, and the break-up kernel $B(x|y,z)$, which describes the rate at which x-mers are generated as result of the break-up of a y-mer in a collision between a y-mer and a z-mer. The rate equations for $c(x,t)$ now are,[16,17]

$$\frac{\partial c(x,t)}{\partial t} = -\, c(x,t) \int_0^\infty K(x,y) c(y,t)\, dy$$

$$+ \int_0^\infty dz \int_x^\infty K(y,z) B(x|y,z) c(y,t) c(z,t)\, dy \tag{2}$$

The two terms on the right-hand side account for the loss and gain of x-mers by fragmentation, respectively. Loss of x-mers arises because of collisions of x-mers with any of the particles in the system, while gain of x-mers arises from collisions in which the mass of y is larger than x. In this non-linear model, the collision kernel is symmetric, *i.e.* , $K(x,y) = K(y,x)$, while mass conservation implies that $y = \int_0^y x\, B(x|y,z)\, dy$.

In analogy with both linear fragmentation and with aggregation, one generally considers only systems with homogeneous collision kernels, *i.e.*

$$K(ax, ay) = a^\lambda K(x, y). \tag{3}$$

For the break-up kernel, there is a considerable latitude of possibilities, as many dependences of $B(x|y, z)$ and $K(x, y)$ on their arguments lead to physically reasonable microscopic processes. I will concentrate on a general class of "splitting" models which retain the essential nonlinearity of collision-induced fragmentation, but are simple enough to be studied analytically. They are:
MODEL I: Both particles split exactly in two in a collision.
MODEL II: Only the larger particle splits in two.
MODEL III: Only the smaller particle splits in two.

Models II and III represent the two extreme and opposite situations in which the larger, or the smaller clusters is more susceptible to breakup, respectively.

3. Scaling

I will mainly focus on solutions to the rate equations that can be cast into a scaling form. Scaling solutions are emphasized for several reasons. First, the kernels for which scaling holds represent most physically relevant systems. Second, the scaling ansatz reduces a two variable problem to a single variable problem, thus providing a simpler description of the problem. Third, a scaling solution is universal in the sense that it is independent of the initial condition. Thus scaling provides a robust classification scheme for the solutions.

To analyze the rate equations, I use the familiar scaling ansatz for the cluster size distribution,[18–21]

$$c(x, t) = s(t)^{-2} \phi(x/s(t)), \tag{4}$$

where $s(t)$ is a typical particle mass. According to this scaling ansatz, the distribution of fragment sizes is not a function of mass and time separately, but rather is a function only of the ratio of the mass to the typical mass, $s(t)$. The prefactor of $s(t)^{-2}$ is required by mass conservation. This scaling form also implies that the moments of the cluster size distribution are all related. Define the αth-moment of the cluster-size distribution and the αth-moment of the scaling function, by

$$M_\alpha(t) = \int_0^\infty x^\alpha c(x, t)\, dx, \qquad m_\alpha = \int_0^\infty x^\alpha \phi(x)\, dx. \tag{5}$$

These are just the Mellin transforms of $c(x, t)$ of $\phi(x)$, respectively (except for a trivial shift of 1 in the definition of α). From this, the "bare" and "scaled" moments are related by

$$M_\alpha(t) = m_\alpha\, s(t)^{\alpha-1}, \tag{6}$$

and in particular,

$$s(t) = m_0/M_0(t). \tag{7}$$

This states that the typical size varies as the inverse of the cluster concentration. These moment relations follow solely from the ansatz (4); therefore they are common to both fragmentation *and* aggregation. Notice that in the scaling ansatz there are two free parameters, namely the amplitude of $s(t)$ and the amplitude of $\phi(x)$. Without loss of generality, one may normalize both m_1 and m_0 to 1 in order to fix these two parameters. Let us now exploit this scaling formulation to obtain asymptotic solutions to the rate equations.

4. Solutions to the Rate Equations for Linear Fragmentation

4.1 Moment relations

For the linear model, substituting Eq. (4) into (1), and defining $\xi = x/s$, yields the scaling equations

$$\omega[2\phi(\xi) + \xi\,\phi'(\xi)] = -\xi^\lambda\,\phi(\xi) + \int_\xi^\infty \phi(\eta)\,\eta^{\lambda-1}\,b(\tfrac{\xi}{\eta})\,d\eta, \tag{8}$$

$$\dot{s}\,s^{-(1+\lambda)} = -\omega, \tag{9}$$

where $\omega > 0$ is the separation constant, and the overdot denotes the time derivative. From Eq. (9), the typical fragment size has the time dependence,

$$s(t) \sim \begin{cases} t^{-1/\lambda}, & \text{for } \lambda > 0 \text{ and } t \to \infty, \\[2mm] e^{-\omega t}, & \text{for } \lambda = 0 \text{ and } t \to \infty, \\[2mm] (t_c - t)^{1/|\lambda|}, & \text{for } \lambda < 0 \text{ and } t < t_c. \end{cases} \tag{10}$$

To find the solution to Eq. (8), one converts it to a relation involving the moments of $\phi(\xi)$ by multiplying both sides by ξ^α and integrating over all ξ. An important point in this derivation is that the second term on the right-hand side of Eq. (8) leads to the integral

$$\int_0^\infty d\xi \int_\xi^\infty \xi^\alpha\,\eta^{\lambda-1}\,\phi(\eta)\,g(\xi/\eta)\,d\eta, \tag{11a}$$

and the order of integration can be interchanged to yield

$$\int_0^\eta d\xi \int_0^\infty \xi^\alpha\,\eta^{\lambda-1}\,\phi(\eta)\,g(\xi/\eta)\,d\eta. \tag{11b}$$

Now by introducing the moments of the reduced break-up kernel

$$L_\alpha = \int_0^1 x^\alpha\,b(x)\,dx, \tag{12}$$

and in terms of the moments of the scaling function m_α, one obtains the linear recursion relation

$$m_{\alpha+\lambda} = \omega \, \frac{1-\alpha}{L_\alpha - 1} \, m_\alpha. \tag{13}$$

Notice that the explicit dependence on the kernel is contained *only* in L_α. This suggests that many of the results that follow from this recursion relation will be universal in nature.

To determine the asymptotic behavior of the cluster size distribution, I make extensive use of the following "correspondence" procedure to relate the moment function, m_α, and its inverse Mellin transform, $\phi(x)$. First, Eq. (13) is used to compute the reduced moments for a discrete set of equidistant α values. Next, "smoothness" is invoked, in which the form of the moments from the discrete set $\{\alpha\}$ is extended to continuous values of α. Finally, one computes the inverse Mellin transform to reconstruct the functional form of the scaling function. Although such a procedure is not mathematically precise, this correspondence is supported by available exact solutions, and it is reasonable to assume that it will be correct for physically realizable fragmentation processes.

4.2 Cluster size distribution in the large x limit

The large x behavior of $\phi(x)$ corresponds to m_α for large values of α. To obtain this latter quantity, substitute $\alpha = k\lambda$, with k a positive integer, iterate Eq. (13), and use the fact $m_0 = 1$. This leads to

$$m_{k\lambda} = \omega^{k-1} \prod_{n=1}^{k-1} \frac{n\lambda - 1}{1 - L_{n\lambda}}. \tag{14}$$

For large k, $m_{k\lambda}$ is dominated the large-n factors in the product. Concomitantly, the large n behavior of $L_{n\lambda}$ depends only on the limiting form of $b(x)$ at x near 1 (*i.e.*, in the   limit of production of large fragments). Thus the form of $m_{k\lambda}$ is universal in that the only details of a specific model that appear are the homogeneity index and the limiting form of $b(x)$ for $x \to 1$; other aspects of the breakup kernel are irrelevant.

For illustration purposes, consider the general class of kernels which have the form for x near 1

$$b(x) = b(1) + O((1-x)^\mu) \tag{15}$$

where $b(1) \geq 0$ and $\mu > 0$ are constants. For this form of $b(x)$, one has

$$L_\alpha = b(1)/\alpha + O(\alpha^{-(\mu+1)}), \tag{16}$$

for large α. Using this in Eq. (14) eventually leads to

$$m_{k\lambda} \propto (k-1)! \, (\omega\lambda)^{k-1} \, (k-1)^{(b(1)-1)/\lambda}. \tag{17a}$$

Employing Stirling's approximation now yields

$$m_\alpha \to c\left(\frac{\omega}{e}\right)^{\alpha/\lambda} \, \alpha^{(b(1)-1)/\lambda-1/2} \, \alpha^{\alpha/\lambda}, \qquad \text{for} \ \ \alpha \to \infty, \tag{17b}$$

where c is a constant and where k has been replaced by α/λ. Although Eq. (17a) is valid only for a discrete set of equidistant values of α, one expects that Eq. (17b) is generally true, since the moment function is a smooth function of α in the large α limit.

Owing to this universality of m_α, it follows that the corresponding functional form of $\phi(x)$ is also universal. One can now deduce the general form of $\phi(x)$ by computing the inverse Mellin transform of m_α. This gives

$$\phi(x) \sim x^{b(1)-2} \, \exp(-a\,x^\lambda) \qquad (x \to \infty). \tag{18}$$

It can be verified directly that this form for $\phi(x)$ for large x indeed yields the large-α form for the moments m_α given in Eq. (17).

More generally, the above result for $\phi(x)$ can be extended to general homogeneous kernels by noticing that the controlling factor, $\alpha^{\alpha/\lambda}$ in Eq. (17b), is responsible for the quasi-exponential form, $\exp(-a\,x^\lambda)$, in $\phi(x)$. This controlling factor arises from the kernel-independent quantity, $\prod_{n=1}^{k-1}[n\lambda - 1]$. Consequently, the scaling function has the *universal* controlling factor $\exp(-a\,x^\lambda)$ for large x, for kernels with homogeneity index λ.

4.3 Cluster size distribution in the small x limit

In the small-x limit, there is a lesser degree of universality since the small mass tail is not influenced by particles of the typical size. One does find, however, that there are only two possible generic scaling forms for $\phi(x)$ at small x, whose applicability depends on whether or not the moments $L_{-\alpha}$ exist for all α. To obtain $\phi(x)$ in the small-x limit, one requires the behavior of $m_{-\alpha}$ as $\alpha \to \infty$. Accordingly, choose $\alpha = 1 - k\lambda$ in Eq. (13) and iterate to arrive at the analog of Eq. (14), namely,

$$m_{1-k\lambda} = \omega^{-k} \, \frac{\prod_{n=1}^{k}[L_{1-n\lambda} - 1]}{\prod_{n=1}^{k} n\lambda}. \tag{19}$$

In analogy with the case of large positive α, the dependence of $m_{1-k\lambda}$ on k for large k is controlled by the limiting form of $b(x)$ for x near 0.

To appreciate the consequences of this fact, let us first treat the general case of kernels that are cut off at small fragment sizes, *i.e.* , "flaking off" of infinitesimal size pieces in a breakup event is not allowed. For example, consider $b(x) = 0$ for $x < x_0$, with $0 < x_0 < 1$ and $b(x) = b_1(x - x_0)^\mu$ for $x \to x_0^+$. That is, the smallest fragment that can be created from breaking a piece of size y is $x_0 y$. For this case, $L_{1-\alpha} \sim c \cdot x_0^{-\alpha}/\alpha^{1+\mu}$, for large α, where c is a constant independent of α. Substituting this leading behavior into Eq. (19), yields

$$m_{1-k\lambda} \sim \left(\frac{c}{\omega\lambda^{\mu+2}}\right)^k \cdot (k!)^{-\mu-2} \cdot x_0^{-\lambda(k+1)k/2}. \tag{20}$$

In terms of $\alpha = 1 - k\lambda$, the controlling factor of $m_{-\alpha}$ has a universal form,

$$\exp\left[-\alpha^2(\ln x_0)/2\lambda\right], \tag{21}$$

as it is only in the lower order contributions in Eq. (20) that dependence on μ appears. By taking the inverse Mellin transform, the controlling factor of $\phi(x)$ has the classical log-normal form[24,25]

$$\phi(x) \sim \exp\left[-\frac{\lambda}{2\ln x_0}(\ln^2 x)\right] \qquad (x \to 0). \tag{22}$$

More generally, for an arbitrary kernel $b(x)$ with a small-size cutoff, one can verify that the controlling factor in Eq. (21) arises from $\prod_{n=1}^{k}[L_{1-n\lambda}-1]$ in Eq. (19). Since for any $b(x)$ with a cutoff, there always exists some $0 < x_0 < 1$ such that $L_{-\alpha} > x_0^{-\alpha}$, this product leads to a factor which diverges at least as fast as $\exp(c\,\alpha^2)$ as $\alpha \to \infty$. Thus one concludes that, in general, $\phi(x)$ decays as $\exp[-(\ln^2 x)/4c]$, or slower, as $x \to 0$.

The log-normal form can be obtained from a simple multiplicative argument which appears to contain the essence of a repeated fragmentation process with a small size cutoff.[25,26] According to this process, the mass of a given fragment schematically evolves as

$$x_0 \to x_1 \to x_2 \to \cdots \to x_N, \tag{23}$$

where the successive reduction factor $r_k = x_k/x_{k-1}$ is a random variable with a well-behaved distribution, for a kernel with a small-mass cutoff. Consequently, the central limit theorem states that $\log x_N = \sum_{k=0}^{N} \log r_k$ will be normally distributed, so that x_N will be distributed log-normally.

A second general universality class for the small-x behavior of $b(x)$ is the situations where "flaking off" of infinitesimal size pieces in a single fragmentation is allowed. This corresponds to a kernel with no cutoff at the small-size limit. For example, for the kernel $b(x) \sim x^{\nu}$ for small x, it follows, by solving for m_α in terms of $m_{\alpha+\lambda}$ in Eq. (13), that m_α diverges whenever L_α diverges. This occurs for α less than a critical value α_c, which is less than 0, since m_0 is finite. For α close to α_c one keeps only the leading term in Eq. (13) to give

$$m_\alpha = L_\alpha \frac{m_{\alpha_c+\lambda}}{\omega(1-\alpha_c)} \propto L_\alpha = \int x^\alpha b(x)\,dx. \tag{24}$$

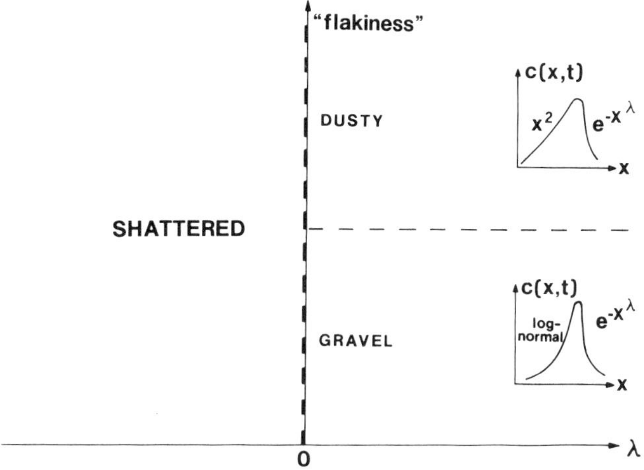

Figure 1 "Phase diagram" for linear fragmentation in the plane defined by the homogeneity index λ and a loosely-defined parameter, the "flakiness" of the relative breakup rate. Small flakiness corresponds to a vanishingly small probability of small flakes being produced in a single breakup event, $i.e.$, a small-size cutoff in $b(x)$, while large flakiness corresponds to the opposite limit of a power law tail in $b(x)$. In the phase plane there is a "shattered" phase for $\lambda < 0$, while for $\lambda \geq 0$ there is a "gravel" phase for small flakiness and a "dusty" phase for large flakiness. The fragment size distributions corresponding to these latter two phases are also shown.

This implies that $\phi(x)$ coincides with $b(x)$, so that

$$\phi(x) \sim x^{\nu} \qquad (x \to 0). \tag{25}$$

Eqs. (22) and (25) provide the asymptotic small x behavior of $\phi(x)$ for an encompassingly wide class of breakup kernels. The range of possibilities for the system can be conveniently summarized by the "phase diagram" of Fig. 1.

4.4 Existence of scaling

Thus far, I have merely assumed that scaling holds for $\lambda > 0$. I now show that $\lambda > 0$ is a necessary condition for scaling, and give a plausibility argument that $\lambda > 0$ is also a sufficient condition for scaling. First notice that $\lambda < 0$ implies no scaling. If scaling existed for $\lambda < 0$, then Eq. (10) predicts that $s(t)$ vanishes in a finite time, and from Eq. (7), $M_0(t)$ is singular. This contradicts the fact, from Eq. (1), that $c(x,t)$ must decay exponentially in time, or slower, for any value of x. Hence scaling solutions do not exist when $\lambda < 0$.

To argue that $\lambda > 0$ is a sufficient condition for scaling, first write the "bare" moment relation which is derived directly from the rate equation (1),

$$\dot{M}_\alpha = (L_\alpha - 1) M_{\alpha+\lambda}. \tag{26}$$

For $\alpha = 1 - k\lambda$, it is possible to solve for the moments iteratively. Starting with $M_1(t) = 1$, Eq. (26) gives $M_{1-\lambda}(t) = (L_{1-\lambda} - 1)t + \text{const.}$, where $L_{1-\lambda} - 1$ is a positive constant. Continuing this process, one finds the asymptotic solution,

$$M_{1-k\lambda} \sim \frac{1}{k!} \prod_{j=1}^{k} (L_{1-j\lambda} - 1)\, t^k \tag{27}$$

for a discrete set of index values $1 - k\lambda$. This is the same time dependence as in Eq. (6) with $\alpha = 1 - k\lambda$. Invoking our smoothness postulate, wherein the form of M_α for arbitrary α interpolates smoothly between the moments defined on the discrete set, then Eq. (26) reproduces the temporal behavior of the moments from the scaling ansatz. It is important to note that for a diminishment process such as fragmentation, the *negative* moments of the cluster size distribution are the fundamental quantities that characterize the evolution of the cluster size distribution. They play the corresponding role of the positive moments of the cluster size distribution in growth phenomena such as aggregation.

4.5 Shattering transition

An intriguing aspect of fragmentation processes is the possibility of a "shattering" transition, in which mass is "lost" to a phase of zero mass particles. This represents an inverse to the phenomenon of gelation, in which the mass of the "sol" phase is lost to the phase consisting of the infinite gel molecule. The potential existence of shattering depends on the mass dependence of the overall cluster breakup rate. If very small clusters break up sufficiently rapidly, then the entire fragmentation process is dominated by a cascade of very rapid breakup of smaller and smaller fragments. Mathematically, both gelation and shattering are signaled by the condition $\dot{M}_1 < 0$. As discussed first by McGrady and Ziff[15] in the context of a particular fragmentation model, both shattering and breakdown of scaling were found to occur when $\lambda < 0$, *i.e.* , whenever the breakup rate favors the fragmentation of small clusters over large clusters. I now show generally that $\lambda < 0$ is a necessary and sufficient condition for the existence of shattering.

To locate the shattering transition, consider Eq. (26) for $\alpha = 1 + \epsilon$, in the limit $\epsilon \to 0$,

$$\dot{M}_1 = (L_{1+\epsilon} - 1)\, M_{\lambda+1+\epsilon} . \tag{28}$$

Since L_α approaches 1 as $\alpha \to 1$ (mass conservation), $\dot{M}_1$ can be nonzero only if $M_{\lambda+1+\epsilon}$ diverges as $\epsilon \to 0$. Without loss of generality, suppose that initially there is no particle with a mass larger than unity, so that will be no particle with a mass larger than unity at any later time. Thus, M_α is non-decreasing as α decreases, at any fixed time. This fact, together with $M_1 < \infty$, imply that M_α can diverge only for $\alpha < 1$. Thus a necessary condition for shattering is $\lambda < 0$.

To show that $\lambda < 0$ is also a sufficient condition for shattering, let us assume the converse and derive a contradiction. For $\lambda < 0$, Eq. (26) gives

$$\dot{M}_{1+|\lambda|} = (L_{1+|\lambda|} - 1)\, M_1 . \tag{29}$$

Since it has been assumed that there is no shattering, M_1 remains fixed, so that the right hand side is a negative constant. This implies that $M_{1+|\lambda|}$ and hence $c(x,t)$ would vanish at a finite time. On the other hand, from Eq. (1), $c(x,t)$ decays exponentially in time, or slower, for any value of x. This contradiction implies that a sufficient condition for shattering is $\lambda > 0$.

An important feature of shattering in linear fragmentation is that, from Eq. (29), mass loss starts right at $t = 0$. This is in contrast to the corresponding situation of the gelation transition in coagulation, where mass loss does not occur until a finite gel time has been reached.

5. Solutions to the Rate Equations for Collision-Induced Fragmentation

For collision-induced fragmentation, one anticipates that the cluster-size distribution will evolve into a scaling form in the long-time limit for a certain range of parameter values that characterize the reaction kernel. Thus using the scaling ansatz in Eq. (2), and assuming that the collision kernel has a homogeneity index λ, one arrives at the integro-differential equation for the scaling function

$$\omega[\xi\phi'(\xi) + 2\phi(\xi)] = -\phi(\xi)\int_0^\infty K(\xi,\eta)\phi(\eta)\,d\eta$$
$$+ \int_0^\infty d\zeta \int_\xi^\infty K(\eta,\zeta)B(\xi|\eta,\zeta)\phi(\eta)\phi(\zeta)\,d\eta \qquad (30)$$

and for the time dependence one finds

$$\dot{s}\,s^{-\lambda} = \omega^{-1}, \qquad (31)$$

with solution,

$$s(t) \sim \begin{cases} t^{\frac{1}{1-\lambda}}, & \text{for } \lambda > 1 \text{ and } t \to \infty, \\[2ex] e^{-\omega t}, & \text{for } \lambda = 1 \text{ and } t \to \infty, \\[2ex] (t_c - t)^{\frac{1}{\lambda-1}}, & \text{for } \lambda < 1 \text{ and } t < t_c. \end{cases} \qquad (32)$$

From Eq. (32), a shattering transition now occurs at $\lambda_s = 1$. For $\lambda < \lambda_s$, the total number of clusters diverges in a finite time and mass is not conserved. Interestingly the critical value of λ is unity, corresponding to a situation where smaller particles are relatively less likely to collide. However, this inhibitory feature is compensated by the relatively rapid production of small clusters by two-body collisions. In linear fragmentation, there is no amplification coming from a collision process, so that shattering first appears only when the break-up of clusters of all sizes contribute equally to the fragmentation process, $i.e.$, $\lambda_s = 0$.

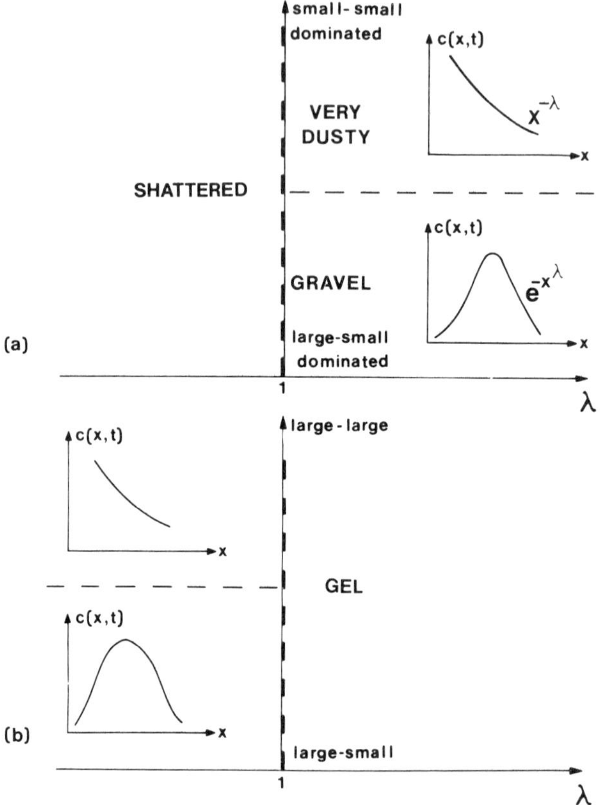

Figure 2 (a) Phase diagram for collision-induced fragmentation in the plane defined by λ and a second parameter which quantifies the relative importance of large-small versus small-small reactions. The shattered phase now occurs for $\lambda < 1$, while for $\lambda \geq 1$, there is a gravel phase when large-small reactions dominate, and a "very dusty" phase when small-small reactions dominate. For comparison, the phase diagram for a general aggregation processes is shown in (b). The mechanisms that lead to either a monotonic or non-monotonic cluster size distribution appear to be common to both aggregation and fragmentation.

To understand the overall features of nonlinear fragmentation, let us now turn to the three splitting models discussed earlier. Because clusters split exactly in half, the mass x can be written in the discrete form $x = 2^{-n}$ with integer n. This discretization greatly simplifies the analysis of the rate equations. The asymptotic solutions to the rate equations for the splitting models will be obtained in the regime where scaling holds. Numerically, this occurs when $\lambda > \lambda_c$, where λ_c is model dependent and less than one in general.

Before delving into mathematical details, it is worthwhile to summarize the basic results graphically. For the splitting models, the basic classifying feature is whether reactions of small fragments with large fragments dominates over, or is dominated by, small-small reactions. (In any case, large-large reactions are irrelevant asymptotically.) In Models I and II, the larger of the two colliding fragments always splits. Therefore large-small reactions eventually dominate, since large clusters become effectively the most reactive as the fragment size distribution evolves. As I will show in the next section, this gives rise to a peaked fragment size distribution, as illustrated in the phase diagram for collision-induced fragmentation (Fig. 2(a)). On the other hand, in model III, small-small reactions eventually dominate. Consequently large clusters become "frozen out" of the fragmentation process at long times, leaving behind an appreciable "residue" of large clusters. These qualitative features of the fragment size distributions have strong parallels with the corresponding phase diagram of aggregation processes, as indicated in Fig. 2(b). Now I turn to a quantitative analysis of the behavior in the splitting models.

5.1 Model I: Both particles split upon collision

For concreteness, let us assume that the collision rate has the product form, $K(x, y) = (xy)^{\lambda/2}$. Then the rate equations for this model can be written in the form,

$$\frac{dc_n}{dt} = [-2^{-\lambda n/2} c_n + 2 \cdot 2^{-\lambda(n-1)/2} c_{n-1}] [\sum_{j=0}^{\infty} 2^{-\lambda j/2} c_j]. \tag{33}$$

The first term accounts for the loss of 2^{-n}-mers due to their collision with all other clusters in the system. Similarly, the second term accounts for the gain of 2^{-n}-mers due to the collision of a $2^{-(n-1)}$-mer with other clusters. The additional factor of 2 arises because two 2^{-n}-mers are produced from the breakup of a $2^{-(n-1)}$-mer.

Of some pedagogical interest is the case $\lambda = 0$, *i.e.* , the rate of fragmentation is mass independent. With the initial condition $c_n(0) = M \delta_{n,0}$, the exact solution, found by elementary methods, is

$$c_n(t) = \frac{1}{n!} \frac{t_c - t}{t_c^2} (2 \ln \frac{t_c}{t_c - t})^n, \tag{34}$$

where the shattering time $t_c = 1/M$. Mass is conserved for times $t < t_c$, but then shattering occurs in which the cluster concentrations of all finite sizes vanish.

For general λ, Model I can be recast as a linear fragmentation model by introducing a rescaled time $t' = \int^t d\tau \, M_{\lambda/2}(\tau)$. In this effective linear model, the parameters that characterize the breakup kernel are related to those of the corresponding non-linear model by $\lambda' = \lambda/2$, $b(x) = 2\delta(x - 1/2)$. From our results for the linear model, one therefore concludes that the scaling is valid for $\lambda > 0$, but that a shattering transition of an intrinsically non-linear nature, as embodied, for example, by the solution given in Eq. (34), occurs for $\lambda < 1$.

In the scaling regime ($\lambda > 0$), the mapping to the linear model, together with the results quoted in Eqs. (18) and (22) yields the asymptotic behaviors for the non-linear model,

$$\phi(x) \sim \begin{cases} \exp[-x^{\lambda/2}]\,/x^2, & (x \to \infty); \\[2mm] \exp[\frac{\lambda}{4\ln 2}(\ln^2 x)], & (x \to 0), \end{cases} \tag{35a}$$

where $x = 2^{-n}/s(t)$ and $s(t)$ is determined by Eq. (32). From the scaling ansatz, this can be rewritten in term of $c_n(t)$ as,

$$c_n(t) \sim \begin{cases} 2^{2n}\exp[-(2^{-n}/s(t))^{\lambda/2}], & (s(t) \ll 2^{-n}); \\[2mm] s(t)^{-2}\exp[\frac{\lambda}{4\ln 2}(\ln^2 2^{-n}/s(t))], & (s(t) \gg 2^{-n}). \end{cases} \tag{35b}$$

In summary, the behavior of Model I closely resembles that of linear fragmentation with a small-size cutoff in the relative breakup rate.

5.2 Model II: Larger particle splits upon collision

For this model, let us take for the collision rate $K(x,y) = x^\lambda$, if $x \geq y$, and $K(x,y) = y^\lambda$ otherwise. This represents the appropriate adaptation of the product form for the collision rate in Model I, for the case where only the larger cluster splits. The rate equations for this process are,

$$\begin{aligned} \frac{dc_n}{dt} &= -2^{-\lambda n}c_n \sum_{j=n}^{\infty} c_j + 2\cdot 2^{-\lambda(n-1)}c_{n-1}\sum_{j=n-1}^{\infty} c_j, \\ &\equiv -\gamma^n c_n \sum_{j=n}^{\infty} c_j + 2\gamma^{n-1}c_{n-1}\sum_{j=n-1}^{\infty} c_j. \end{aligned} \tag{36}$$

In the long-time limit, one expects that c_n will decay to zero for any fixed n, while the total number of clusters increases indefinitely. Therefore the lower limits on both sums can be extended to 0, so that one arrives at rate equations very similar to Eq. (33) of Model I. Accordingly, let us introduce the rescaled time variable $\tau = \int^t dt'\, M_0(t')$ which linearizes the rate Eqs. (36) to the form

$$\frac{dc_n(\tau)}{d\tau} = -\gamma^n c_n(\tau) + 2\gamma^{n-1}c_{n-1}(\tau). \tag{37}$$

One can again appeal to the general results for linear fragmentation to infer that the behavior of Model II is qualitatively similar to that of Model I.

However, a more direct solution will be given, which also serves as a check of the scaling ansatz for Model II. Using elementary methods, the solution to Eq. (37) can be written as

$$c_n(\tau) = e^{-\gamma^n \tau}\left[a_n - 2\gamma^{n-1}\int_\tau^\infty d\tau'\, e^{\gamma^n \tau'} c_{n-1}(\tau')\right], \tag{38a}$$

where

$$a_n = c_n(0) + 2\gamma^{n-1} \int_0^\infty d\tau' \, e^{\gamma^n \tau'} c_{n-1}(\tau') \qquad (38b)$$

is positive. In this solution, one makes use of the boundary condition that all the cluster concentrations of finite mass are equal to zero at infinite time. Another reason for choosing the infinite limit for the integral is that the contribution of this integral becomes vanishingly small as $\tau \to \infty$, and this feature permits the asymptotic solution to be found. Since $c_n(\tau)$ is non-negative for any n, Eq. (38a) provides the bound,

$$c_n(\tau) \le a_n e^{-\gamma^n \tau}. \qquad (39a)$$

Substituting this bound for c_{n-1} in Eq. (37) and performing the integral, one deduces that $a_n e^{-\gamma^n \tau}$ also represents a lower bound for $c_n(\tau)$. Thus for $\lambda > 0$ and large τ,

$$c_n(\tau) \sim a_n e^{-\gamma^n \tau}. \qquad (39b)$$

The determination of the amplitude a_n cannot be performed by simply substituting Eq. (39b) into (35b), since Eq. (39b) is valid only for large τ. However, one can proceed by Laplace transforming Eqs. (37) and (38b), thus reducing them to linear recursion relations. Defining the Laplace transform $\tilde{c}_n(\sigma) = \int_0^\infty d\tau \, c_n(\tau) e^{-\sigma\tau}$, Eq. (37) becomes

$$\tilde{c}_n(\sigma) = \frac{c_n(0) + 2\gamma^{n-1}\tilde{c}_{n-1}(\sigma)}{\sigma + \gamma^n}, \qquad (40a)$$

while the recursion relation for a_n takes the form,

$$a_n = c_n(0) + 2 \cdot \gamma^{n-1}\tilde{c}_{n-1}(-\gamma^n). \qquad (40b)$$

Upon iterating Eq. (37a), the dominant contribution to $\tilde{c}_{n-1}(-\gamma^n)$, is proportional to $(2/\gamma)^{n-1}$ as $n \to \infty$, for $\lambda > 0$ and for a monomer-only initial condition. Consequently Eq. (40b) gives $a_n \propto 2^n$, and this then yields the asymptotic solution,

$$c_n(t) \sim 2^{2n} \exp[-2^{-\lambda n}\tau(t)], \qquad (\tau \to \infty). \qquad (41)$$

Up to an overall constant factor this asymptotic solution should be independent of the initial condition.

To express c_n explicitly as a function of the time, one first needs to determine the time dependence of τ. By invoking scaling, one deduces that $M_0(t) \sim 1/s(t)$. Consequently, $\tau = \int^t M_0(t')\,dt'$ can be identified with $s(t)^{-\lambda}$. As a result, $c_n(t)$ can be expressed in the scaling form in the long-time limit,

$$c_n(t) \sim 2^{2n} \exp[-(2^{-n}/s(t))^\lambda]. \qquad (42a)$$

In the scaling regime ($\lambda > 0$), Eq. (42a) can be recast in the scaling form

$$\phi(x) \sim \exp[-x^\lambda] \, /x^2, \qquad (x \to \infty), \qquad (42b)$$

where $x = 2^{-n}/s(t)$ and $s(t)$ is determined by Eq. (32). Since Eq. (42b) is obtained independent of the scaling assumption, the validity of the scaling ansatz for $\lambda > 0$ has thus been demonstrated.

5.3 Model III: Smaller particle splits upon collision

In close analogy with Model II, let us now take for the collision rate $K(x, y) = x^\lambda$, if $x \leq y$, and $K(x, y) = y^\lambda$ otherwise. The rate equations for this process are,

$$\frac{dc_n(t)}{dt} = -\gamma^n c_n \sum_{j=0}^{n} c_j + 2\gamma^{n-1} c_{n-1} \sum_{j=0}^{n-1} c_j. \tag{43}$$

To obtain the asymptotic behavior, one first notes that the rate equation for $c_0(t)$ may be easily solved to yield the long-time solution

$$c_0(t) \sim 1/t. \tag{44}$$

Let us now show that c_n has the same time dependence as c_0 for all n. First, define the new variables,

$$R_n(\tau) = c_n(t)/c_0(t),$$
$$\tau = \int^t dt' \, c_0(t') \sim \ln t. \tag{45}$$

In terms of these rescaled variables, Eq. (43) becomes,

$$\frac{dR_n(\tau)}{d\tau} = -\gamma^n [R_n^2(\tau) - A_n(\tau)R_n(\tau) - B_n(\tau)] \qquad n > 0 \tag{46}$$

where

$$A_n(\tau) = \gamma^{-n} - \sum_{j=0}^{n-1} R_j(\tau)$$
$$B_n(\tau) = 2\gamma^{-1} R_{n-1}(\tau) \sum_{j=0}^{n-1} R_j(\tau). \tag{47}$$

By noticing that $A_n(\tau)$ and $B_n(\tau)$ involve only the $R_j(\tau)'$s with $j < n$, it is not difficult to show inductively that the ratio $R_n(\tau)$ converges to a positive constant r_n, for any n, in the long-time limit. Thus one concludes that

$$c_n(t) \sim r_n/t. \tag{48}$$

From a detailed analysis of the recursion relations that determine r_n, one ultimately finds the asymptotic solution

$$c_n(t) \sim \begin{cases} 2^{n\lambda}/t, & \lambda \geq 1, \\ 2^{n(1+\lambda)/2}/t, & 1 > \lambda \geq -1, \\ 2^{n(1+\lambda)}/t, & -1 > \lambda. \end{cases} \tag{49}$$

A salient feature of this solution is that the cluster size distribution at fixed time has a power-law dependence of the mass. This gives rise to the anomalously large residue of large clusters which persists at long times, as illustrated in Fig. 2. Notice also that only the first line can be expressed as a function of the scaling variable $2^{-n}/s(t)$. Thus scaling evidently breaks down for $\lambda < 1$.

6. Summary and Discussion

A general treatment of linear and collision-induced fragmentation processes has been given within the framework of the rate equations. This is an approach of a mean-field character, as fluctuations in the spatial positions of the clusters, as well the variability in cluster shape are ignored. In the rate equations for linear fragmentation, the microscopic details of the breakup events are accounted for by two kernels: an overall breakup rate which that specifies the rate at which a fragment of a given mass breaks, and the relative breakup rate, which gives the size distribution of products from a single breakup event. Within the scaling regime, the homogeneity index λ of the overall breakup rate is the crucial parameter which characterizes the properties of the cluster size distribution. For linear fragmentation models, the asymptotic scaling solutions to the rate equations were obtained both in the large and small mass limits. The conditions for the existence of scaling solutions and also the basic features of the shattering transition were also discussed.

To gain some insight into the behavior that can occur when non-linear effects play a role in particle breakup, a class of collision-induced fragmentation models was considered. In order to treat models which exhibit the essential non-linear nature of collision-induced breakup, yet are simple enough to analyze in detail, three types of "splitting" models were introduced, in which a fragmenting cluster always breaks into two equal-sized pieces. In Model I, both colliding fragments split in two upon collision, while in Models II and III, only the larger or only the smaller of the two colliding fragment splits in two, respectively. For these models, scaling is valid for different ranges of λ. Model III is rather unique in that large clusters are "frozen out" of the fragmentation process at long time. This leads to a monotonic fragment size distribution, with a power law tail at large masses, while the distribution in Models I and II is non-monotonic and decays quasi-exponentially for large mass. Moreover, in Model III the concentration of fragments of any finite size decays as a power law in time, and in Models I and II it decays exponentially.

While the rate equations approach provides a useful and comprehensive account of a wide range of fragmentation phenomena, there are many interesting questions of potential experimental relevance that are worth addressing. Most importantly, it is necessary to develop approximations which can take inhomogeneities of various types into account. While the assumption of perfect mixing is probably appropriate for ball milling types of comminution processes, it is clearly inadequate in many geophysical situations, for example, where fragments remain in fixed spatial positions throughout the breakup process. Attempts to account for this type of spatial inhomogeneity have been attempted, but primarily at a qualitative level. It is also clear that the rate of fragmentation processes should generally depend on the cluster shape. Whether this shape dependence can be accounted for by averaging over all possible fragment shapes has yet to be understood. Finally, in many fragmentation processes there are various non-linear and non-local effects. When breakup is driven by high pressures, there can be a transfer of stress across many fragments. When fragments possess considerable kinetic energy, non-linear effects, induced by collisions may play a substantial role. The understanding of the kinetics in these types of systems should pose rich areas for new research.

Acknowledgements

This work has been supported in part by grant #DAAL03-89-K-0025 from the Army Research Office. I would like to gratefully acknowledge this financial support.

References

[1] E. W. Montroll and R. Simha, *J. Chem Phys.* **8**, 721 (1940)

[2] P. J. Blatz and A. V. Tobolsky, *J. Phys. C* **49**, 77 (1945); H. H. Jellinek and G. White, *J. Polymer. Sci.* **6**, 745 (1951); O. Saito, *J. Phys. Soc. Jpn.* **13**, 198 (1958)

[3] A. M. Basedow, K. H. Ebert, and H. J. Ederer, *Macromolecules* **11**, 774 (1978)

[4] R. M. Ziff and E. D. McGrady, *Macromolecules* **19**, 2513 (1986)

[5] R. Shinnar, *J. Fluid Mech.* **10**, 259 (1961)

[6] J. J. Gilvarry, *J. Appl. Phys.* **32**, 391 (1961)

[7] L. Austin, K. Shoji, V. Bhatio, K. Savage, and R. Klimpel, *Ind. Eng. Chem. Process Des. Dev.* **15**, 187 (1976)

[8] M. Sahimi and T. T. Tsotsis, *Phys. Rev. Lett.* **59**, 888 (1987)

[9] A. F. Filippov, *Theory Probab. Its Appl.* (Engl. Transl.) **4**, 275 (1961)

[10] T. W. Peterson, M. V. Scotto, and A. F. Sarofim, *Powder Tech.* **45**, 87 (1985)

[11] T. W. Peterson, *Aerosol Sci. Tech.* **5**, 93 (1986)

[12] A. R. Kerstein, preprint (1986)

[13] R. M. Ziff and E. D. McGrady, *J. Phys. A* **18**, 3027 (1985)

[14] R. M. Ziff and E. D. McGrady, *Macromolecules* **19**, 2513 (1986)

[15] E. D. McGrady and R. M. Ziff, *Phys. Rev. Lett.* **58**, 892 (1987)

[16] Z. Cheng and S. Redner, *Phys. Rev. Lett.* **60**, 2450 (1988)

[17] Z. Cheng and S. Redner, in preparation

[18] S. K. Friedlander and C. S. Wang, *J. Colloid Interface Sci.* **22**, 126 (1966)

[19] T. Vicsek and F. Family, *Phys. Rev. Lett.* **52**, 1669 (1984)

[20] P. G. D. van Dongen and M. H. Ernst, *Phys. Rev. Lett.* **54**, 1396 (1985)

[21] K. Kang, S. Redner, P. Meakin, and F. Leyvraz, *Phys. Rev. A* **33**, 1171 (1986)

[22] A. P. Siebesma, R. R. Tremblay, A. Erzan, and L. Pietronero, *Physica A* **156**, 613 (1989)

[23] See, *e.g.* , R. C. Srivastava, *J. Atmos. Sci.* **39**, 1317 (1982), for a particular limit of nonlinear fragmentation

[24] A. N. Kolmogorov, *Doklady Akad. Nauk SSSR* **31**, 99 (1941)

[25] See *e.g.* , J. Aitcheson, *The lognormal distribution* (Cambridge University Press, London, England, 1957), and references therein

[26] See *e.g.* , E. W. Montroll and M. F. Shlesinger, in *Nonequilibrium Phenomena II: From Stochastics to Hydrodynamics*, eds. J. L. Lebowitz and E. W. Montroll (North-Holland, The Netherlands, 1984)

II Diffusion-Limited Aggregation Model

In the past few years, there has been a large effort spent on the structure of Diffusion-Limited Aggregation (DLA) clusters. This model and extensions of it, such as the Dielectric Breakdown model, share some similarities with fracture phenomena. Since this subject is rather new, and exhibits a lot of complex features, it is worth explaining it in some details. The first lecture of this part, by Kertész, introduces the model, and some physical situations where it is encountered. In particular, the linear stability analysis of the local growth is detailed. A similar problem is encountered in the development of a single crack. This parallel is made explicit, and differences are underlined.

Feder *et al.*, in the second lecture, discuss various cases of fluid injection where DLA as well as other models (viscous fingering, invasion percolation, dendritic growth, and so on) are encountered. This sheds light on the connections between these models and their respective field of relevance. Some other situations are also mentioned, such as the dispersion front separating two miscible fluids. In each cases, an emphasis is put on the comparison between the experimental cases, and the numerical modelling of these cases.

The final lecture of this section, given by Van Damme *et al.*, is divided in two parts. The first one shows on an experimental example a continuous evolution from the previous cases of viscous fingering, and DLA type growth, toward a real fracture case. This example uses a non–Newtonian fluid injected in porous media experiments. The non–Newtonian character of the fluid makes the problem much more complex but also much richer. The analysis of the patterns which are formed in those circumstances reveals an impressive collection of different behaviors. The second part of the lecture contains a discussion of some properties of packing of crushed aggregates and reports an experimental analysis of self–similarity properties, using various tools.

Chapter 4

Theory and Simulation of Diffusion-Limited Growth

János Kertész[†]

Institute for Theoretical Physics, University of Cologne
Zülpicher. Str 77, D-5000 Köln 41, FRG

1. Introduction

Crack patterns show often apparent similarity to structures grown in diffusion-limited aggregation (DLA). The basis of this similarity is that the equations governing fracture can be considered in some cases as the vectorial versions of those describing DLA-type growth processes. During the last years experiments, computer simulations and theoretical investigations have led to a great deal of information about DLA and related (scalar) processes which may be of relevance for understanding crack patterns. The purpose of this lecture is to give a short introduction to these phenomena; for comprehensive treatment the reader is referred to extensive recent reviews.[1-5]

2. Interfacial growth: instability and patterns

Viscous fingering in a Hele-Shaw cell[1-3,5] is a demonstrative example of pattern formation via interfacial growth. The Hele-Shaw cell consists of two glass plates of width L separated by the distance $b \ll L$ and filled with oil (or some other viscous liquid). The driving force is the applied pressure by which air (as a liquid with low viscosity) is pressed into the cell from one of the edges which pushes the oil out (Fig. 1).

In the theoretical description the starting point is the linearized Navier-Stokes-equation of fluid motion:

$$\eta \nabla^2 \mathbf{v} = \nabla p \tag{1}$$

where η is the viscosity, $\mathbf{v}$ the velocity of the liquid and p the pressure. According to the geometry, the solution is assumed to have the form:

$$\mathbf{v} = \frac{3}{2} \left(1 - \frac{4z^2}{b^2} \right) \mathbf{v}_{\|} \tag{2}$$

[†] On leave from Institute for Technical Physics, H-1325 Budapest, Hungary

Disorder and Fracture, Edited by J. C. Charmet *et al.*
Plenum Press, New York, 1990

i.e., the problem can be considered as two-dimensional and the third direction, perpendicular to the plates, can be integrated out. Eq.(2) expresses the sticking boundary conditions on the plates ($\mathbf{v} = 0$) and the coefficient $3/2$ is given by the constraint of the average velocity. Substituting Eq.(2) into Eq.(1) and taking into account that b is small one gets an equation similar to *Darcy's law* of fluid motion in porous media:

$$\mathbf{v}_\| = -\frac{b^2}{12\eta}\nabla p. \tag{3}$$

Assuming incompressibility, we get the *Laplace equation*

$$\nabla^2 p = 0. \tag{4a}$$

The boundary condition at the liquid-gas interface is

$$p_\Gamma = p_{gas} - \gamma\kappa \tag{4b}$$

where the viscosity of the gas is neglected. γ is here the surface tension and κ is the curvature of the interface. The normal velocity of the motion of the interface is given by the conservation of fluid; according to Eq.(3) this leads to

$$v_n = -\frac{b^2}{12\eta}\mathbf{n}\nabla p. \tag{4c}$$

Introducing dimensionless variables the full set of equations is:

$$\nabla^2 u = 0, \tag{5a}$$

$$u_\Gamma = d_0\kappa, \tag{5b}$$

$$v_n = \mathbf{n}\nabla u, \tag{5c}$$

with d_0 being a constant of dimension of a length and κ is in this case the curvature parallel to the plates. Eq.(5) represents a moving boundary problem. The moving boundary Eq.(5c) introduces the nonlinearity which is essential in order to obtain pattern formation.

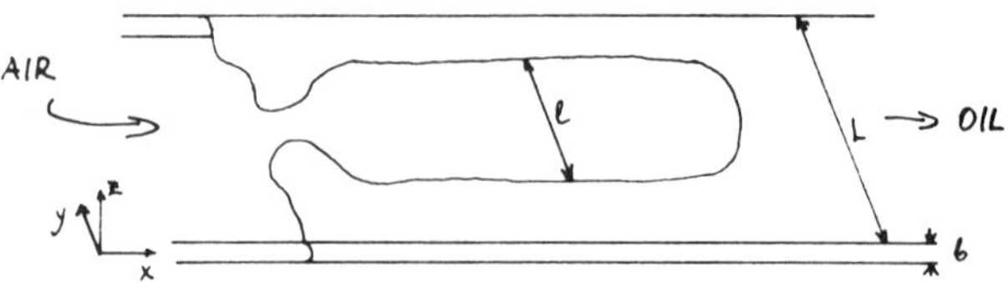

Figure 1   Schematic picture of a linear Hele-Shaw cell.

It seems that *gradient governed Laplacian growth* as formulated in Eq.(5) describes a broad class of pattern forming phenomena, at least in some approximation. (Sometimes the diffusion operator $(\partial_t - const\ \nabla^2)$ is more precise than the Laplacian ∇^2 in Eq.(5a), and the boundary conditions may also be more complicated.) In addition to viscous fingering, the following physical systems are worth mentioning:

• *Dendritic crystal* growth from the melt where the driving force is the undercooling. The equation corresponding to Eq.(5a) describes heat diffusion and the boundary conditions are given by the Gibbs-Thomson relation and the energy conservation law. Dendritic growth from oversaturated solutions can be treated in analogous terms.

• *Electrodeposition* of metal ions from solution. The Laplace equation is valid here for the electric potential and the driving force is the applied voltage. The boundary conditions are determined in part again by particle conservation and are more complex than in Eq.(5).

Figure 2 contains some characteristic patterns grown in Laplacian processes. We see that there are three types of patterns: *dense branching* (2 a, d, g), *open branching* (2 b, e, h) and *dendritic* (2 c, f, i) structures. The former two are characterized by splitting tips and the last one by stable propagating tips. It depends on the physical conditions which of the possible patterns will emerge and the *morphological transitions* between them constitute a current field of research.[6]

Why does Eq.(5) lead to pattern formation? Due to the nonlinearity introduced by the moving boundary condition Eq.(5c), the originally symmetric solutions become unstable. The qualitative explanation for the instability is that in gradient governed growth a protuberance due to fluctuations will feel larger gradient resulting in faster growth i.e. increased gradients and so on. This is the so-called *Mullins-Sekerka instability*. (Clearly, changing the sign of the driving force removes the instability.) The surface tension is a stabilizing factor; it introduces a lower cutoff *i.e.* perturbations of shorter wavelengths than this cutoff will day out.

The *linear stability analysis* makes these considerations more quantitative. Let us take a Hele-Shaw cell with the geometry of Fig. 1. The solution of Eq.(5) in the absence of perturbations is a front moving with constant velocity $\Gamma(y,t) = x_0 + t$ and a field $u_0 = x - \Gamma$. Now a small amplitude sinusoidal perturbation is put onto the interface:

$$\Gamma(y,t) = x_0 + t + \delta_k(t)e^{iky} \tag{6a}$$

and the field is expected to have the form $u = u_0 + u_k$ with

$$u_k = \epsilon_k(t)e^{iky - kx} \tag{6b}$$

Substituting this Ansatz into Eq.(5b) and using $\kappa \approx -k^2$ one gets

$$\epsilon_k = \delta_k(1 - d_0 k^2). \tag{7}$$

Using Eq.(5c) the following equation for δ follows:

$$\partial_t \delta_k(t) = |k|(1 - d_0 k^2)\delta_k(t) \tag{8}$$

with the solution $\delta_k(t) = \exp(\omega(k)t)$. $\omega > 0$ means exponential initial growth, *i.e.* instability. The surface tension (d_0) yields to stability at wavenumbers $k > 1/\sqrt{d_0}$.

3. Mode selection

A continuum of modes $0 < k < 1/\sqrt{d_0}$ turns out to be unstable within the framework of linear stability analysis. Which one will win? Is there a new, stable growing mode? These are highly nontrivial questions.[3]

The question of mode selection in the absence of surface tension is mathematically ill posed. Physically, the system is sensitive to any sort of perturbations and a stationary stable state cannot develop. It was of some surprise when it turned out that surface tension alone is not enough to repair the situation (see Ref.[3]). As an illustration viscous fingering in the radial Hele-Shaw cell (Fig. 3) is shown where the gas is introduced into the liquid through a central hole.

The pattern develops in the following way: First a symmetric, circular bubble grows. When the radius becomes large enough (as it can be calculated from linear stability analysis), bumps start growing and become fingers. These increase and

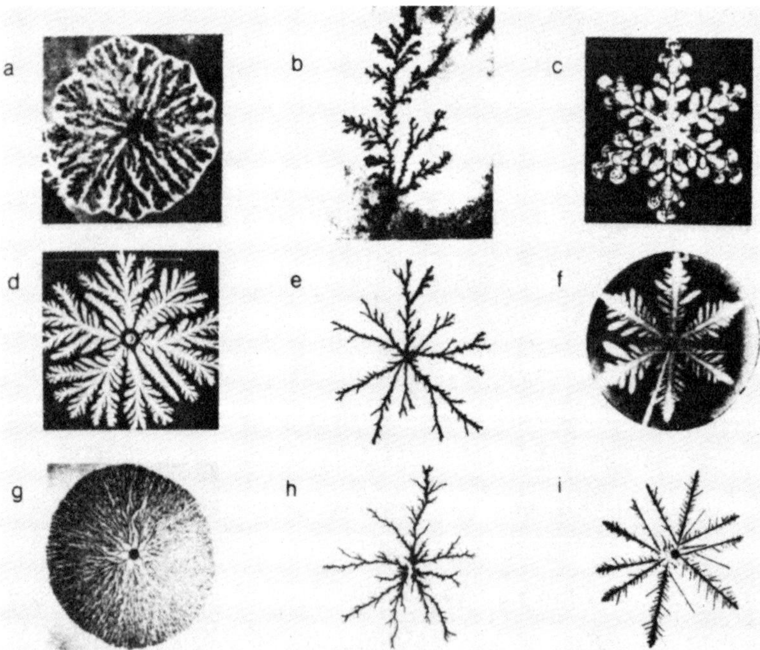

Figure 2 Pattern formation in crystallization (a:[7], b:[8], c:[9]), viscous fingering (d:[10], e:[11], f:[12]) and electrodeposition (g:[13], h:[14], i:[13]). This set of pictures is reproduced from Ref.[15].

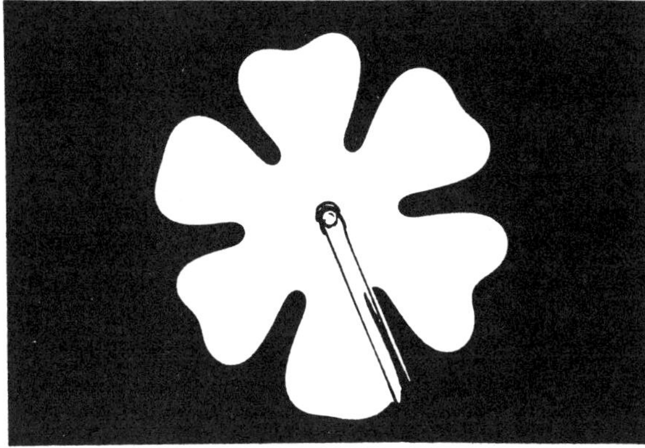

Figure 3 Viscous fingering in a radial Hele-Shaw cell (after Ref.[16]).

thicken until the radius of curvature at their tips will again be too large – then the tip of the finger splits. One of the new branches will eventually screen the other one. Thus, the pattern is essentially determined by the sequence of these tip splitting and screening instabilities.

The situation is different if in addition to the surface tension some kind of anisotropy is present in the system. In the linear Hele-Show cell a stationary mode emerges after some transient time (c.f. Fig. 1) which is stabilized by the edges of the cell. The ratio $\lambda = \ell/L$ depends on the surface tension and is constant in time for fixed driving force. Similarly, in crystal growth the propagating tip of a dendrite has a constant radius and velocity: The crystalline *anisotropy stabilizes the tip*.

The necessity of anisotropy in tip stabilization was shown mathematically by transforming the problem of mode selection to a nonlinear eigenvalue problem (see Ref.[3]). While there are no solutions for the shape of finger without anisotropy, a discrete set of solutions appears in the presence of anisotropy (in contrast to the continuum of unstable modes obtained in the linear stability analysis without anisotropy). Physically, the discrete eigenvalues correspond to situations where the shape of the tip becomes smooth. The finally selected mode is chosen from the possible solutions by stability criteria.

In the presence of tip splitting and screening the disordered open structure of Figure 2 (b, e, h) is expected. Under some circumstances a stable propagating front of branches can be observed (Fig. 2 a, d, g). By now it is not clear what are the selection principles between these two morphologies, but randomness definitely promotes the former one.

4. Diffusion-limited aggregation

So far we have not take into account noise explicitly (except for the initial perturbation in the stability analysis). However, noise must play an important role in the growth of patterns, especially in the case of small surface tension. The famous computer model of *diffusion-limited aggregation* (DLA)[17] is very suitable to investigate this limit.

Let us consider the two-dimensional square lattice version of Eq.(5a):

$$u(\mathbf{x}) = 1/4 \sum_{\mathbf{a}} u(\mathbf{x} + \mathbf{a}) \tag{9}$$

where the summation goes over the first lattice neighbors of $\mathbf{x}$. Equation (9) expresses also the conservation of probability if u is considered as the probability of finding a *random walker* at $\mathbf{x}$. This analogy is utilized in DLA which is defined as follows: The cluster or aggregate is initially a single particle at the origin of the lattice. Another particle is launched at "infinity" and undertakes random walk until it hits a site next to the aggregate where it sticks to the cluster. Then a new particle is started and the procedure is repeated, in principle ad infinitum. Here we do not want to go into details of programing the model.[1,4]

The sticking of the particle corresponds to the moving boundary condition in Eq.(5c): $v_n = 1/4 \sum_{\mathbf{a}} u(\mathbf{x} + \mathbf{a}) \approx \nabla u_\Gamma$ if $u_\Gamma = 0$ (zero surface tension) and $\mathbf{x}$ is a perimeter site of the cluster. Even in the absence of surface tension there is a lower cutoff in the system introduced by the lattice constant. Of course, the rules can be applied to other lattices or even without an underlying lattice — in the latter case the lower cutoff can be identified with the size of the diffusing particles.

A closely related computer model is the so-called *dielectric breakdown model* (DBM). Here the Laplace-equation is solved numerically on a lattice, the ∇u_i-s are determined (i is the index of a perimeter site). Then one of the perimeters is occupied with probability

$$p_i = |\nabla u_i| / \sum_{j} |\nabla u_j|, \tag{10}$$

according to Eq.(5c). Now the Laplacian field has to be calculated again according to the new boundary and the procedure is repeated many times until a large aggregate is grown. The boundary condition expressed in Eq.(10) is a special case in the "eta-model" where $p_i = |\nabla u_i|^\eta / \sum_j |\nabla u_j|^\eta$ is taken. Even for $\eta = 1$ there is a slight difference between the boundary conditions of DLA and DBM: while $u = 0$ is taken on the aggregate for DBM, $u = 0$ already on the perimeter for DLA.

Figure 4 shows a typical off-lattice DLA cluster. The above-mentioned sequence of tip-splitting and screening instabilities are clearly determining the shape of the DLA clusters. It is also apparent that the noise, which is an inherent part of the algorithms plays a crucial role in the growth. However, the patterns show some regularity: they are *statistically self-similar* and can therefore be described by *fractal geometry* (see Refs.[1,2]).

If the number of particles M is measured as a function of the distance R from

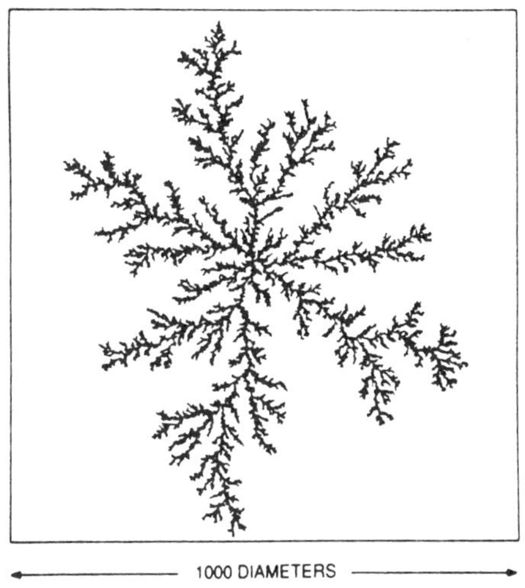

Figure 4 Off-lattice DLA cluster of size $M = 50000$.[18]

the origin, the following power law behavior is observed:

$$M \sim R^D \tag{11}$$

where D is called the fractal dimension of the aggregate. (The name dimension is clear if we take into account that for nonfractal objects D is just the Euclidean dimension.)

Equation (11) is a result of the statistical self-similarity of the aggregates. This means that the density-density correlation $c(R)$ function is scale invariant: $c(bR) = b^{-\zeta}c(r)$. This function equation has the solution $c(R) = r^{-\zeta}$ and by integration and comparison with Eq.(11) we get $\zeta = D - d$ where d is the Euclidean dimension of the space the aggregate is embedded in.

Much computer time has been devoted to determine D and to understand the structure of the DLA clusters. The most accurate value in two-dimensions is $D = 1.715 \pm 0.004$.[19] This is close to the values which were measured on patterns grown in systems discussed in the section 2 (see Refs.[8,11,14]). In higher dimensions the formula[20] $D = (d^2 + 1)/(d + 1)$ seems to be a good approximation.[19] (Unphysical dimensions $d > 3$ are of special theoretical interest.)

The single number D is not capable to characterize the full complexity of DLA. As it is apparent from visual observation, the origin of the cluster plays a special role in the cluster: It is this point from which the dilatation symmetry applies. This observation is reflected in the tangential correlation function which measures the decay of correlations at fixed distance from the origin as a function of an angle. In contrast to isotropic fractals (like the infinite percolation cluster at threshold) the radial correlations decay faster than the radial ones.[21] This is due to the special tree-like structure of the aggregate.[22]

Another interesting aspect of the structure of aggregates is related to the width ξ of the *active zone*, *i.e.*, the part of the cluster where particles arrive.[23] It is defined as the mean deviation of the growth probability distribution $p(\mathbf{r})$ and $\xi \sim M^{\nu'}$ where M is the size of the cluster. For a simple fractal one would expect $\nu' = 1/D$ which expresses that all lengths scale in the same way. At small to medium sizes this does not seem to be the case, although the exponent ν' turns out to depend slightly on the M and probably approaches $1/D$ from below in the limit of $M \to \infty$.[19]

The parts behind the active zone are hardly reached by the randomly walking particles. The reason is well known from classical electrostatics: it is the screening due to the elongated parts which makes the field small in the fjords. It turns out that a finer subdivison than active zone and dead parts of the cluster is of interest. Let us consider the set of sites with growth probability p. The strength of the screening is characterized by the exponent α defined by

$$p \sim R^{-\alpha}. \tag{12a}$$

The number of sites with the same p_i is denoted by $n(p)$ and is expected to behave like:

$$n(p) \sim R^{f(\alpha)}. \tag{12b}$$

The last equation expresses that the set of points with a given p (or α) is itself a fractal. The function $f(\alpha)$ is nontrivial: there is an infinity of fractal dimensions which are needed to characterize the system (Fig. 5).

This phenomenon is called *multifractality* (see Refs.[1,2]) and has already been discussed at this school.[25] Of course, the whole multifractal apparatus (Legendre

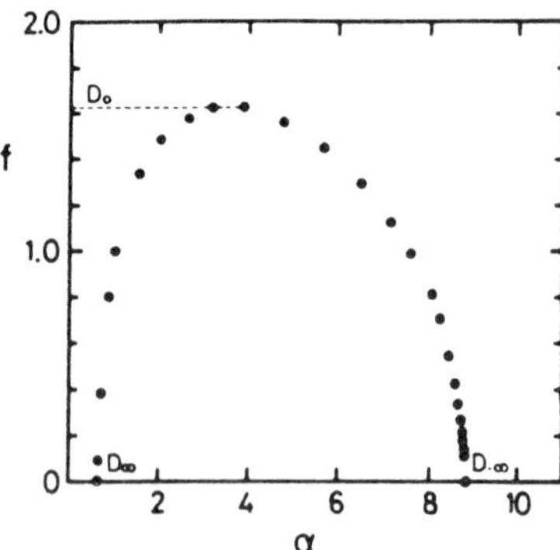

Figure 5 The multifractal exponents $f(\alpha)$ for two-dimensional DLA clusters.[24]

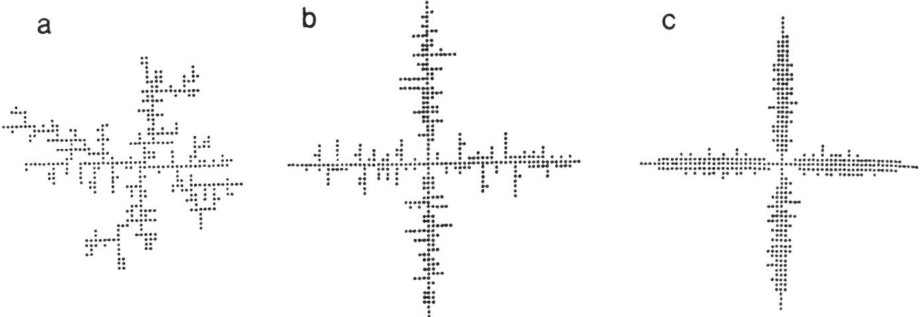

Figure 6 The effect of noise reduction on DLA clusters[30] with $M = 400$ and $m = 2$(a), $m = 20$(b) and $m = 400$(c).

transforms, momenta, D_q-s) can also be applied to DLA: $\sum p_i^q \sim R^{(q-1)D_q}$; D_q, α and f are interrelated by $(q - 1)D_q = q\alpha_q - f_q$ and $\alpha_q = \frac{d}{dq}\{(q - 1)D_q\}$ with $f(\alpha) = f_q(\alpha_q)$. In addition to the generally valid relations $(D = D_0 = f_0 = f_{max}, D_{\pm\infty} = \alpha_{\pm\infty}, D_1 = \alpha_1 = f_1)$ we have $2D_3 = D$[26] and in two dimensions $D_1 = 1$.[27]

Another interesting aspect of DLA is its reaction to external anisotropy which can be either a direction dependent sticking rule or just the underlying lattice structure. In the view of Section 3 it is not surprising that DLA is sensitive to anisotropy. In fact, anisotropic sticking rule changes entirely the geometry of the clusters: It leads to elongated cigar-shaped patterns which are not fractals.[28]

The situation is more subtle in the case of lattice anisotropy. Lattice DLA-clusters of small size are very similar to the off-lattice ones and effect of the underlying lattice becomes apparent only for larger sizes (*e.g.* for the square lattice $M \approx 10^5$ particles.[29] Why does the effect of the lattice show up so slowly? The reason is that the fluctuations due to the random character of the walks act against the manifestation of anisotropy. The actual shape of lattice-DLA clusters of a given size emerges from an interplay of between fluctuations and anisotropy.

The effect of this interplay can be investigated by the so-called noise reduction method which controls the fluctuations in DLA.[30,31] Instead of adding the random walker to the cluster when it hits a growth site, one keeps counting trajectories terminating at a given site. Only those sites are added to the clusters at which the counter reaches a prescribed value m. Large m corresponds to low level of noise, $m = 1$ to ordinary DLA.

For fixed number of particles an interesting morphological transition can be observed as a function of m (Fig. 6). Not too large noisy DLA clusters grown on the square lattice are open branching structures with unstable tips like off-lattice aggregates. If noise is reduced, the anisotropy due to the grid breaks through and the clusters become cross-shaped with stable tips reminiscent of dendritic crystals. For very large values of m even the sidebranches vanish and one recovers needles growing out of a center. The transition from tip-splitting to stable tip can also be seen in a multifractal analysis[32]: The localized propagating stable tips appear as single peaks in the $n(\ln p)$ spectrum and the corresponding $f(\alpha)$ becomes zero according to dimension of the set with maximal growth probablility.

What happens if the size M of the aggregate is increased for a fixed value of m? There is theoretical and numerical evidence[33] that the asymptotic shape will be the (fractal) dendrite (and not the needle). Ordinary lattice DLA clusters ($m = 1$) are expected to behave asymptotically like noise reduced ones: the lattice affects the fractal dimension ($D \approx 1.53$ for the square lattice[29]). This means that both anisotropy and noise are relevant parameters to the problem of asymptotic shape of DLA clusters.

5. Discussion: relation to fracture

The relevance of diffusion-limited aggregation to fracture is apparent if the equation of symmetric elasticity (Lamé equation) is considered[34] :

$$(\lambda + \mu)\nabla(\nabla \mathbf{u}) + \mu\nabla^2\mathbf{u} = 0 \tag{13}$$

which is a sort of generalized "vectorial Laplace equation." In Eq.(13) the displacement $\mathbf{u}$ is a vector field (in contrast to the scalar field of Eq.(5)), λ and μ are the Lamé coefficients. In the presence of a propagating crack we have also the moving boundary problem. Unfortunately, in contrast to Eq.(5b), the equation of motion of the interface is obtained by phenomenological conditions: Crack propagation is not simply gradient governed.

In spite of the differences to diffusion-limited growth many analogue questions arise and some methods can be transplanted. A real bridge between these two areas is built by viscoelastic media in which the crossover from fingering to cracking can be observed.[35] The problem of mode selection in crack propagation in such a medium turned out to be simpler than the corresponding diffusion-limited case.[36] However, this is certainly not the whole story and we are probably just starting to make use of the analogies.

In computer simulations mainly the dielectric breakdown approach is applied. Instead of the Laplace-equation, Eq.(12) (or its simplified version) is solved numerically on a lattice. Than the p_i-s are calculated according to the applied criterion. Here p_i means the probability of breaking a bond instead of adding a particle. Than one of the bonds is broken with the calculated probability and the procedure is strarted again with the new boundary. (This approach is suitable even for the study of deterministic crack patterns[37]). Recently much effort has been devoted to the investigations of such models which are reviewed in other lectures of this school.[25,34]

I would like to finish with the remark that this brief survey is by no means intended to be complete neither with respect to the content nor to the references. The relation between laplacian growth and fracture is discussed in more details in Ref.[38].

Acknowledgements

Collaboration with T. Vicsek and H.J. Herrmann helped my understanding of the presented material in many ways. Thanks are due to the Humboldt Foundation for support.

References

[1] T. Vicsek, *Fractal Growth Phenomena*, (World Scientific, Singapore, 1989)

[2] J. Feder, *Fractals*, (Plenum, New York, 1988)

[3] D.A. Kessler, J. Koplik and H. Levine, Adv. Phys. **37**, 255, (1988)

[4] P. Meakin, *in Phase Transitions and Critical Phenomena*, ed. by C. Domb and J.L. Lebowitz (Academic, New York, 1988) Vol. 12, p. 335

[5] D. Bensimon, L.P. Kadanoff, S. Liang, B. Shraiman and L. Tang, Rev. Mod. Phys. **58**, 977, (1986)

[6] J. Kertész, *in Random fluctuation and pattern growth: Experiments and models*, ed. H.E. Stanley and N. Ostrowsky (Kluwer, Dordrecht, 1988) p. 42

[7] E. Ben-Jacob, Y. Godbey, N. Goldenfeld, J. Koplik, H. Levine, T. Mueller and L.M. Sander, Phys. Rev. Lett. **55**, 1315, (1985)

[8] G. Radnóczy, T. Vicsek, L.M. Sander and D. Grier, Phys. Rev. A **35**, 4012, (1987)

[9] W.A. Bentley and W.J. Humphreys, *Snow Crystals*, (Dover, New York, 1962)

[10] A. Buka, J. Kertész and T. Vicsek, Nature **343**, 424, (1986)

[11] G. Daccord, J. Nittmann and H.E. Stanley, Phys. Rev. Lett. **56**, 336, (1986)

[12] E. Ben-Jacob, G. Deutscher, P.Garik, N. Goldenfeld and Y. Lereah, Phys. Rev. Lett. **57**, 1903, (1986)

[13] Y. Sawada, A. Dougherty and J.P. Gollub, Phys. Rev. Lett. **56**, 1260, (1986)

[14] M. Matsushita, M. Sano, Y. Hayakawa, H. Honjo and Y. Sawada, Phys. Rev. Lett. **53**, 286, (1986)

[15] T. Vicsek and J. Kertész, Eurphys. News **19**, 24, (1988)

[16] L. Paterson, J. Fluid. Mech. **113**, 513, (1981)

[17] T.A. Witten and L.M. Sander, Phys. Rev. Lett. **47**, 1400, (1981)

[18] P. Meakin, J. Colloid Interface Sci. **105**, 240, (1985)

[19] S. Tolman and P. Meakin, Phys. Rev. A **40**, 428, (1989)

[20] M. Muthukumar, Phys. Rev. Lett. **50**, 839, (1983)

[21] P. Meakin and T. Vicsek, Phys. Rev. A **32**, 685, (1985)

[22] B. Mandelbrot and T. Vicsek, J. Phys. A **22**, L377, (1989)

[23] M. Plischke and Z. Rácz, Phys. Rev. Lett. **95**, 415, (1984)

[24] Y. Hayakawa, S. Sato and M. Matsushita, Phys. Rev. A **36**, 1963, (1987)

[25] S. Roux, this volume

[26] T. Halsey, Phys. Rev. Lett. **59**, 2067, (1987)

[27] N.G. Makarov, Proc. London. Math. **51**, 1119, (1985)

[28] R.C. Ball, R.M. Brady, G. Rossi and B.R. Thompson, Phys. Rev. Lett. **55**, 1406, (1985)
[29] P. Meakin, R.C. Ball, P. Ramanlal and. L. Sander, Phys. Rev. A **35**, 5233, (1987)
[30] J. Kertész and T. Vicsek, J. Phys. A **19**, L257, (1986)
[31] J.Nittmann and H.E. Stanley, Nature **321**, 663 (1986)
[32] C. Amitrano, L. de Arcangelis, A. Coniglio and J. Kertész, J. Phys. A **21**, L15, (1987)
[33] J.P. Eckmann, P. Meakin, I. Procaccia and R. Zeitak, Phys. Rev. A **39**, 3185, (1989)
[34] H.J. Herrmann, this volume
[35] H. Van Damme, this volume
[36] M. Barber, J. Donley and J.S. Langer, Phys. Rev. A **40**, 366, (1989)
[37] H.J. Herrmann, J. Kertész and L. de Arcangelis, Europhys. Lett. **10**, 147, (1989)
[38] J. Kertész, *in Statistical models for the fracture in disordered media*, H.J. Herrmann and S. Roux, eds., North Holland, Amsterdam, (1990)

Chapter 5

Growth Patterns and Fronts:
Fluid Flow Experiments

Jens Feder, Finn Boger, Liv Furuberg,
Einar Hinrichsen, Torstein Jøssang,
Knut Jørgen Måløy, and Unni Oxaal

Department of Physics, University of Oslo
Box 1048 Blindern, 0316 Oslo 3, Norway

1. Introduction

Patterns and fronts arise in most fluid flow situations. Waves, clouds, convection patterns and turbulence are well known examples. In porous media the displacement of one fluid by another fluid leads to many new, often fractal,[1,2] fronts and patterns. The disorder of the porous matrix plays a key role that is not well understood. Depending on the displacement rates, viscosity ratios, miscibility, interfacial tensions and pore geometry a bewildering variety of displacement fronts arise. Lenormand[3] has studied many of the regimes observed under various conditions during two-fluid displacement processes in micromodels of porous media.

Here we discuss simple experiments that lead to patterns and fronts where disorder is an important factor. We first discuss the flow in a Hele-Shaw channel[4] where a single viscous finger is formed when air displaces glycerol. The flow in a Hele-Shaw cell has often been used to model flow in porous media since the (simplified) equations are the same for the two problems. The shape and stability of the finger has been extensively discussed.[5] We find that small particles in the fluid lead to instabilities in the flow. We also find new hydrodynamic instabilities in the fluid remaining after breakthrough, which show interesting patterns.

The growth of the invading fluid generates spatial structures that are regular patterns in ordered porous media, see Fig.(2) and fractal structures in disordered porous media, see Fig.(6). At high displacement rates the front structure is now well understood and is best characterized as a *fractal* fingering structure. Chen and Wilkinson[6] and Måløy et al.[7] demonstrated that viscous fingering (VF) in a random porous medium at high capillary numbers, $Ca >> 10^{-4}$, generates structures with a fractal[1,2] geometry. This fractal structure closely resembles that obtained from the diffusion limited aggregation (DLA) model of Witten and Sander.[8,9] Sim-

Disorder and Fracture, Edited by J. C. Charmet *et al.*
Plenum Press, New York, 1990

ilar structures have also been obtained by fluid-fluid displacement in radial Hele-Shaw cells using non-Newtonian viscous fluids.[10,11] The relationship between fluid-fluid displacement in porous media and DLA was first discussed by Paterson.[12] The original DLA model of Witten and Sander specifies a time ordered sequence of events but is not explicitly time dependent. While the time dependent aspects of DLA and closely related processes have been extensively explored theoretically[13,14] and by means of computer simulations,[15,16] experimentally this aspect of DLA has been explored only recently.[7,17] The dynamics of the viscous fingering structures have also been discussed in terms of multifractal measures.[17,18] Recently a crossover function has been shown to describe the structure of DLA and viscous fingers, which are not self-similar.[19]

Fractal structures are also observed for the *invasion percolation*[20] process at very low displacement rates. The *dynamics* of the processes that arise at low displacement rates has been much less studied than static properties. We have shown[21] that the growth of simulated planar invasion percolation clusters exhibits a novel dynamic scaling. This dynamical scaling structure has so far not been analyzed in a quantitative fashion for experimental structures.

In the last part we discuss the movement of tracers in a fluid flowing through the two-dimensional porous model. In this case there is *no* hydrodynamic instability in the fluid flow. However, we found that tracer *dispersion*, which maps the fluid flow through the porous medium, moves in such a way that equal concentration contours are self-affine fractal curves.[1,22,23] Apart from the flow of a single component fluid, the dispersion of tracers is the simplest flow problem in porous media and the motion and spreading of the front is well understood in terms of the diffusion-convection equation. Our observation of a fractal front structure was surprising and it remains a theoretical challenge to explain these results.

2. Experimental

The fluid-fluid displacement experiments described here have been done in several geometries. We discuss the two-dimensional case and the fronts and patterns that arise by displacement processes.

The Hele-Shaw cells were constructed by using glass plates 15 mm thick, 40 cm long and 8 cm wide. The top plate had a 5 mm hole 10 mm from the end where fluids could be injected. As spacers we used monofilament nylon thread 0.5 mm in diameter. The nylon threads were stretched with a separation W and clamped between the two glass plates, and the end of the channel near the injection hole was then closed. Experiments were done by first filling the model with glycerol and then injecting air at constant pressure.

We constructed porous models with a controlled pore geometry using a method developed by Bonnet and Lenormand:[24] thereby obtaining an epoxy resin plate where the pores and connecting channels formed a recessed pattern.[25]

Disordered models consisted of a single random layer of 1 mm glass beads glued onto a thin support. Both types of models were closed by sandwiching the model between a stiff glass disc, and a thin plastic sheet held in contact with the spheres by compressed air. This assembly was supported from above and below by 10 mm thick glass discs, and clamped in an aluminum ring. Fluids were injected at the center and left the model at the rim. The models had a porosity $\phi \simeq 0.7$. In a typical experiment air injected at the center displaced glycerol filling the pore space of the model.

In a typical experiment we used a black glycerol-water-Negrosine solution, having the viscosity $\mu \simeq 700 - 50$ cP, and density $\rho = 1.21 \pm 0.03$ g/cm^3 at 25°C. To inject air or low viscosity fluids at a constant rate we used a piston type pump. In experiments were air was injected at a constant pressure we used a pressure control system described in Ref.[26].

For quantitative analysis the pictures of the experiments were digitized with a resolution of 4000 × 4000, using a Data Copy scanning digitizing camera interfaced to an Apollo DN3000 workstation. The resolution used corresponds to 0.2 mm per pixel. In the digitizing process we selected a logarithmic response curve so that the resulting gray levels (0–255) are roughly proportional to the concentration of Negrosine in the volume imaged onto a pixel.

3. Hele-Shaw Cell Experiments

A Hele-Shaw cell[4] consists of two transparent plates separated a distance b. The equation for the flow velocity U, derived from the Navier-Stokes equations governing flow in the Hele-Shaw cell is[27]

$$U = -\frac{k}{\mu}\nabla(p + \rho g z) = -M\nabla\phi, \tag{1}$$

where p is the pressure, ρ the density and g the component of the acceleration of gravity along the z-coordinate of the cell. The mobility is $M = k/\mu$ and the flow potential is $\phi = (p + \rho g z)$. For a cell placed in the horizontal position we have $g = 0$. The viscosity of the fluid is μ and the *permeability* of the Hele-Shaw cell is

$$k = \frac{b^2}{12}, \tag{2}$$

Note that the velocity in Eq.(1) is the *average* velocity over the thickness of the cell. For incompressible fluids the equation of continuity gives

$$\nabla \cdot U = -\frac{k}{\mu}\nabla^2(p + \rho g z) = M\nabla^2\phi = 0. \tag{3}$$

This is the Laplace equation and it is characteristic of potential problems encountered in electrostatics, in diffusion problems and in many other fields; consequently we call flows controlled by Eq.(3) potential flows.

We will discuss the situation where a fluid (index 1) displaces another fluid (index 2). The interface between the two fluids is controlled by capillary forces when the fluids are at rest and there is a pressure difference between the two fluids

$$(p_1 - p_2) = \sigma \left(\frac{1}{R_x} + \frac{1}{R_y} \right). \tag{4}$$

Here σ is the interfacial tension between the two fluids. The two principal radii of curvature, R_x and R_y, describe the interface locally. The radius of curvature R_y is controlled by the contact angle, θ, describing how the two-fluid interface contacts the plates that define the cell geometry. Typically one finds that $R_y \sim b/2$, and $R_x \gg R_y$. We have that $p_1 > p_2$ when the fluids are at rest and (2) is the wetting fluid.

Now let us inject the fluid (1) at a constant rate U at $z = -\infty$ and withdraw fluid (2) at the same rate at $z = \infty$. The interface between the two fluids then moves with a velocity $U = (0,0,U)$ along the z-axis. However, the interface is *unstable* if the viscosity of the driving fluid is smaller than the viscosity of the fluid being driven. Flow in porous media also follows equations (1) and (3), and therefore the flow in Hele-Shaw cells is often used to model the flow in porous media. However, there are important differences and the validity of using the Hele-Shaw cell as a model of flow in porous media is questionable.

The theory of viscous fingering was developed and compared to experiments independently by Saffman and Taylor[27] and by Chuoke et al.[28] Recently there has been a growing interest in the field and many new theoretical and experimental results have been published.[29−33]

When gravity effects can be ignored, *i.e.* for a horizontal cell, linear stability analysis[27,28] gives the result that the front is *stable* if the driving fluid has the higher viscosity and *unstable* if the driving fluid has the lower viscosity. For $\mu_1 < \mu_2$, the front is unstable with respect to perturbations that have a wavelength λ that is longer than a *critical wavelength* λ_c and have the largest growth rate for a wavelength λ_m given by:

$$\lambda_m = \sqrt{3}\lambda_c = \pi b \sqrt{\frac{\sigma}{U\mu}} = \frac{\pi b}{\sqrt{Ca}}, \tag{5}$$

in the limit of negligible viscosity of the driving fluid ($\mu_1 \ll \mu_2$), as is the case when glycerol is displaced by air. Here we have introduced the dimensionless *capillary number Ca* defined by

$$Ca = \frac{U\mu}{\sigma}, \tag{6}$$

which measures the ratio of viscous to capillary forces.

After the initial linear instability, nonlinear growth usually selects a single dominating finger as shown in top of Fig.(1). As the capillary number increases the width of the finger approaches $0.5W$. In the figure one sees that the width of the finger decreases along the channel as expected in experiments performed at constant pressure. Why the width $0.5W$ is selected has been a difficult theoretical problem.[5,33] Also as the capillary number becomes very large one finds that the single viscous finger becomes unstable with respect to *tip-splitting*.[10,27,34,35]

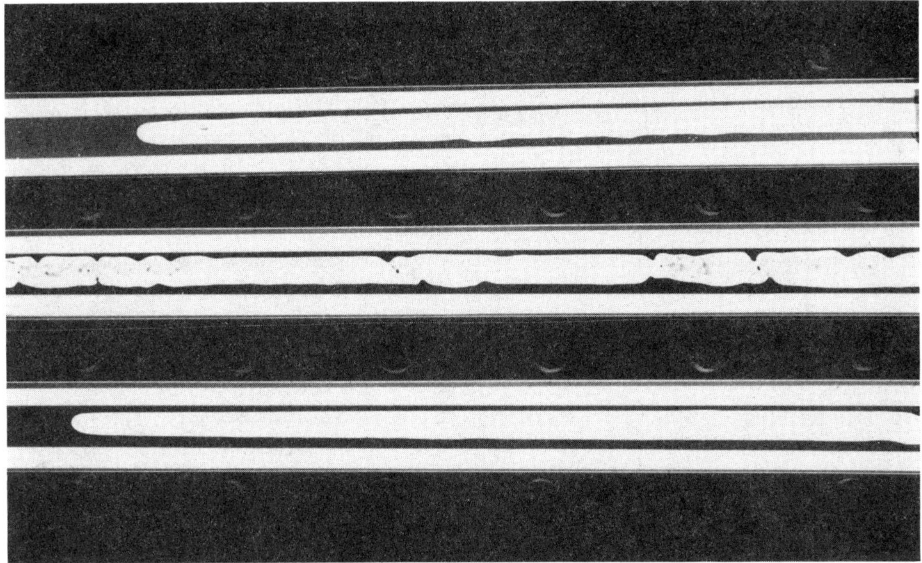

Figure 1 Air displacing glycerol (black) in a Hele-Shaw channel. The channel dimensions are $b = 0.5$ mm, $W = 1.5$ cm, and $L = 40$ cm. The air is injected at a constant pressure, and the capillary number is $Ca \simeq 0.02$. (top) – The single finger selected by nonlinear growth from the initial (linear) instability. (middle) – Instabilities in the flow of the fluid left after the finger has reached the end of the cell. (bottom) – The viscous finger is perturbed by the presence of 0.1 mm latex spheres suspended in the glycerol.

We wish to point out that many interesting patterns and fronts arise in the simple geometry discussed here. For instance, if the experiment is continued beyond the point of breakthrough, one finds that the air (injected at constant pressure) will flow more rapidly and cause *multiphase flow* as illustrated in the middle of Fig.(1). The fluid remaining at the sides of the channel is being dragged along by the air stream and the patterns observed are formed by the Kelvin-Helmholtz instability of the interface. The nonlinear growth of this instability then leads to deviations from the sinusoidal pertubations of the two fluid layers, which finally grow into tips that break up into droplets that also are dragged along.

The effect of disorder, may be seen in the bottom of Fig.(1), where small latex particles in the fluid perturb the fluid and the meniscus separating the air from the glycerol. The stochastic perturbations are clearly difficult to treat theoretically and these results are presented only to show that even the simple Hele-Shaw geometry allows the study of patterns and nonlinear behavior, and we hope that progress in the understanding of nonlinear phenomena will permit the analysis of experiments of the type illustrated here.

4. Flow in Micromodels

The displacement of glycerol by air in regular micromodels at different rates is illustrated in Figs.(1) and (2). The model consists of a regular lattice of cylinders blocking the flow. There is a striking difference in the patterns generated at low and high capillary numbers. At low Ca we find compact growth of the invaded region, with a pattern of trapped menisci when the displaced fluid wets the model. In the non-wetting case a little fluid is trapped in pores. At high displacement rates snow-flake like dendritic patterns are generated clearly showing the effect of the quadratic symmetry of the lattice. In the non-wetting case the branches cover several pores and the general structure is dominated by the four main branches.

We expect a circular front at very low displacement rates and dendritic structures with quadratic symmetry at high displacement rates. However, imperfections in the model due to small variations in pore size and in the wetability (contact angle) of the matrix lead to deviations from the idealized structures. This is indeed an indication that *disorder* is a relevant perturbation that changes the growth pattern in a significant way.

The effect of perturbing the model by adding small glass particles so that there on the average is one per pore, is shown in Fig.(4). The glass particles are trapped in the pores, some have one, some none and others several. Therefore the micromodel is no longer a regular lattice model but a random porous medium with an overall quadratic symmetry. The flow in such models thus exhibits the competition between lattice symmetry and disorder. The fingering structure in Fig.(4) is similar to the fractal viscous fingering structures observed in the random porous models.

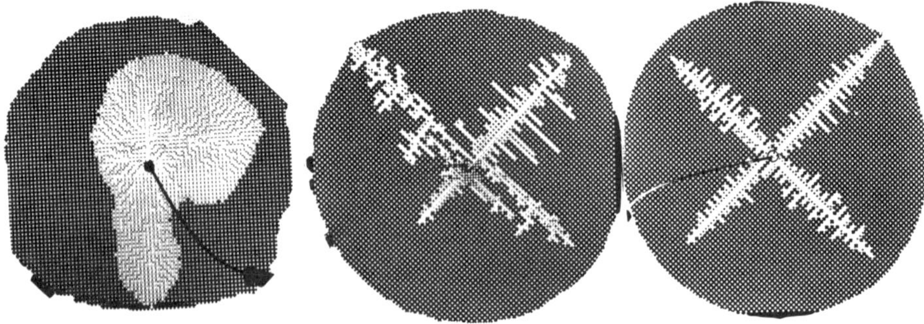

Figure 2 Displacement of glycerol (black) by air in lattice micromodel. Glycerol wets the model. (a) $Ca = 0.0004$; (b) $Ca = 0.04$; (c) $Ca = 0.1$.

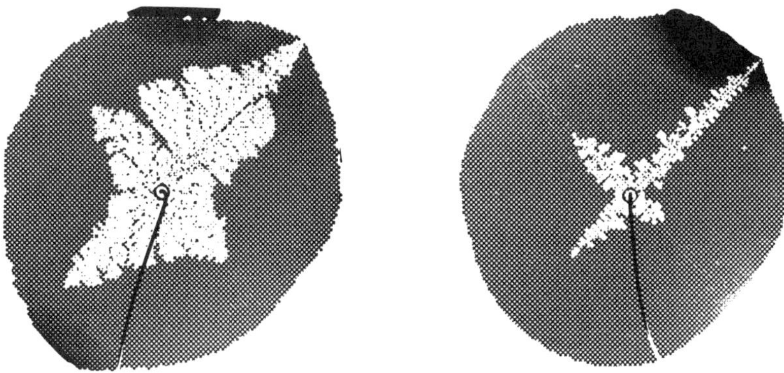

Figure 3 Displacement of glycerol (black) by air in lattice micromodel. Glycerol does not wet the model. (a) $Ca = 0.04$; (b) $Ca = 0.1$.

In Fig.(5) we see the result of displacing glycerol by air at high rates on a percolation cluster *i.e.* when the porous model is a self-similar fractal structure.[25] In this case the flow is confined to the backbone of the percolation cluster. The geometry of the backbone requires the flow to go through single bonds (the red bonds) and therefore the flow is geometry dominated, and the disorder of the model dominates the growing pattern both for high and low Ca. The fractal dimension of the growing front is $D \simeq 1.3$ for high Ca, and $D \simeq 1.5$ at low Ca. Numerical simulations[25] describe the experimental results very well.

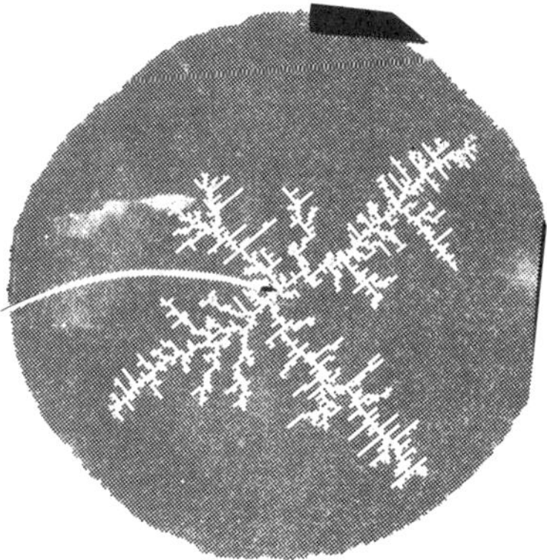

Figure 4 Displacement of glycerol (black) by air in lattice micromodel. Glycerol wets the model. Glass beads at a concentration of 1 per pore introduces disorder in the model. $Ca \simeq 0.1$.

5. Flow in Disordered Models

The lattice symmetry of the micromodels is absent in the disordered models consisting of a random layer of spheres. In these models the pore sizes fluctuate. Glycerol, which wets the model, was displaced by air or oils. At low displacement rates we find that the flow can be described by *invasion percolation* (see Fig.6a). At high displacement rates we find *fractal viscous fingering*[7] (see Fig.6b).

5.1 Fractal Viscous Fingering

When a high viscosity fluid is displaced at high capillary numbers by a low viscosity fluid in a porous medium fractal viscous fingering (VF) results. We analyzed fingering structures by digitizing pictures such as the one shown in Fig.(6b). The number $N(r)$ of black pixels was measured as a function of distance, r, from a point near the center of injection. Fractal structures exhibit scaling and we expect $N(r)$ to have the form

$$N(r) \sim N_0(r/R_g)^D. \tag{7}$$

Here D is the fractal dimension of the structure and R_g is the its radius of gyration. N_0 is the total number of black pixels, which correspond to porespace from which

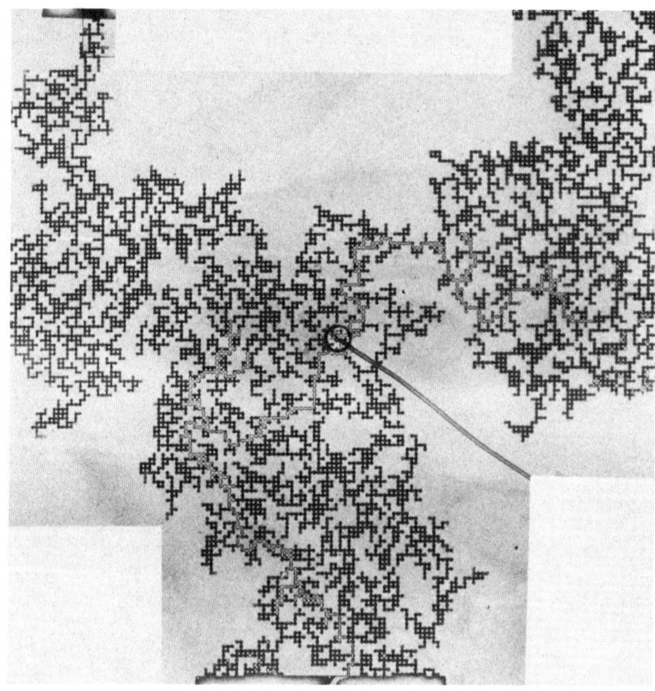

Figure 5 Displacement of glycerol (black) by air (white) at high capillary numbers in a micromodel representing a realization of a percolation cluster.

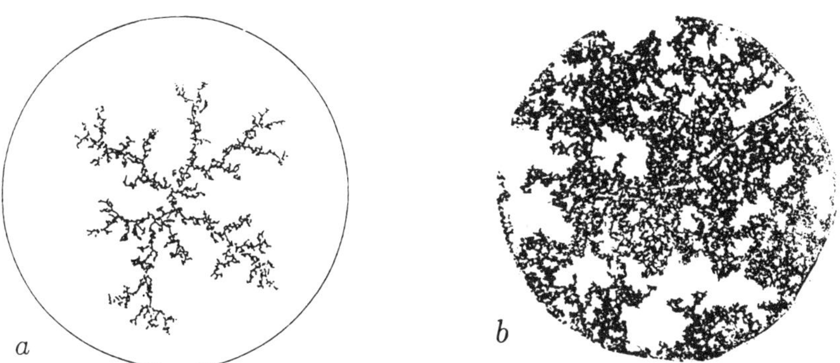

Figure 6 Displacement of glycerol (black) by air in a two-dimensional model consisting of a random layer of 1 mm spheres. (a) Invasion percolation at low displacement rates $D \simeq 1.8$; (b) Fractal viscous fingering at high displacement rates, $D = 1.64$.

glycerol has been displaced. The scaling power-law Eq.(7) is expected only for the range $a < r < R_0$, where a is a typical pore dimension, and R_0 is the radius of the model.

In Fig.(7) we plot $\log\left[N(r)/N_0\right]$ as function of $\log(r/R_g)$ for the structure shown in Fig.(6b). By fitting Eq.(7) in the range $2a < r < R_g$, we find a fractal dimension $D = 1.64 \pm 0.04$.

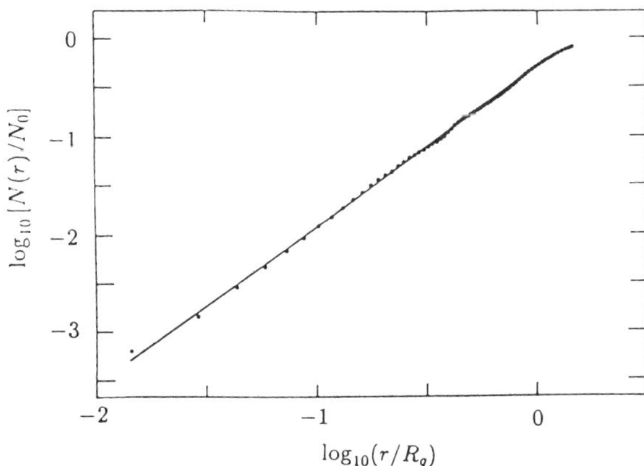

Figure 7 The normalized finger structure area $N(r)/N_0$ as function of the reduced radius r/R_g. The solid line obtained from a fit to the experimental data has a slope of 1.64

5.2 Fractal Growth Measures

The structure and the dynamics of irreversible growth processes that generate fractal structures depends on the local growth probabilities, which are fractal measures.[36,1] The use of fractal measures as a tool to analyze experimental results were strongly stimulated by the recent work on convection[37] which allowed a compact representation of the dynamics by a spectrum of fractal dimensions — a so called $f(\alpha)$ curve. For a recent discussion of multifractals see Ref.[38]. We have discussed the *observed growth measure*[17] for viscous fingering using these ideas.

The growing finger structure develops in the active zone. In our experimental observations of the active zone, we measure the "mass" m_i of the growth "islands" shown in Fig.(8). Index $i = 1, 2, \ldots, N_I$, where N_I is the number of sites at which we observe growth.

The normalized mass distribution $\mu_i = m_i/m_0$, where m_0 is the total mass of the growth zone, characterizes the growth dynamics. The set of observed normalized island masses $\mathcal{M} = \{\mu_i\}$ is a *measure* on the fractal finger structure.

The number of points in the set of root points of the growth is N_I. We find that N_I increases with the size of the viscous fingering structure. For a fractal structure we expect N_I to be given by:

$$N_I = a \left(\frac{R_g}{\delta} \right)^{D_I} . \tag{8}$$

Here D_I is the fractal dimension of the growing interface, δ is the pixel size at which

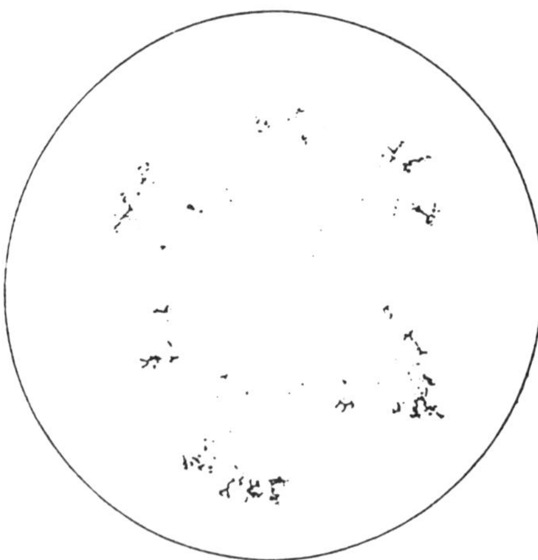

Figure 8 Active growth zone of a viscous fingering structure shown in Fig.6b. The time between the two pictures used to construct the growth zone was 2.6s, which corresponds to a relative time increment of $\Delta t/t_0 = 0.1$.

the structure is analyzed, and R_g is the radius of gyration.

We have counted N_I, for three sequences of pictures of our experiments find that the set of points is fractal with a fractal dimension

$$D_I = 1.0 \pm 0.1, \tag{9}$$

and with an amplitude $a = 1.1 \pm 0.5$. Therefore the set of growth sites is a one-dimensional fractal "dust". This result is consistent with DLA simulations[39]

So far we have only discussed the set of points at which $\mu_i > 0$. We may specify a *subset* consisting of all the growth sites for which $\mu \leq \mu_i \leq \mu + \Delta\mu$. If we specify the measure in a scale independent way we may find that such subsets are fractal sets. We therefore choose to specify subsets of the growth sites by the scaling relation

$$\mu = \left(\frac{\delta}{R_g}\right)^{\alpha}. \tag{10}$$

From Eq.(10) we find the range of μ_i corresponding to a range $\Delta\alpha$. The set of growth sites that have islands that give μ_i in the specified range form a set of points $\mathcal{N}_\alpha$ that may be a fractal set just as $\mathcal{N}_I$

$$N_\alpha(R_g, \delta) \sim \left(\frac{R_g}{\delta}\right)^{f(\alpha)}. \tag{11}$$

Where $f(\alpha)$ is the fractal dimension of the subset $\mathcal{N}_\alpha$. We have analyzed the experimental results for VF and find an $f(\alpha)$ curve in agreement with expectations. However, the accuracy is low and experiments in much larger models should be done.

5.3 Structure of DLA and Viscous Fingering

Using a modification of the DLA algorithm we found quantitative agreement with experimental results on the growth dynamics of viscous fingering.[26] However, the DLA clusters are not strictly self-similar, and the fractal dimensions of the aggregates are not well defined. We have shown[19] that it is possible to characterize DLA and VF structures (such as the one seen in Fig.6b) using ideas developed to describe river systems.[40] We assign branch *orders n* as follows: Each branch defines a continuous line, starting at a tip, and ending on another branch of lower order if it is not the "trunk" (zeroth order branch). More than two branches may meet at a single point. The highest order branches are those which have no side branches. The next to highest order branches have side branches of the highest order, and so on. We find a fixed branching ratio r_N and length ratio and r_L for the DLA- and VF-structures. The number N_n and length L_n of branches of order n are given by

$$L_n = L_m r_L^{n-m} \quad \text{and} \quad N_n - N_m r_N^{n-m}, \tag{12}$$

Table 1 Fractal dimensions for DLA and VF. † Estimated by scaling different cluster sizes onto the same curve. ‡ Box-counting only points with $r < R_g$.

		DLA	VF
		branch order ratios	
$r_N = N_n/N_{n-1}$	number	5.2 ± 0.2	4.8 ± 0.5
$r_L = L_n/L_{n-1}$	length	0.35 ± 0.01	0.34 ± 0.04
		dimensions	
$D_s = \frac{\ln(r_N)}{\ln(1/r_L)}$	self-sim.	1.6 ± 0.1	1.5 ± 0.1
$N(\delta) \sim \delta^{-D_b}$	box	1.62 ± 0.02†	1.51 ± 0.06
$N(\delta) \sim \delta^{-D_b}$	box‡	1.67 ± 0.03	$- - -$
$M(R_g) \sim R_g^{D_g}$	growth	1.710 ± 0.005	$- - -$
$M(r) \sim r^{D_c}$	cluster	1.69 ± 0.01†	1.62 ± 0.05†

where and $m = 1$ is a lower cut-off for the validity of the scaling relations. We have shown that the similarity-dimension is given by $D_s = \ln(r_N)/\ln(1/r_L)$ for these tree structures. The results obtained for DLA and VF are summarized in Table 1.

For the DLA clusters, we have analyzed each branch order separately in terms of the box counting algorithm. The number $N_n(\delta)$ of filled boxes of order n as function of the box size δ has the scaling form

$$N_n(\delta) = M \cdot \delta^{-D} g(\delta/L_n). \tag{13}$$

The mass M is the total cluster mass. The branches of DLA are linear on length scales less than the average branch length, and fractally distributed on larger scales. This feature is characterized by the scaling function $g(\delta/L_n)$, valid for all but the largest and smallest branches. The crossover function $g(x)$ is constant for $\delta > L_n$ and tends to x^{D-1} for $\delta < L_n$. We have also shown that the lengths L_n all scale with R_g.

5.4 Invasion Percolation

At very low displacement rates the invading fluid (air) displaces the defending fluid (glycerol) by moving through the widest pore-neck at the perimeter of the invaded region. Of course, regions that are trapped cannot be invaded since the glycerol is incompressible. This process has been modeled by a simple computer algorithm, where on a lattice random numbers are assigned to bonds, and the invader advances through the bond with the highest random number. Interestingly this "greedy" algorithm where one always enters the "best" pore results in the trapping of domains of all sizes. The invaded region is fractal so that the mass

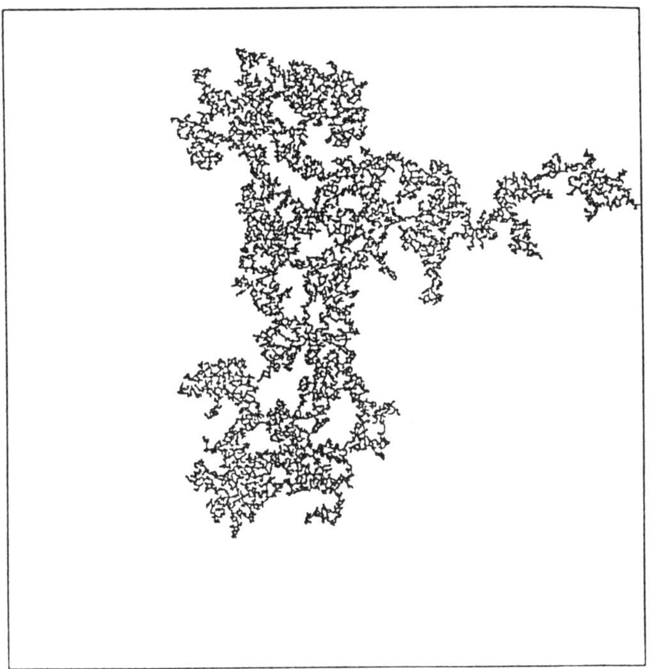

Figure 9 Invasion percolation with trapping on a 500×500 lattice. The invader (black) enters from the center and the defender (white) escapes through the square rim. At break-through the invader reaches the edge and has invaded 20368 sites.

$M(L)$ increases with the size L of the invaded region:

$$M(L) \sim L^D, \tag{14}$$

where the fractal dimension of invasion percolation with trapping is $D \simeq 1.82$. The structure of percolation and invasion percolation processes are relatively well understood (for a brief review see Ref.[2]).

The *dynamics* of the invasion percolation process has not been studied experimentally. However, we have recently shown that this process also has an interesting dynamic structure, which describes burstlike growth of the front.[21] Qualitatively, this can be understood: The invading front exhausts all easily accessible pores, then by forcing the invader through a difficult pore a new region, which may have new easily invaded areas is made available to the front. The front then moves into this area until it gets stuck again having exhausted the new easily invaded pores.

To obtain a *quantitative* measurement of these time correlations, we considered the pair correlation function $N(r,t)$, giving the probability that a site a distance r from a reference site is invaded a time t later than the reference site. Thus $N(r,t)drdt$ is the conditional probability that if a site at the position $\mathbf{r}_0$ was invaded

 Jens Feder *et al.*

at time t_0, another site at a distance between r and $r+dr$ away from $\mathbf{r}_0$ is invaded in an interval dt around t_0+t. During the simulations the reference site is successively chosen to be the last invaded site. After some initial transient effects, the function $N(r,t)$ is found to be independent of t_0. This is expected since the invasion process is governed by local rules, and the surroundings of every new invaded site are statistically similar to the surroundings of any other invaded site independent of how far the process has developed.

We find that the correlation function obeys the *dynamic scaling* form

$$N(r,t) = r^{-1} f(r^z/t), \tag{15}$$

with the new dynamic exponent

$$z = D. \tag{16}$$

We also find that the scaling function $f(u)$ is peaked at $u \sim 1$, and has an unusual limiting power-law behavior at both limits,

$$f(u) \sim \begin{cases} u^a, & u \ll 1 \\ u^{-b}, & u \gg 1 \end{cases} \quad \text{with} \quad a \simeq 1.4,\ b \simeq 0.6. \tag{17}$$

This implies that the most probable growth occurs at $r_t \sim t^{1/D}$. At time t, most of the region within distance r_t has already been invaded, the rest of the region contains trapped defender fluid, and new growth there is rare. On the other hand, the probability to select a new growth site at $r \gg r_t$ decays rapidly with r. Roux and Guyon[41] have recently shown that $b = 1/D$ consistent with our results.

In Fig.(10) we show the correlation function $N(r,t)$ obtained in extensive simulations. First we note that for $t = 1$ growth is most probably close to the previously invaded site, and that the growth decreases as a power law r^{-2} as r increases. For $t > 1$, we find that after a time t has passed, growth is most probable at some distance r_t away. We find that $r_t \sim t^{1/z}$ with $z \simeq 1.8$. These results suggest the scaling form given in Eq.(15), which leads to the satisfying data collapse shown in Fig.(10) with $z = 1.82$.

Experiments are in progress to measure the dynamics of the invasion percolation process in two-dimensional disordered models.

6. Fractal Dispersion in Porous Models

When a tracer is added to a fluid flowing in a porous medium it disperses because of molecular diffusion and convection. We recently discovered that the dispersion front of miscible displacement has a fractal structure.[42] As shown in Fig.(12) our experiments on dispersion in miscible flow at high Péclet numbers are *on the average* accurately described by convective diffusion equation.[43,44] The *new feature* presented here appears when we analyze the dispersion front itself: Contours of constant tracer concentration are *self-affine fractal curves*[1,2,23] confined by the width of the dispersion front.

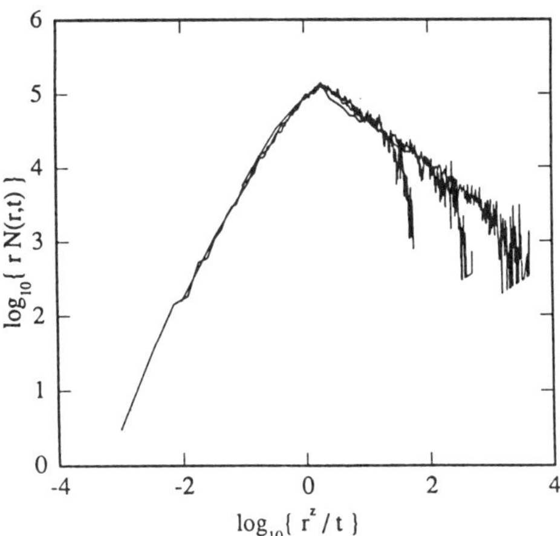

Figure 10 Log-log plot of the scaling function $f(u) = r \cdot N(r,t)$ as a function of $u = r^z/t$ and with $z = 1.82$ The correlation function $N(r,t)$ is a function of the separation r between sites invaded with a time separation t, and the figures show results for time separations $t = 1$, 10, 100 and 1000. The data in the figure were generated in 8 simulations of the invasion percolation process on a lattice of size 300×600. Finite size effects result in a violation of the data collapse for $r \sim L$.

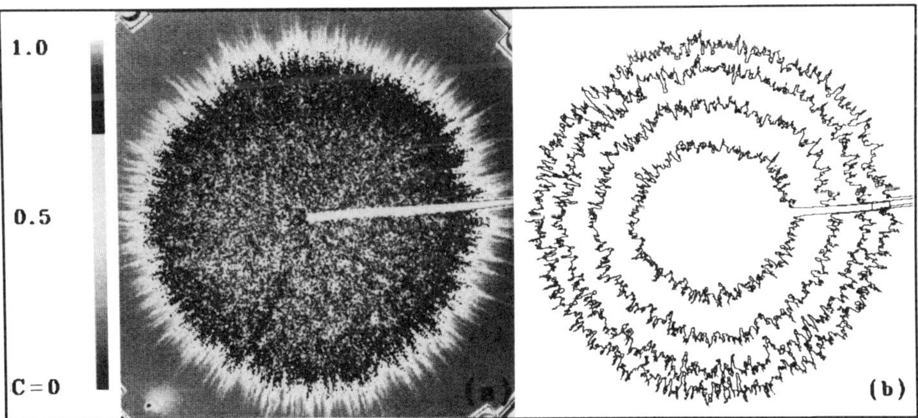

Figure 11 The dispersion obtained by injecting glycerol colored by Negrosine into glycerol of the same viscosity and density in a two-dimensional porous model. (a) The logarithm of the transmitted light intensity is coded with a grayscale shown in the insert to emphasize the front. For a color version of this figure see Ref.[42]. Time $t = 160$ s. (b) The time development of the fractal contour where the concentration is $1/2$ the injected tracer concentration. The four contours correspond to increasing times: $t = 36$ s; $t = 77$ s; $t = 122$ s; $t = 160$ s.

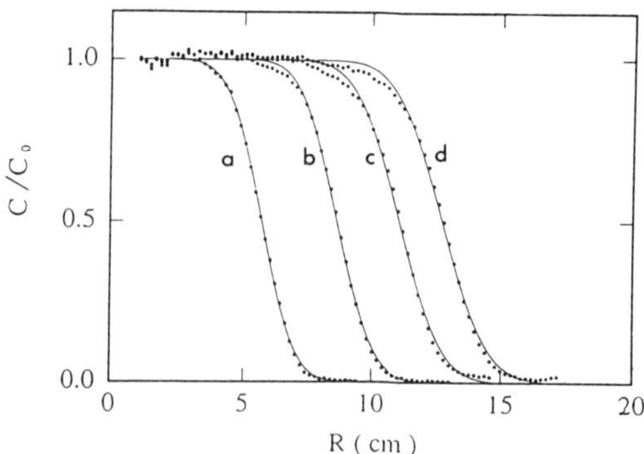

Figure 12 The average concentration $C(r,t)/C_0$ as a function of r for different times. The fitted functions $C/C_0 = \frac{1}{2}(1 - \text{erf}\,[(r - R_0)/\lambda])$ are shown as solid lines in the figure. R_0 and λ are the position and width of the front respectively. (a): $t = 36$ s; (b): $t = 77$ s; (c): $t = 122$ s; (d): $t = 160$ s.

An example of a digitized picture is shown in Fig.(11). The figure illustrates the strong fluctuations of the fractal dispersion front caused by geometrical dispersion, which generates concentration "spikes" in the radial direction.

The structure of the dispersion front was studied in terms of the fractal dimension of the contour of the interface where the local concentration is $C \in$

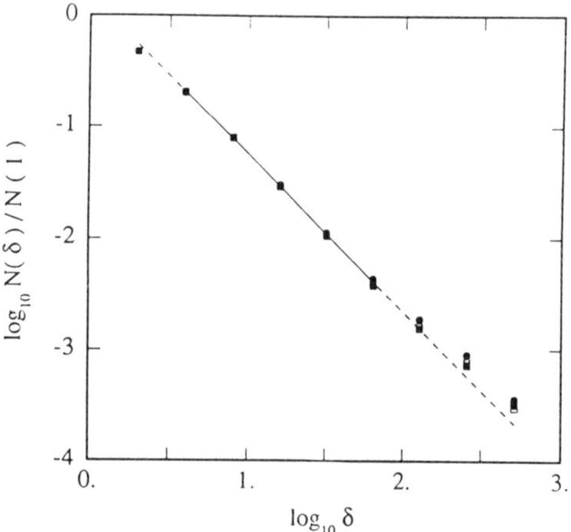

Figure 13 The number of "boxes" $N(\delta)$ needed to cover the contours where the tracer concentration equals $\frac{1}{2}$ the injected concentration as a function of the box size δ given in pixels. $N(1)$ is the total number of pixels in the contour. The fractal dimension is $D = 1.42 \pm 0.05$ is estimated from the fit in the region drawn with a full curve.

[0.500, 0.504]. The fractal coastlines for four time stages are shown in Fig.(11b). Any practical fractal structure will have a lower and an upper length scale cut-off. In the present case the particle dimension $d = 1$ mm, which corresponds to five pixels, is the lower cut-off. The upper cut-off is given by the width λ of the dispersion front. In Fig.(13) we show the result of using the "box-counting" method to estimate the fractal dimension of the front. The number of boxes $N(\delta)$ needed to cover the front with a box size δ (in pixels) is $N(\delta) \sim \delta^{-D}$. We find that the dispersion front is fractal with $D = 1.42 \pm 0.05$.

In Fig.(13) we see that for $\delta > \lambda$ there is a crossover to a lower fractal dimension. This is to be expected since the front on a scale $\delta \gg \lambda$ is a circle (with $D = 1$) and therefore is a self-affine[1,2,23] curve for which one must distinguish between *local* and *global* fractal dimensions.

Discussion of Fractal Dispersion

We have shown that the detailed structure of a dispersion front in a two dimensional porous medium is fractal and that the contour of the concentration where $C = 0.5$ has a fractal dimension of $D = 1.42 \pm 0.05$. It is tempting to relate the fractal contours we observe to the fractal diffusion front discovered by Sapoval et al.[45]. The diffusion of particles from a source is controlled by the diffusion equation *without* the convection term. The front of the region of diffusers *connected* to the source is a fractal curve corresponding to the hull of a percolation cluster and a fractal dimension of $D = 1.75$. However, the diffusion front is defined in terms of the *connectivity* between diffusing particles whereas the fractal structure of the dispersion front reflects the probability distribution of where diffusing tracers end after time t. This probability distribution is the result of the combined diffusion and random convection in the porous medium. The physics of the two fronts is therefore different. The dispersion front contains different physics and we know of no model of fractal physics onto which the dispersion front may be directly related. We therefore suggest that dispersion contours in fact belong to a universality class that is distinct from that of the diffusion front.

Why is the dispersion front fractal? Since the dispersion on the average is well described by the convective diffusion equation[43,44] we know that the front is characterized by only two length scales that both depend on time: the average position of the front, and the width of the dispersion front. This leads to the conclusion that dispersion has no intrinsic length scale independent of time in the range between the pore size and the width of the front. It is by now well established both from the study of phase transitions and from the applications of fractals that for problems which have no intrinsic length scale we expect *scaling* behavior, and that the geometry of the system is *fractal*. Unfortunately, we can say little in general about the expected fractal dimension for the dispersion process. A proper understanding of the fractal structure of the dispersion front remains a challenge for future theories.

Acknowledgements

Many interesting discussions with Amnon Aharony and Paul Meakin have been very helpful. This work has been supported by VISTA, a research cooperation between the Norwegian Academy of Science and Letters and Den norske stats oljeselskap *a.s.* (STATOIL), and by NAVF.

References

[1] B.B. Mandelbrot, *The Fractal Geometry of Nature*, W.H. Freeman, New York, (1982).

[2] J. Feder, *Fractals*, Plenum, New York, (1988).

[3] R. Lenormand, J. Fluid Mech., **189**, 165, (1988).

[4] H.S. Hele-Shaw, Nature, **58**, 34, (1988).

[5] D. Bensimon, L.P. Kadanoff, S. Liang, B. Shraiman and C. Tang, Rev. Mod. Phys., **58**, 977, (1986).

[6] J.D. Chen and D. Wilkinson, Phys. Rev. Lett., **55**, 1892, (1985).

[7] K.J. Måløy, J. Feder and T. Jøssang, Phys. Rev. Lett., **55**, 2688, (1985).

[8] T.A. Witten and L.M. Sander, Phys. Rev. Lett., **47**, 1400, (1981).

[9] T. Vicsek, *Fractal Growth Phenomena*, World Scientific, Singapore, (1989).

[10] J. Nittmann, G. Daccord and H.E. Stanley, Nature, **314**, 141, (1985).

[11] H. VanDamme, F. Obrecht, P. Levitz, L. Gatineau and C. Laroche, Nature, **320**, 731, (1986).

[12] L. Paterson, Phys. Rev. Lett., **52**, 1621, (1984).

[13] J.M. Deutch and P. Meakin, J. Chem. Phys., **78**, 2093, (1983).

[14] H.G.E. Hentschel, J.M. Deutch and P. Meakin, J. Chem. Phys., **81**, 2496, (1984).

[15] P. Meakin and J.M. Deutch, J. Chem. Phys., **80**, 2115, (1984).

[16] R.F. Voss, J. Phys. A, **17**, L373, (1984).

[17] K.J. Måløy, F. Boger, J. Feder and T. Jøssang, in *Time-Dependent Effects in Disordered Materials*, R. Pynn and T. Riste eds., Plenum, New York, 111, (1987).

[18] P. Meakin, *Fractal aggregates and their fractal measures* in *Phase Transitions and Critical Phenomena*, C. Domb and J.L. Lebowitz eds., Academic press, New York, 336, (1987).

[19] E.L. Hinrichsen, K.J. Måløy, J. Feder and T. Jøssang, J. Phys. A: Math. Gen., **22**, L271, (1989).

[20] D. Wilkinson and J.F. Willemsen, J. Phys. A, **16**, 3365, (1983).

[21] L. Furubeg, J. Feder, A. Aharony and T. Jøssang, Phys. Rev. Lett., **61**, 2117, (1988).

[22] B.B. Mandelbrot, Phys. Scr., **32**, 257, (1985).

[23] B.B. Mandelbrot in *Fractals in Physics*, L. Pietronero and E. Toatti eds., North-Holland, Amsterdam, 3, (1986).

[24] J. Bonnet and R. Lenormand, Rev. Inst. Fr. Pet., **42**, 477, (1977).

[25] U. Oxaal, M. Murat, F. Boger, A. Aharony, J. Feder and T. Jøssang, Nature, **329**, 32, (1987).

[26] K.J. Måløy, F. Boger, J. Feder, T. Jøssang and P. Meakin, Phys. Rev. A, **36**, 318, (1987).

[27] P.G. Saffman and G.I. Taylor, Proc. R. Soc. Lond., **245**, 312, (1958).

[28] R.L. Chuoke, P. van Meurs and C. van der Poel, Trans. Metall. Soc. of AIME, **216**, 188, (1959).

[29] L. Paterson, J. Fluid Mech., **113**, 513, (1981).

[30] G. Tryggvason and H. Aref, J. Fluid Mech., **136**, 1, (1983).

[31] J.V. Maher, Phys. Rev. Lett., **54**, 1498, (1985).

[32] D.A. Kessler and H. Levine, Phys. Rev. A, **32**, 1930, (1985).

[33] M.H. Jensen, A. Libchaber, P. Pelcé and G. Zocchi, Phys. Rev. A, **35**, 2221, (1987).

[34] J. Nittmann and H.E. Stanley, Nature, **321**, 663, (1986).

[35] G. Daccord, J. Nittmann and H.E. Stanley in *On Growth and Form*, H.E. Stanley and N. Ostrowsky eds., Martinus Nijhoff, Dordrecht, 203, (1986).

[36] B.B. Mandelbrot, J. Fluid Mech., **62**, 331, (1974).

[37] M.H. Jensen, L.P. Kadanoff, A. Libchaber, I. Procaccia and J. Stavans, Phys. Rev. Lett., **55**, 2798, (1985).

[38] B.B. Mandelbrot in *Fluctuations and Pattern Formation*, H.E. Stanley and N. Ostrowsky eds., Kluwer, Boston, 279, (1988).

[39] P. Meakin and T.A. Witten, Phys. Rev. A, **28**, 2985, (1983).

[40] R.E. Horton, Geol. Soc. Am. Bull., **56**, 275, (1945).

[41] S. Roux and E. Guyon, J. Phys. A, to appear.

[42] K.J. Måløy, J. Feder, F. Boger and T. Jøssang, Phys. Rev. Lett., **61**, 2925, (1988).

[43] A.L. Dullien, *Fluid Transport and Pore Structure*, Academic Press, New York, (1979).

[44] E. Guyon, J. Nadal and Y. Pomeau eds, *Disorder and Mixing*, Kluwer Academic Publishers, Dordrecht, Nato ASI Series E:152, (1988).

[45] B. Sapoval, M. Rosso and J.F. Gouyet, J. Phys. Lett., **46**, L149, (1985).

Chapter 6.1

From Flow to Fracture and Fragmentation in Colloidal Media

1: Non-Newtonian Fingering and Visco-Elastic Fracturing

H. Van Damme and E. Lemaire

C.N.R.S.-C.R.S.O.C.I.
F-45071 Orléans Cédex 02, France

1. Introduction

Colloidal materials can show themselves in a wide range of physical states of matter. When stabilized as dilute suspensions in simple liquids, they form weakly non-newtonian fluids. At higher concentrations, larger departure from newtonian behavior is observed and they often form viscoelastic, shear-thinning pastes. At still higher concentration they can form physical gels. Finally, when all the dispersion liquid is removed, by evaporation for instance, they may form coherent and fragile solids.

An ubiquitous sample family of this type of material in our environment is the family of clay minerals,[1,2] which we encounter as muds, soils, or sedimentary rocks. Mud is generally an unpleasant fluid, but it can also be used for medical purposes or as drilling fluid in petroleum engineering. Soil can be a fine-grained powdery solid, a collection of well-structured aggregates (clods), or a dangerous pseudo-plastic fluid which undergoes landslides. Sedimentary rocks start forming as soft solids on the bottom of oceans or river deltas and end up as compact solids where petroleum genesis takes place. Artificial sedimentary rocks can also be prepared by compaction for embedding nuclear wastes. Cement hydrates are artificial materials which have much in common with clays.[3]

Because they are coherent granular media, colloidal solids may exhibit a strong tendency to branching during crack propagation and applying a shear- or compression stress on a piece of material often leads to extensive fragmentation (*i.e.* to the generation of a large number of fragments).

On the other hand, because they are non-newtonian viscoelastic fluids, colloidal suspensions and pastes exhibit a wide range of mechanical behaviours, depending

Disorder and Fracture, Edited by J. C. Charmet *et al.*
Plenum Press, New York, 1990

on the class of flow in which they are involved and the time scale of the mechanical excitations.[4] Because the elementary particles they contain are of nanometric size (which makes them sensitive to brownian motion) and often of non-spherical shape, their rheology is very sensitive to surface physico-chemical interactions and their structure may be liquid crystal-like.

In simple shear flow, colloidal fluids often display shear-thinning behaviour with a yield stress, which is described by the Herschel-Bulkley equation[4]:

$$\tau = \tau_y + \beta \dot{\gamma}^n \tag{1}$$

$\tau = |\tau_{xy}|$, $\dot{\gamma} = |\dot{\gamma}_{xy}|$, τ_y is the yield stress and $n < 1$ is the shear-thinning exponent. β has dimensions Pa s^n. Thus, the non-newtonian viscosity, $\eta = \tau/\dot{\gamma}$, decreases as $\dot{\gamma}$ increases. For $n = 1$, Eq.(1) describes a Bingham fluid, whereas for $\tau_y = 0$ and $n < 1$, it describes a simple shear-thinning fluid.

In addition, colloidal fluids often display viscoelastic properties, with an elastic character which shows increasingly up as the characteristic time of the flow event, τ_{fl}, approaches (from above) the time taken by the microscopic reorganization processes in the fluid, τ_{str}. This can be rationalized using the concept of Deborah number:[4]

$$De = \frac{\tau_{str}}{\tau_{fl}} \tag{2}$$

For a characteristic experimental length l and a velocity U, $\tau_{fl} = l/U$. The structural relaxation time is less obvious and is related to the reciprocal frequencies of the colloidal particles motion in the fluid. De may be interpreted as the ratio of elastic to viscous forces. For $De \ll 1$, viscous effects dominate, whereas for $De \gg 1$, the system will essentially behave as an elastic solid.

Our purpose is twofold. In part 1 of this contribution, we will show on experimental grounds how one can go continuously from interfacial *flow* instabilities to a complex *fracture* process in colloidal fluids. Our approach is based on a simple experimental device — the Hele-Shaw cell (two transparent plates separated by a narrow gap) — and on a simple experimental process: the injection of a low viscosity fluid (air, water, or a light hydrocarbon) into a colloidal medium filling the cell. As shown in figures 1 and 2, dramatically different intrusion patterns can be obtained, some of which are clearly reminiscent of flow patterns, whereas others are close to crack networks. Part 2 will be devoted to colloidal solids and we will analyse, also on experimental grounds, the relationship between the local order within these materials and their fragmentation morphology.

2. Experimental

The colloidal fluids that we used were all obtained by dispersion of commercial bentonite clays in water: either purified Wyoming bentonite from N.L. Industries, or the FTP3 grade from CECA S.A. At microscopic level, the bentonite particles

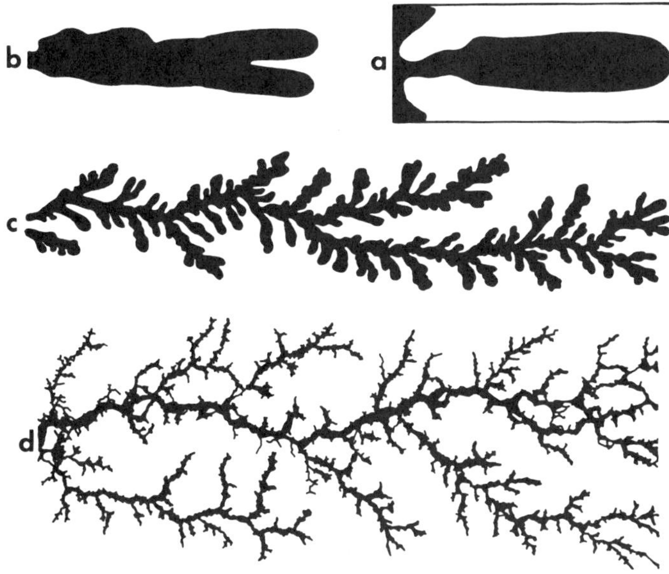

Figure 1 Patterns obtained in rectilinear Hele-Shaw cells, by injecting:
(a) air in glycerine (adapted from Ref.[17];
(b) water in oil (adapted from Ref. [31]);
(c) air in a relatively fluid water-based clay mud (from Ref.[21]);
(d) water at high injection rate in a strong water-based clay mud.

(in fact, bentonite is the commercial name for clays which contain essentially the mineral montmorillonite; more details on the classification of clays will be given in part 2) are extremely anisotropic single crystal sheets, with an aspect ratio of the order of 500. The thickness of the sheets is close to 1 nm, whereas their size in the a,b plane is of the order of 500 nm. In a dry montmorillonite powder, the sheets are strongly aggregated, whereas the ultimate dispersion state which can be reached in very diluted colloidal suspensions (typically less than 1% of clay in water) is that of individual sheets surrounded by water. In more concentrated suspensions and pastes, the sheets start aggregating and the aggregates form a connected and entangled network which generates a lenticular porosity, as shown in figure 3. As the solid/liquid ratio is increased, the average number of sheets within the aggregates increases, as well as the overal connectivity of the system.[5]

In our experiments, the range of solid to liquid ratio, S/L (weight by weight), went from 0.03 to 0.60. This corresponds to huge modifications of the rheological properties. For instance, going from $S/L = 0.03$ to $S/L = 0.10$, boosts the yield stress (measured in simple cylindrical shear flow) from 0.4 to 137 Pa and depresses the shear-thinning exponent from 0.73 to 0.42. Going from $S/L = 0.15$ to $S/L = 0.60$ increases the real part of the complex modulus, G' (the storage modulus), from 10^3 to 10^5 Pa, at a strain of 10^{-2} and an angular frequency of 50 rad s^{-1} (in dynamic cone-plate testing).

The experimental devices were centrosymmetric ("radial") or axisymmetric ("linear") Hele-Shaw cells.[6]

3. Viscous fingering and crack propagation compared

The intrusion under pressure of one phase into another is a process which can be observed both in fluids and in solids. "Viscous fingering" is the phenomenon one refers to when a high-viscosity fluid is displaced by a low viscosity fluid within a medium where flow is dominated by viscous friction, such as a porous medium or a Hele-Shaw cell. On the other hand, "hydraulic fracturing" is the process one refers to when a solid (generally a rock) is broken open by injection of a liquid at very high pressure.[7] Both processes may lead to simple or complex intrusion patterns which one refers to as "viscous fingers" in the former case or simply as "cracks" in the later case.

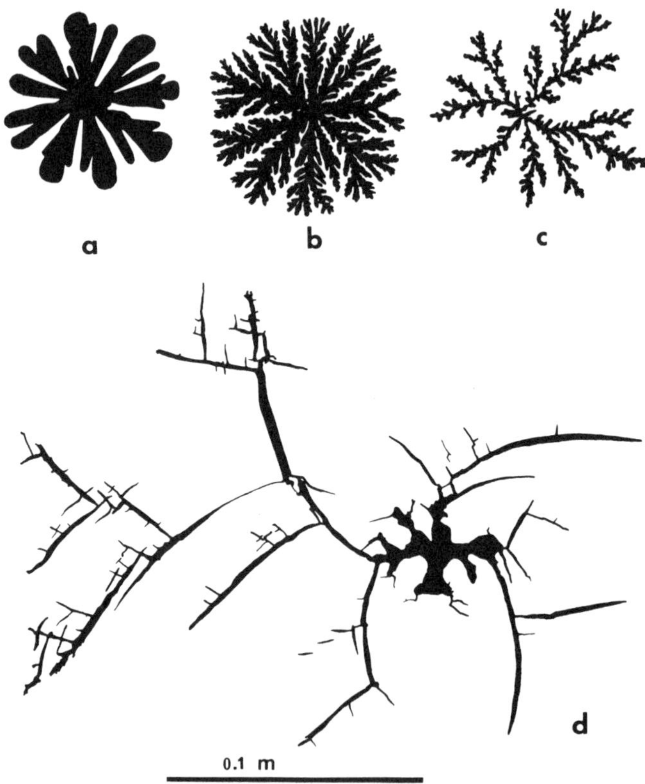

Figure 2 Patterns obtained in radial Hele-Shaw cells, by injecting:
(a) and (b) air into glycerine, at small and large injection pressure (from Refs. [10] and [33], respectively);
(c) and (d) water in a water-based clay mud, at relatively low (c) and large (d) clay concentration and injection rate.

Figure 3 Digitized transmission electron micrograph of a water-based montmorillonite mud (by courtesy of D. Tessier). Note the connectivity of the solid network due to face-to-face associations between particles, and the lenticular porosity. The pore walls are in fact aggregates of elementary lamellae, 1 nm thick, separated by layers of water molecules.

Although viscous fingering and hydraulic fracturing are both failure processes (of the more rigid medium under the action of the less rigid one), the mechanism by which a viscous finger grows in a newtonian viscous liquid has, *a priori*, little in common with the mechanism by which a crack grows in an elastic solid. Viscous fingering is an interfacial instability which stems from the destabilizing action of the viscous forces, whereas hydraulic fracturing is a structural instability which is driven by the release of elastic stress. However, as we shall see below, they display several analogies.

In the absence of net gravitational forces (horizontal cell), viscous fingering between immiscible newtonian fluids is controlled by the balance between viscous and capillary forces.[8] In each fluid, the flow velocity is governed by Darcy's equation and Laplace's equation:

$$U_i = -(b^2/12\mu_i)\nabla P \text{ and } \nabla U_i = \nabla^2 P = 0 \tag{3}$$

U_i is the velocity in fluid i averaged over the thickness of the cell, b; μ_i is the fluid viscosity. Equation 3 implies that any bump generated by fluctuations of the interface increases the pressure gradient in front of it, thereby accelerating further the more viscous fluid and increasing the size of the bump. Capillary forces enter in the game as stabilizing forces. The pressure jump across the interface between fluid 1 pushing fluid 2 is given by:

$$P_1 - P_2 = \Gamma_{12}K \tag{4}$$

where Γ_{12} is the interfacial tension and K the local curvature of the interface.

As a result, each bump (positive K) produces a pressure drop in front of it which decreases the gradient induced by the viscous forces. In terms of dimensionless group, the stability of the interface is controlled by two dimensionless numbers, the capillary number, Ca, and the viscosity contrast, A:

$$Ca = U\mu/\Gamma_{12} \qquad A = \frac{\mu_2 - \mu_1}{\mu_2 + \mu_1} \tag{5}$$

For an infinite viscosity contrast ($A = 1$), the interface may be considered as an equipotential surface (P constant) and Ca is the only relevant parameter. In these conditions, the linear stability analysis performed by Chuoke et al[9] for the boundary conditions of rectilinear flow shows that there is a critical wavelength, λ_c, below which all fluctuations are damped out by the interface tension, and a wavelength of fastest growing rate, λ_m, related to λ_c by:

$$\lambda_c = (\pi/\sqrt{3})bCa^{-1/2} \quad \lambda_m = \sqrt{3}\lambda_c \tag{6}$$

Radial source flow requires a slightly more complex analysis, but a modified capillary number still determines the cutoff scale.[10]

Little theoretical work has been done on viscous fingering with non-newtonian fluids. For simple shear-thinning (or thickening) fluids (eq. 1, with $\tau_y = 0$), Darcy's law and Laplace's equation have to be replaced by:[11]

$$U \sim \nabla P^{1/n} \text{ and } \mathrm{div}(|\nabla P|^{\frac{1}{n}-1}\nabla P) = 0 \tag{7}$$

However, no stability analysis is available yet.

On the other hand, crack propagation[12] stems essentially from the release of elastic energy which accompanies the growth of the crack in a solid under load. The driving force is the strain energy release rate, g, which describes the variation of elastic energy with area. When the crack wall area extends by dA, a mechanical energy $g\,dA$ is released. Note that hydraulic fracturing is a particularly unstable process since the total load applied by the fluid on the crack walls increases when the crack extends.

The primary stabilizing factor is the variation of surface energy with crack wall area. When the crack extends by dA, a surface energy equal to $2\Gamma_{12}dA$ is required. If the material is an ideal elastic medium, the balance between elastic energy and interface energy leads to Griffith's criterion for the equilibrium of the crack:

$$g = 2\Gamma_{12} \tag{8}$$

If g exceeds $2\Gamma_{12}$, the excess energy is converted into kinetic energy for crack propagation.

Griffith's and Chuoke's original analyses miss an important point. Griffith's analysis predicts that, in the limit of vanishing interface energy ($\Gamma_{12} \to 0$), even the smallest pressure in the fracturing fluid would be able to open any infinitesimal flaw at the solid surface and to force it to propagate. On the other hand, Chuoke's analysis predicts that fluctuations of any wavelength can grow when the interface tension between the two fluids vanishes. Actually, such an instability is never observed, for one essential reason: *entropy production*, which tends to be minimized in any irreversible process,[13] limits the velocity of cracks and the width of fingers.

A first source of entropy production is viscous energy dissipation. In viscous fingering, minimizing the rate of frictional energy dissipation led Paterson[14] to a wavelength of maximum growth of the order of 4b, but no cutoff was obtained. In crack propagation, energy dissipation occurs by internal friction or viscoelastic losses at the crack tip. As a result, the crack submitted to the extension force $g - 2\Gamma_{12} > 0$ does no longer accelerate continuously, but reaches a constant velocity. If the kinetic energy is neglected, the following equation applies:[15]

$$g - 2\Gamma_{12} = 2\Gamma_{12}\Phi(U) \tag{9}$$

U is here the crack tip velocity. The dimensionless function $\Phi(U)$, which takes into account the viscoelastic properties, is generally a monotonous function of U at low velocity, so that stable (subcritical) crack growth can be observed most of the time. However, this is not a priori guaranteed in hydraulic fracturing since g may increase faster than $\Phi(U)$.

It is also clear from what precedes that crack propagation becomes unstable if, for some reason, $\Phi(U)$ decreases as U increases. This happens in most viscoelastic solids at very high velocity and leads either to catastrophic crack propagation or to stick-slip motion.[15]

To some extent, a negative branch in the $\Phi(U)$ *versus* U diagram for crack propagation may be compared to shear-thinning behavior for viscous finger growth. Both properties are strongly destabilizing and allow for larger tip or front velocities under a given load or pressure.

A second source of entropy production is *mixing*, either by *dispersion*, in viscous fingering, or by *dissolution*, in hydraulic fracturing. Both processes make the stability analysis difficult, because the interface properties are now time-dependent. Dispersion between two (partially or totally) miscible fluids, such as a pure solvent and the same solvent containing a solute or colloidal particles, is controlled by the Peclet number:

$$Pe = UL/D \tag{10}$$

L is a macroscopic characteristic length and D is the diffusion coefficient. As shown by Chuoke for rectilinear flow[16], Pe plays, in miscible displacement, a role similar to that of Ca in immiscible displacement. It introduces a cutoff wavelength and maximum growth rate wavelengths which are, respectively

$$\lambda_c = 8\pi/Pe \text{ and } \lambda_m = 2\lambda_c \tag{11}$$

We are not aware of any stability analysis for hydraulic fracturing of a solid by a liquid able to dissolve the solid. It is clear from Eq.9 that the current models break down in the case of vanishing interface tension. A possible solution to the problem would be to consider a *dynamic interface tension*, applicable to non-equilibrium conditions. Thermodynamically, Γ is defined as the partial derivative of the system free enthalpy, F, with respect to interface area, A, at constant pressure (P), temperature (T) and composition (c), at thermodynamic equilibrium. This later condition can obviously not be met with miscible fluids or with soluble solids without destroying the interface. This contradictory situation could be resolved by considering a dynamic (or transient) surface tension, Γ_d, which is derived from the free enthalpy (or free energy) by replacing the thermodynamic equilibrium condition by a constant interface thickness (d) condition:

$$\Gamma_d = (dF/dA)_{P,T,c,d} \tag{12}$$

Thus, Γ_d is an answer to the question: "How much free enthalpy does it cost to increase the interface area between two miscible liquids or between   liquid and a soluble solid at constant temperature, pressure and composition, while keeping the microscopic structure of the interface unchanged?" Keeping the structure of the interface unchanged involves a work against the entropy or free enthalpy production associated with dispersion or dissolution. This work is:

$$dF = T \; d_i S \; dA \tag{13}$$

in which $d_i S$ is the entropy production per unit area which would stem from dispersion or dissolution. The entropy production *rate*, $d_i S/dt$, can be expressed in terms of a flux, J, and a conjugated generalized force, X. J is the diffusion flux given by Fick's first law, $J = -D(dc/dx)$, whilst the generalized force is the concentration gradient, dc/dx).[13] This leads to:

$$\Gamma_d = TD(dc/dx)^2 dt \tag{14}$$

which, as expected, is a time-dependent quantity. The main problem is then to evaluate the time during which the entropy production process has to be considered. A natural figure seems to be the time taken by the viscous finger or the crack to travel over a distance equal to its width.

4. The morphogenesis of fractal viscous fingers

The initial phase of a rectilinear viscous fingering experiment with immiscible newtonian fluids is characterized by the formation of ripples at the interface. Very soon the wavelength selection mechanism operates and a steady state regime settles in, in which one single finger emerges (the famous "Saffman-Taylor finger"[17]), with smooth walls and a relative width λ (λ = finger width, l, over channel width, w) between 1 and 0.5 (figure 1a). In order to take into account the aspect ratio of the cell, a modified capillary number, Ca' or $1/B$, is often used as control parameter:

$$Ca' \text{ or } 1/B = 12(w/b)^2 Ca \tag{15}$$

Saffman-Taylor finger develop for $1/B$ below, say, one thousand. Thus, λ starts from 1 at $1/B = 0$ and drops to 0.5 when $1/B$ reaches about 2×10^3.[18] The

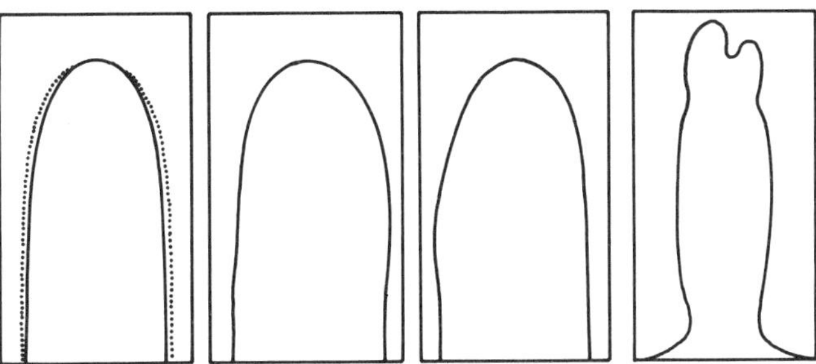

Figure 4 Unstable modes in newtonian fingering. From left to right: the symmetric "breathing" mode; the asymmetric "hump" mode; the asymmetric "tip wobbling" mode; tip-splitting (Refs.[19] and [20]).

emergence of one single finger after the initial ripples of the interface can be readily understood in terms of the *shielding effect* which favors any bump which is slightly ahead of the others and has the largest pressure gradient at his tip.

At large velocities or low interface tension, ST fingers become themselves unstable, for the same reasons as the initial interface. What has been predicted, either by random-walk-type simulations[19] or by direct solution of the flow equations[20] is a sequence of a few instability modes including a "breathing" mode, a "tip wobbling" mode, a "hump" mode and, finally, tip-splitting (figure 4). However, in rectilinear cells (channels), the boundary conditions imposed by the closed walls of the cell are so stabilizing that one can hardly observe these instabilities experimentally with immiscible newtonian fluids. What has actually been observed is the hump mode[18] and the onset of tip-splitting, for $1/B$ values ranging from 600 to 3×10^3 [8] (Figure 1).

Owing to its diverging geometry, radial source flow is much more favourable to tip-splitting. Using the same type of fluids as above (immiscible, newtonian), the same fundamental instabilities as in rectilinear flow were observed , with a larger number of tip splittings (figure 2a and 2b). The use of miscible newtonian liquids allows to push further the cascade of tip splitting.

The rather robust behaviour of viscous fingers in newtonian fluids contrasts with their extremely unstable behaviour in shear-thinning fluids. As soon as a non-newtonian shear-thinning fluid is displaced, finger branching becomes extremely difficult to control, even at very low velocity, and one is rapidly led to a quasi-explosive branching cascade, even in rectilinear cells (figures 1c and 2c). The capillary number, Ca, or its modified form, $1/B$, can no longer be used as control parameter since the viscosity is shear-rate dependent, but modified "effective" forms, $\overline{Ca}$ and $\overline{1/B}$, can be defined, using the non-newtonian viscosity at a shear rate, averaged over the gap of the cell, of the order of $2U/b$.[21] By doing so, one notices that shear-thinning properties allow for the exploration of a broad range of extremely high values of

$\overline{1/B}$, typically of the order of 10^6,[21] in a regime where all kinds of instabilities are indeed expected. The patterns obtained in this regime (figures 1c and 2c) are fractals without loops, with a DLA-type structure. The absence of loops makes their "mass" fractal dimension (which describes the scaling of the pattern area) equal to their "surface" fractal dimension (which describes the scaling of the perimeter). That a mass fractal structure can be obtained is not obvious *a priori* since viscous fingering is primarily an interfacial instability. Mass fractality stems from the fact that a statistically uniform thickness is associated to the interface.

The onset of fractal behaviour can be investigated in detail, starting from weakly shear-thinning fluids and low front velocities, and increasing progressively either the velocity or the shear-thinning properties, or both. Figure 5 is a sequence of patterns obtained in that way, by injecting air at increasing pressure in a dilute bentonite clay suspension.[22] One first encounters a tip-wobbling mode, which gives the finger its tottering aspect. This is soon relayed by what seems to be the asymmetric hump mode which is rapidly quenched instead of receding and sliding along the finger as in newtonian fluids[18], probably due to the existence of a threshold shear stress for flow, and gives rise to side ripples. At still higher velocity, the ripples turn into individualized bilateral buds. Tip-splitting can be detected at the forefront of the pattern, but the final structure is a very disrupted chaotic-like pattern in which no clear branching cascade can be detected.

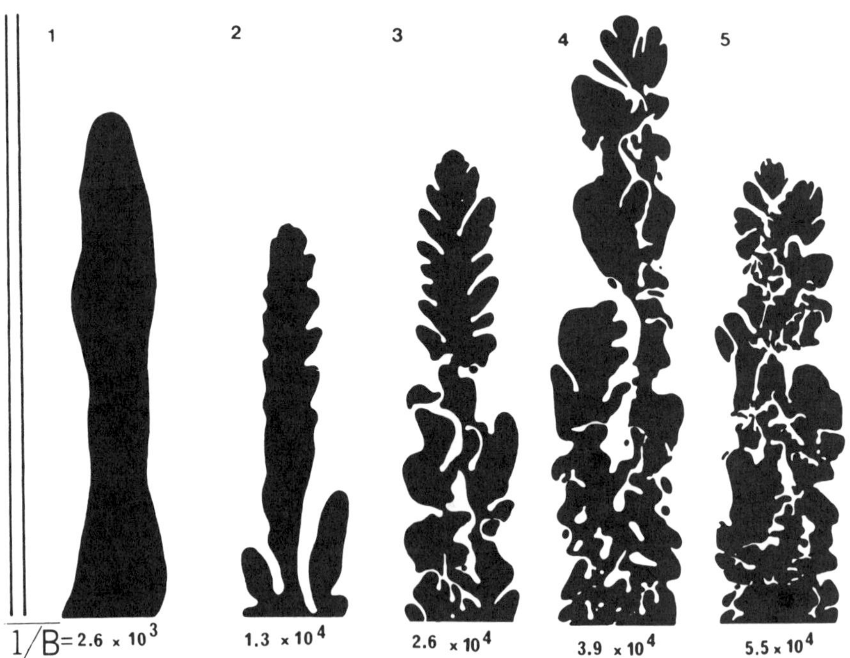

Figure 5 Patterns obtained by injecting air in a weakly non-newtonian water-based clay mud. The driving pressure increases from 1 to 6.

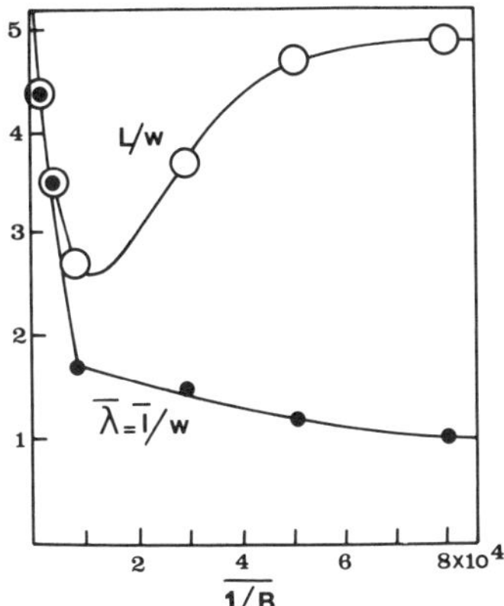

Figure 6 Opening of the fractal window between the reduced averaged finger — or branch — width, $\overline{l}/w$, and the reduced pattern width, L/w, in the sequence of Fig.5.

Remarkably, in spite of the fact that no clear structure can be detected in the "mass" distribution within such disrupted patterns, a simple coastline analysis reveals that the interface is already fractal over half a decade.[22] The lower and upper cutoffs of this "surface" fractal domain are the average finger- or branch width, $\overline{l}$ (which can be measured by randomly intersecting the patterns with straight lines), and the pattern width, L, respectively. Thus, the development of the instability has opened a window for surface fractal behaviour by pulling apart two cutoffs which in ST fingering are merging: $\overline{l}$ and L. As shown in figure 6, the opening of the window takes place rather abruptly around $\overline{1/B} = 10^3$. This corresponds to the development of bilateral buds.

In order to escape this chaotic-type regime, one has to increase the concentration of solid particles in the colloidal fluid, in order to strengthen its non-newtonian and viscoelastic character (the two effects cannot be controlled separately). Keeping a low concentration (with, concomitantly, a weak shear-thinning and viscoelastic character) and increasing further the flow velocity does not at all lead to the same evolution. It introduces important inertial effects which flush most of the colloidal fluid out of the cell. Conversely, as shown in Fig.7, increasing the colloid concentration allows for more shielding and initiates a branch width selection process which progressively narrows the size distribution of all the buds, leaves, fragments,... obtained in the chaotic-like regime. This process ends up in patterns with a well-defined and homogeneous branch width and a clear branching cascade. At this point, the patterns are loop-free mass fractals, with a mass fractal dimension equal to their surface (boundary) fractal dimension, over a decade.

In terms of dimensionless control parameter, the landmarks of the settling of fractal behaviour are as follows: $\overline{1/B} = 10^3$ for the first perturbation of the Saffman-Taylor fingers; $\overline{1/B} = 10^4$ for the development of the hump mode and for the emergence of a self-similarity window in the shape of the interface; $\overline{1/B} = 10^5$ for the development of tip-splitting and for the development of a fractal interface over about a decade; $\overline{1/B} = 10^6$ for the selection of an homogeneous branch width and, simultaneously, for the emergence of a DLA-type fractal behaviour.

The morphogenetic process which turns a smooth ST finger into an essentially compact surface fractal and then into a loop-free mass fractal can also be followed by applying an image analysis operation known as skeletisation.[23] The skeleton of a pattern is an ensemble of connected lines formed by the points located midway from the boundaries of each branch, fragment,... Thus, branches are replaced by single lines whereas blobs with a central hole are replaced by loops.

The skeletisation of the patterns of figures 5 and 7 produces the sequence shown in figure 8. One may notice two opposite trends. From skeleton 1 to skeleton 6 or 7, the trend is clearly towards a greater complexity, due to the development of branching first, and to the closure of loops later. On the other hand, beyond patterns 6 or 7, the skeletons show clearly a tendency towards a greater simplicity, due to the opening of loops and to the pruning of branches. Thus, the morphologically "chaotic" intermediate regime corresponds to the maximum number of loops. Whether passing through this regime is a requisite or not for going from ST fingers to fractal arborescences remains to be established.

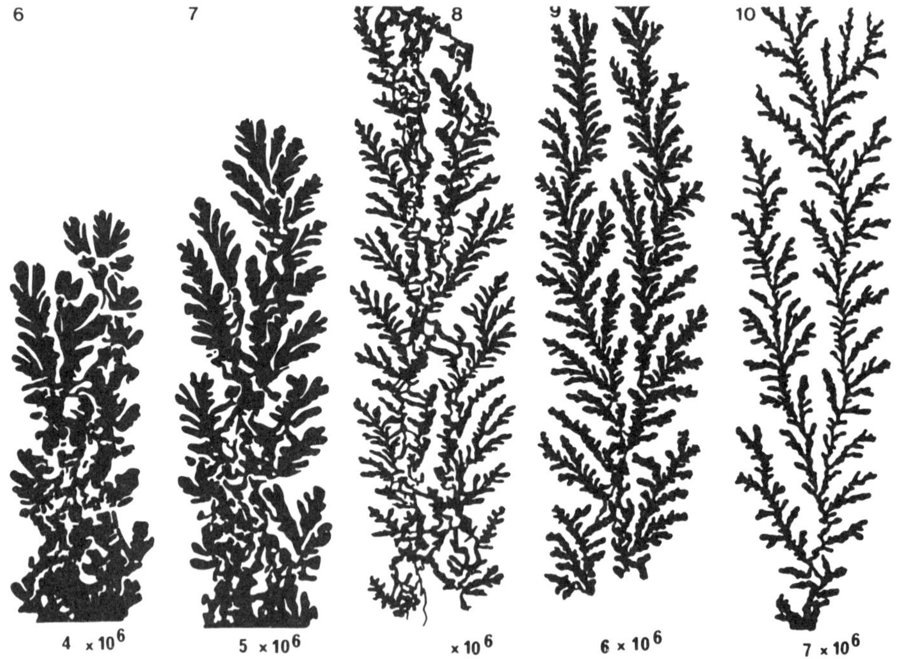

Figure 7 Patterns obtained by injecting air into water-based muds, at constant injection pressure and increasing colloid concentration.

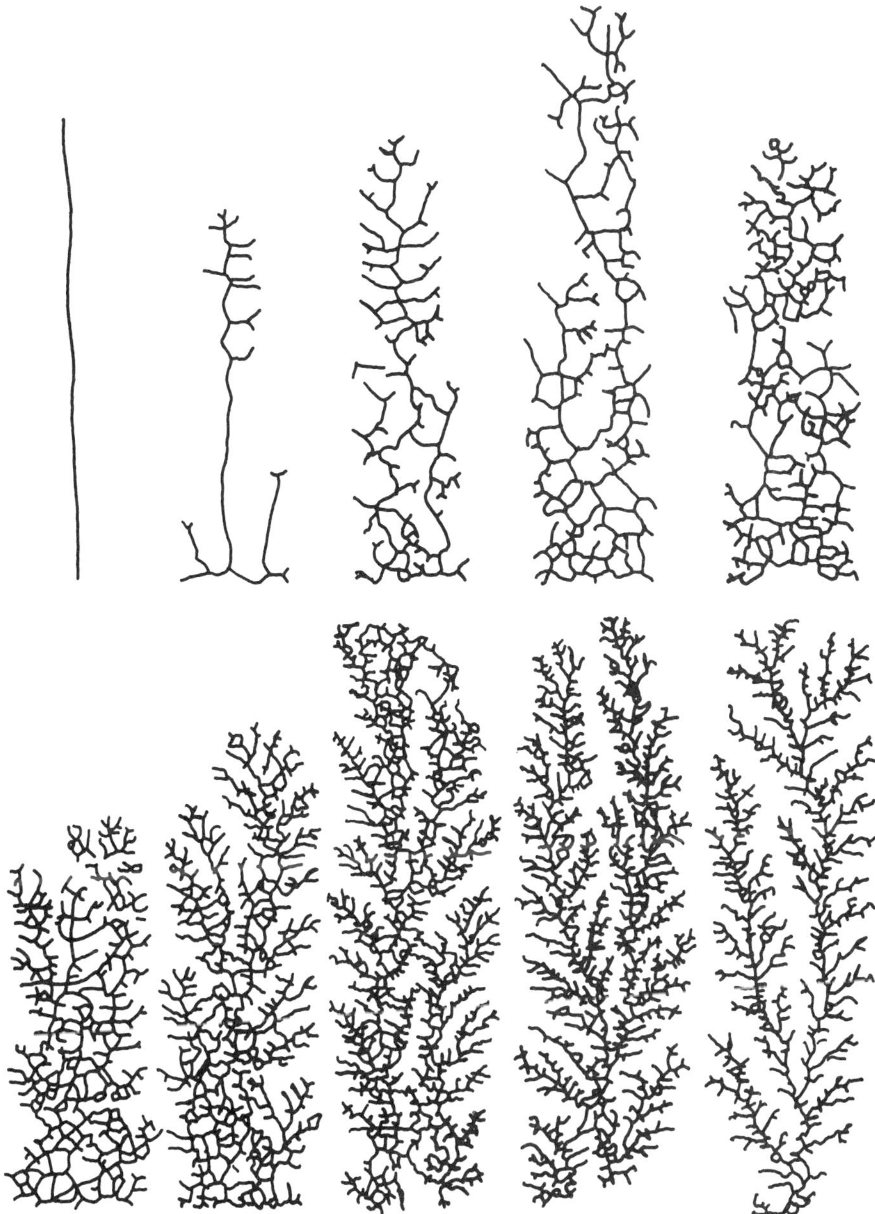

Figure 8 A skeletized view of the morphogenetic process which transforms a smooth Saffman-Taylor finger into a fractal tree.

In addition to this qualitative information, skeletisation permits several quantitative analytical operations, such as the combinatorial (Horton-Strahler) analysis[24,25] from which a mass fractal dimension can be extracted.[26,27]

5. Viscous or visco-elastic?

Fully developed fractal fingering patterns are most easily obtained with colloidal fluids (including polymeric solutions). This is particularly true in rectilinear cells. Since the addition of a colloid to a newtonian liquid introduces simultaneously non-newtonian and viscoelastic properties, the question naturally rises whether viscoelasticity is really required or whether the purely viscous properties of the fluid are sufficient. The best way to try to answer this question is to compare the experimental results with the predictions of models based either on viscous forces or on viscoelastic effects.

The first reference model is the linear stability analysis performed by Chuoke et al[9] for immiscible newtonian fluids, which takes into account viscous and capillary forces (see section 3). Although this analysis is certainly no longer valid for shear-thinning fluids with a yield stress, the wavelength for maximum growth rate derived from it (Eq. 6) remains in order-of-magnitude agreement with the average finger- or branch width all along the sequence of figures 5 and 7, provided one uses the *effective* capillary number. The observed $\bar{l}$ is within a factor of two with respect to λ_m at the onset of instability, and within a factor of five in the developed fractal regime (Fig. 9).

As we already pointed out, there is currently no stability analysis which takes into account the three most important elements of the experiments that we describe, namely the interface tension, the shear-thinning character and the viscoelastic properties. The only attempt to take explicitly into account the viscoelastic properties in fractal fingering is due to de Gennes.[28] Its approach is a direct application of the concept of Deborah number introduced in section 1. The basic assumption is that the characteristic frequency of the structural relaxation process is close to the

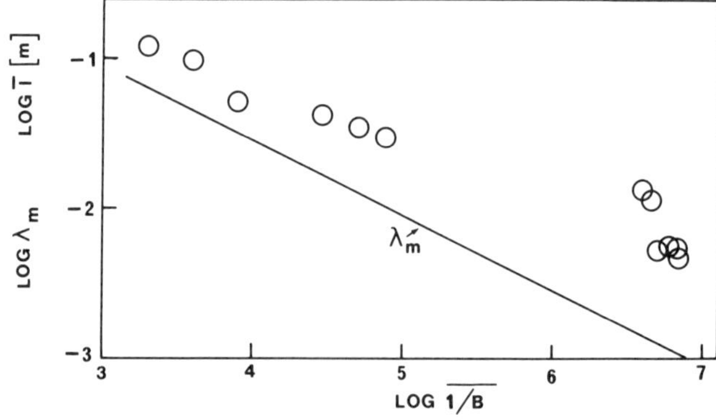

Figure 9 Evolution of the average finger- or branch width in the sequences of figures 5 and 7, compared to the wavelength for maximum growth derived from the effective capillary number and Eq. 6.

frequency for the rise of stress in front of the tips or, in other words, that $De = 1$. Assuming further
(i) that the characteristic frequency of the flow event is U/l;
(ii) that the structural relaxation time of the viscoelastic fluid can be approximated by Maxwell's relaxation time, μ/G, for a   hypothetical body which would be characterized by a newtonian viscosity μ and an elastic modulus G;
(iii) that the pressure gradient is of order P/l;
(iv) that Darcy's law is still valid, one is led to:[28]

$$l = \frac{bP^{1/2}}{(12G)^{1/2}} \qquad (16)$$

The tip velocity does no longer appear explicitly and is replaced by the injection pressure. Another more fundamental difference with respect to the result of the linear stability analysis of Chuoke et al[9] (Eq. 6) is that the elastic modulus is now explicitly controlling the finger width, rather than the viscosity.

Equation 16 accounts correctly for the cell gap dependence of the branch width in the fractal regime and for the decrease of the branch width as the fluid gets stronger ($l \propto G^{-1/2}$ or, more generally, $l \propto G^{-n/(n+1)}$ for a shear-thinning fluid[29]). However, the same type of behaviour can be predicted from Paterson's viscous dissipation model or from Chuoke's analysis.

On the other hand, the viscoelastic model fails to account for a very characteristic feature of fractal viscous fingering, namely that increasing the injection pressure does not at all lead to an increase of the branch width, as predicted by equation 16, but to a *decrease* of l and, concomitantly, either to an increase of the pattern width, L, in rectilinear flow,[29] or to an increase of the branch density, in radial source flow.[30] In other words, increasing the injection pressure does not inflate the branches, but inflates the pattern (the same trend as in patterns 2 to 5 in Fig. 5)! The net result is an increase of the total pattern area.

Actually, the decrease of l with P can be readily understood by realizing that the primary consequence of a larger injection pressure is an increase of the tip velocity, U. Thus, from that point of view, the behavior of fractal viscous fingering patterns is but a continuation of the behavior of stable ST fingers, which extends into the fractal unstable regime. On the other hand, the increase of the pattern width or the branch density can be understood by realizing that the increase of P leads to a general rise of the pressure gradients at the branch tips. This allows the branch tips to approach closer to the walls (in rectilinear cells) or closer to each other (in radial cells).

Thus, when taken as a whole, the elements of our evaluation point to the conclusion that viscoelastic properties are not required to reach "full" development of the fractal fingering regime around $1/B = 10^6$. Viscoelastic effects seem to control neither the width of the branches nor the width, the density or the fractal dimensions of the patterns.

A more quantitative (but still unprecise) confirmation of this can be obtained from estimates of the Deborah number in the sequence leading to fractal fingering (Figs. 5 and 7). Approximating the structural relaxation times by the ratio η/G', where η is the non-newtonian viscosity at a shear stress $= 2U/b$, and where G' is the storage modulus at a frequency U/l, one is led to the following figures: $De = 10^{-3}$ at the onset of instability of the Saffman-Taylor fingers; $De = 10^{-1}$ at the onset of tip-splitting; $De = 0.2$ in the chaotic-like regime and beyond.

However, this does not mean that viscoelastic effects do not play any role. As we shall see in section 6, viscoelasticity seems to control less global features such as the branching angle or the tip profile and, beyond some threshold, introduces deep modifications.

6. Surface tension and branch width

At the time fractal viscous fingering was discovered[31], a low interface tension was believed to be essential for reaching the high capillary number regime were fractal morphology develops. In fact, the destabilizing effect of non-newtonian properties is much more efficient for reaching the fractal regime than lowering the interface tension. As shown in figure 10, the magnitude of the equilibrium interface tension introduces but minor modifications in the fractal pattern morphology. It merely introduces a larger lower cut-off, above which the same fractal dimension is observed.[21]

Even in miscible flow, a finite finger width is obtained. As discussed in section 3, its value is controlled either by viscous dissipation or by dispersion, or both. Viscous dissipation is expected to dominate in highly viscous muds, whereas the dynamical interface tension is expected to control the finger width in the less viscous muds. This is in gross agreements with observations. In DLA-type patterns obtained with muds containing 7% by weight of clay, l was found to be close to $4b$[29], as expected from Paterson's analysis.[14] On the other hand, in more fluid muds containing only

Figure 10 Patterns obtained by injecting air (left) or water (right) in water-based muds. In spite of the vanishing equilibrium interface tension in the latter case, a finite finger width is observed.

Figure 11 Magnifications of patterns obtained at vanishing equilibrium interface tension (water penetrating a water-based mud), at increasing driving pressure and tip velocity. The finger width passes through a maximum.

5% clay, the finger width shows a more complex behavior.[30] At small injection pressures and front velocities, the fingers are extremely narrow (Fig.11). As the velocity increases, the width passes through a maximum and then comes back to lower value.

This behaviour can be understood at least qualitatively from the approach outlined in section 3. Indeed, if one admits that the fractal viscous fingering regime develops in a range were De is close to 1, one may replace the characteristic time for mixing by μ/G. Using Einstein's formula for the diffusion coefficient of a colloidal particle of size a in a medium of viscosity μ, $D = kT/6a\mu$, one obtains:

$$\Gamma_d = (kT/6a)(dc/dx)^2 G^{-1} \tag{17}$$

Thus, since the average concentration gradient across the interface is expected to vary linearly with the concentration in the mud, c_0, whereas the elastic modulus is a much more rapidly increasing function of c_0, the dynamical interface tension and the finger width are expected to pass through a maximum.

7. From flow to fracture

Increasing the effective capillary number beyond the value reached in the last pattern of figure 7, either by increasing the colloid concentration or the injection pressure, leads to the type of pattern which is shown in figure 1d. Although the general branching morphology looks unchanged, the pattern looks much more prickly. A close look to the branches shows that this is due to a significant modification of the *tip profiles*, which were essentially convex up to that point, and which are now essentially concave.

A series of experiments performed at constant injection pressure in pastes of increasing colloid concentration in rectilinear miscible flow[29] shows that the concave-to-convex evolution is continuous (Fig. 12) and follows the development of the yield

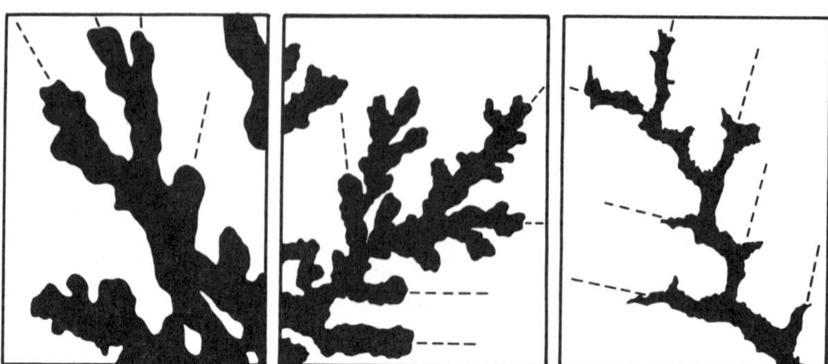

Figure 12 Magnifications of patterns obtained by injecting water in water-based muds at constant driving pressure. The colloid concentration increases from left to right. Note the increasing concavity of the tips and the opening of the branching angle.

stress. In weakly concentrated pastes the tips have a rounded profile, whilst in the more concentrated pastes of the series, they look like crack tips.

The average *branching angle* follows another interesting evolution. As shown in figure 12, the branching geometry goes from almost parallel branching (as in a chandler) in fluid suspensions, to almost perpendicular branching in rigid pastes.

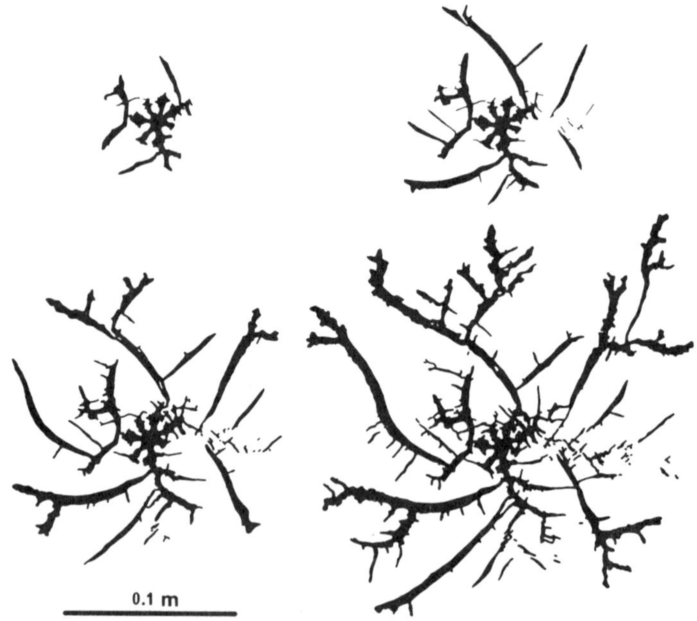

Figure 13 Growth sequence of fracturing patterns. Note the growth of cracks behind the tips.

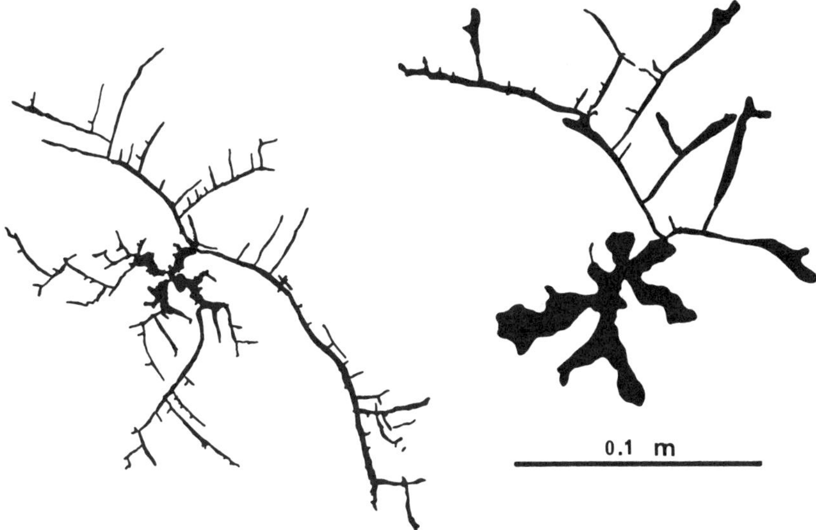

Figure 14 Examples of crossover from the viscous fingering to the fracturing regime in a single experiment, at vanishing (left) and large equilibrium interface tension (right).

Quasi-perpendicular branching and concave tips in concentrated pastes were taken as precursory signs of the onset of fracture by Van Damme et al.[29] They suggested that if a finger tip grows at a velocity larger than l/τ_{str} or b/τ_{str}, a stress would develop *behind* the tip, along the finger. This stress would be oriented parallel to the finger and would grow as the finger moves forward. At some point, a lateral crack could grow to relax the stress, and the orientation of this crack would be naturally at 90° with respect to the main finger. As the tip keeps moving, the lateral stress could grow again and a second lateral crack could form, and so on, leading, eventually, to a quasi-periodicity of lateral branching.

The clear confirmation of this came when even more concentrated pastes (solid/liquid ratio between 0.1 and 0.6, w/w) were studied in radial geometry.[32] The general morphology of the patterns was found to be much closer to what one intuitively expects for fracture (Figure 2d). The growth sequences, like in figure 13 for instance, show clearly that bifurcation may indeed take place behind the tip. The cracks are generally bended and the subcracks grow systematically on the convex side of the main crack, at 90° with respect to the tangent. Growth also takes place by widening of the cracks, essentially on their convex side.

The crossover from the viscous fingering to the fracturing regime sometimes happens within a single experiment, as examplified in figure 14. In this case, the morphological crossover goes along with a velocity jump.

The strong morphological differences between the fingering patterns (figure 2c) and the fracturing patterns (figure 2d and 13) permits, most of the time, to classify a given pattern in either one of these catagories without ambiguity. This in turn allows one to construct a fingering-fracturing "phase" diagram in the plane relating the injection flow rate to the colloid concentration in the paste (which is an indirectly and non linearly related to the cohesion energy and the elastic properties of the paste). As shown in figure 15, the boundary between the two regimes is roughly hyperbolic, which is in agreement with the idea that the crossover occurs when the frequency of the mechanical excitation exceeds the frequency of the internal relaxation processes (the stronger the paste, the smaller the frequency of internal microscopic reorganizations and the smaller the flow rate necessary to exceed that frequency).

The fracturing regime which is revealed by these experiments still raises many questions, particularly on the components of the stress tensor. It cannot be compared directly to type I crack opening or to three-dimensional hydraulic fracturing, where one expects crack bifurcation to occur at the tip, with an angle smaller than $90°$. The major difference that we see is that *friction* with the cell walls is still an important parameter, and in that respect, the process might prove to be of interest for modelling some geotechtonic phenomena.

It may also be involved in the primary migration of petroleum. Source rocks are essentially colloidal rocks and the intrinsically two-dimensional (layered) geometry of sediments makes the fracturing process induced by radial source flow in a Hele-Shaw cell an attractive model for the fracturing process which could be induced by the local generation of hydrocarbons from a piece of kerogen (the organic matter resulting from the embedding of algae or other organisms).

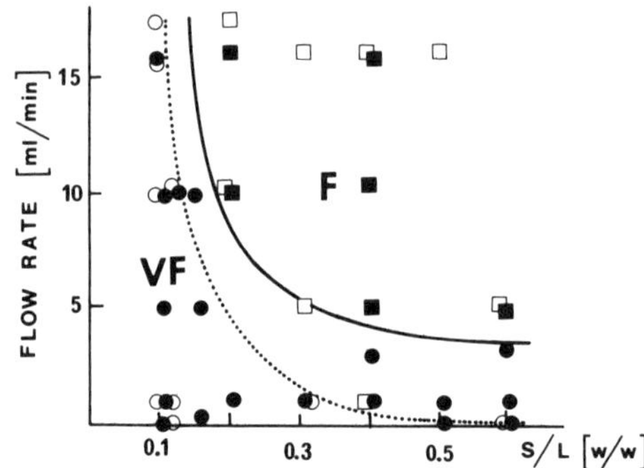

Figure 15 Behavior diagram in a plane "injection flow rate *vs* colloid concentration". Circles and squares correspond to viscous fingering and fracturing patterns, respectively. Open and closed symbols correspond to miscible and immiscible intrusion, respectively. The curves are the estimated boundaries between viscous fingering (VF) and fracturing (F), in the miscible (dotted curve) and immiscible (continuous curve) case.

Acknowledgements

We thank the Institut Français du Pétrole for supporting this research.

References

[1] H. Van Olphen, *Introduction to Clay Colloid Chemistry*, Wiley, New York, (1963).

[2] G. Brown Ed., *Clay Minerals: Their Structure, Behavior and Use*, The Mineralogical Society, London, (1984).

[3] F. Wittman, this volume.

[4] R.B. Bird, R.C. Armstrong and O. Hassager, *Dynamics of Polymeric Liquids*, Vol.1, Wiley, New York, (1987).

[5] C.H. Pons, D. Tessier, H. Ben Rhaiem and D. Tchoubar, Proc. Int. Clay Conf., H. van Olphen and F. Veniale Eds., Elsevier, Amsterdam, p. 165-186, (1982).

[6] J. Kertész, this volume.

[7] D. Mader, *Hydraulic Proppant Fracturing and Gravel Packing*, Elsevier, Amsterdam, (1989).

[8] G.M. Homsy, Ann. Rev. Fluid Mech., **19**, 271, (1987).

[9] R.L. Chuoke, P. Van Meurs and C.J. Van Der Poel, J. Pet. Technol., **11**, 64, (1959).

[10] L. Paterson, J. Fluid Mech., **113**, 513, (1981).

[11] G. Daccord, J. Nittman and H.E. Stanley, Phys. Rev. Lett., **56**, 336, (1986).

[12] T. Halsey, private communication.

[13] I. Prigogine, *Introduction to the Thermodynamics of Irreversible Processes*, Thomas, Springfield, (1955).

[14] L. Paterson, Phys. Fluids, **28**, 26, (1985).

[15] D. Maugis, this volume.

[16] R.L. Chuoke, as quoted in Ref.[8].

[17] P.G. Saffman and G.I. Taylor, Proc. R. Soc. London, A **245**, 312, (1958).

[18] P. Tabeling, G. Zocchi and A. Libchaber, J.Fluid Mech., **177**, 67, (1987).

[19] S. Liang, Phys. Rev. A, **33**, 2663, (1986).

[20] D. Bensimon, Phys. Rev. A, **33**, 1302, (1986).

[21] H. Van Damme, E. Alsac, C. Laroche and L. Gatineau, Europhys. Lett., **5**, 25, (1988).

[22] H. Van Damme, E. Alsac et C. Laroche, C.R. Acad. Sc., Série II, **309**, 11, (1989).

[23] J. Sera, *Image Analysis and Mathematical Morphology*, Academic Press, New York, (1982).

[24] R.E. Horton, Bull. Geol. Soc. America, **56**, 275, (1945) and A.N. Strahler, ibid., **63**, 1117, (1952).

[25] J. Vannimenus and X.G. Viennot, J. Stat. Phys., in press.

[26] E.L. Hinrichsen, K.J. Måløy, J. Feder and T. Jøssang, J. Phys. A, **22**, L271, (1989).

[27] E. Lemaire and H. Van Damme, in preparation.

[28] P.G. de Gennes, Europhys. Lett., **3**, 195, (1987).

[29] H. Van Damme, C. Laroche, L. Gatineau and P. Levitz, J. Physique, **48**, 1121, (1987).

[30] H. Van Damme, C. Laroche and L. Gatineau, Rev. Phys. Appl., **22**, 241, (1987).
[31] J. Nittmann, G. Daccord and H.E. Stanley, Nature, **314**, 141, (1985).
[32] E. Lemaire and H. Van Damme, C.R. Acad. Sc., **309**, Série II, 859, (1989).
[33] T. Vicsek, in *Random Fluctuations and Pattern Growth: Experiments and Models*, H.E. Stanley and N. Ostrowsky Eds., pp 312-319, Kluwer, Dordrecht (1988).

Chapter 6.2

From Flow to Fracture and Fragmentation in Colloidal Media

2: Local Order and Fragmentation Geometry

H. Van Damme and M. Ben Ohoud

C.N.R.S.-C.R.S.O.C.I.
F-45071 Orléans Cédex 02, France

1. Introduction

Colloidal solids are coherent packings of small (say, below one micron) particles which are held together by short-range surface forces.[1] Sedimentary rocks, like clay and carbonate rocks, and hardened cement pastes belong to this family. In spite of their solid character, colloidal solids have to be regarded as granular or divided media in the sense that the iono-covalent or metallic forces within the colloidal particles are much larger than the interparticular forces. Thus, as far as mechanical strength is concerned, there is a clear distinction between the cohesion of the elementary particles and the cohesion of their aggregates.

The central question which will be addressed in this chapter is the following: How does a piece of colloidal solid break, and how are the geometrical properties of the fragments related to the ordering of the elementary particles in the solid. As we shall see, there are strong arguments for saying that, contrary to aggregation of colloidal particles which is known to lead to mass or surface (multi)fractals structures in a wide range of conditions, fragmentation of colloidal packings reveals a scaling behavior which applies to the open porosity.

This evidence comes from volumic mass, surface area and open porosity measurements performed on fragments of artificial colloidal rocks, over almost three decades in particle size. The rocks were obtained by sedimenting and drying aqueous suspensions of several clay minerals, in order to cover a broad range of elementary particle morphology, including needle-like particles, relatively flexible fibers, rigid platelets and flexible sheets, which all belong to the family of clay colloids.

2. Clays[2]

Clay minerals are ubiquitous in nature. They occur in discrete geological de-

Disorder and Fracture, Edited by J. C. Charmet *et al.*
Plenum Press, New York, 1990

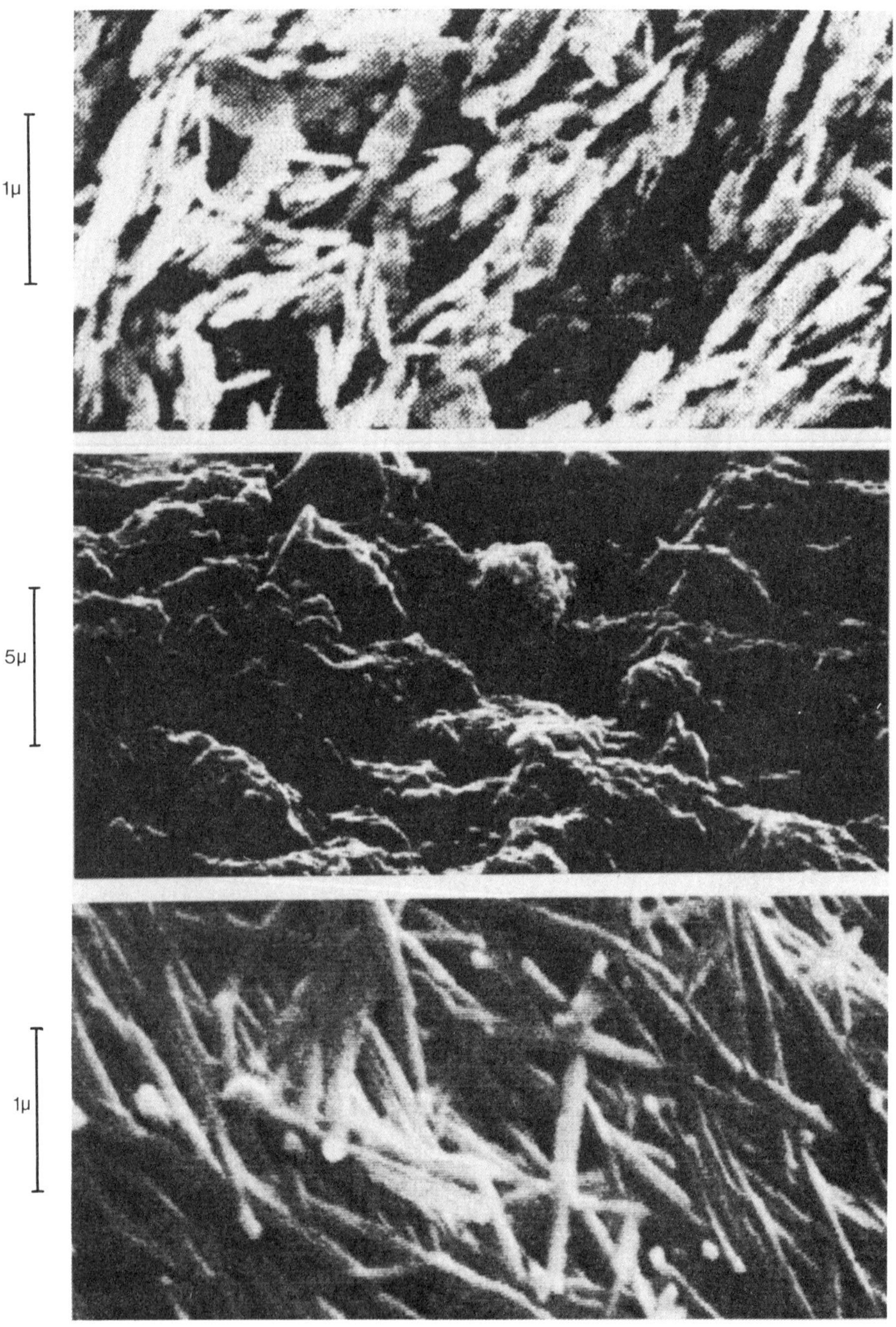

Figure 1 Digitized scanning electron micrographs of a kaolinite packing (top), a montmorillonite packing (middle) and a sepiolite packing (bottom).

posits, in soils and in all products of erosion and breakdown of rocks. They are, by definition, the fine-grained (below, say, one micron) fraction of minerals. It turns out that this purely granulometric definition corresponds closely to a particular class of the silicates with a layer lattice structure: the phyllosilicates, of which micas are probably the most widely known.

The essential feature of phyllosilicates is a continuous two-dimensional sheet of tetrahedra of oxygen ions, coordinating essentially Si, Al or Fe cations. Si is by far the most frequent. This tetrahedral sheet is linked to an adjacent octahedral sheet in which oxygen and hydroxyl (OH) ions coordinate essentially Al, Mg, $Fe(III)$, $Fe(II)$ or Li cations. The unit formed by linking one octahedral sheet to one terahedral sheet is called a 1:1 layer. By linking a tetrahedral sheet on each side of an octahedral sheet, a 2:1 layer is formed.

The 1:1 and 2:1 layers are not necessarily electrically neutral, due to unbalanced ionic substitutions in the cationic lattice which leads to a net negative charge. The total charge balance is maintained by interlayer species, which may be individual cations as in micas, hydrated cations as in smectites, or continuous hydroxyde sheets as in chlorites.

The layer type, the charge on the layer and the nature of the interlayer material provide the basis for the classification of clay minerals. Each of them displays very distinct and typical structural, morphological, colloidal and surface-chemical properties.

Kaolinite is a typical 1:1 clay, with zero layer charge. There is no interlayer material and the elementary layers stack on each other thanks mainly to hydrogen bonds between adjacent layers. The resulting cristallites are *rigid platelets*, typically a few hundreds of nanometers wide and a few tens of nanometers thick. Water does not enter the interlayer space of kaolinite, and kaolinite does not form stable gels in water. In solid deposits (figure 1), the orientation of the platelets is but locally correlated, for the polydispersity of the particle size and the interparticular friction prevents strong liquid crystal order.

Fibrous clays are another important group. *Sepiolite* and *palygorskite* are the most common members of the group. Both have a 2:1 layer structure, but the lateral extension of the layers is restricted to a few tetrahedra, so that the layers turn into parallelepipedic ribbons. Morphologically, fibrous clays appear as more-or-less *rigid needles* (palygorskite) or more-or-less *flexible fibers* (sepiolite). Typical sepiolite fibers have a diameter of 100 nm and are about 2 to 3 μm long. The microtexture of solid deposits (figure 1) is that of a disordered and entangled network.

Smectites have properties almost opposite to those of kaolinite. Their 2:1 layer charge is large and is compensated by hydrated interlayer cations which, in natural clays, are essentially hydrated sodium, calcium or magnesium ions. The hydration energy of the interlayer ions allows water to enter the interlayer space. This makes the original interlayer cations easily exchangeable by other metal cations or any other cationic species. Each 2:1 layer has an extremely high aspect ratio. Its lateral extension is of the order of 500 nm and it is only 1 nm thick.

The most common members of the family are *montmorillonite* and *hectorite* (*Al* and *Mg* in the octahedral layer, respectively). Among the best known properties of smectites is their ability to swell and to shrink upon addition or removal of water. For that reason, smectites are often referred to as *swelling clays*. Swelling is associated with a statistical increase of the average distance between layers.[3] However, even in dilute suspensions (sols), the layers are not totally isolated.[4] The basic units in a slurry of natural sodium-smectite are *aggregates* of a few layers, more or less parallel to each other. At higher concentration in solid, the aggregates start forming a connected solid network (see figure 3 in part 2 of this contribution) with a lenticular porosity.[5] As water is removed, the porosity shrinks and, finally, a compact solid with a wavy surface is obtained, in which there is a strong correlation between the average orientation of the aggregates and the orientation of the supporting solid surface (figure 1).

The size of the aggregates (that is, the average number of stacked layers) is strongly dependent on the nature of the interlayer ions.[5] The general trend is that, the higher the charge of the cation, the larger the number of stacked layers. This is clearly due to short range attractive forces between the interlayer cations, which act as sticking points, and the oxygen ions of the layers surface. The increase of the number of stacked layers has a direct micromechanical consequence: it *stiffens* the particles. This, in turn, restricts the deformation of the particles as they pile up in the drying solid. The final result is a larger orientational disorder and a larger porosity. This can be rationalized in terms of an increase of the excluded volume of the particles as their stiffness increases. Thus, depending on the interlayer ions, the morpho-mechanical properties of smectite particles can go from *highly flexible sheets to rigid platelets*.

This brief description shows the remarkable variety of particle arrangements that one can expect in solids made with clay colloids. As far as porosity and surface area are concerned, it is clear that rigid particles and a large orientational disorder will lead to limited contact areas between particles, hence to brittle systems with a large and accessible void space. Conversely, highly oriented flexible particles will generate compact and hardly penetrable systems.

3. Experimental

One kaolinite, one palygorskite, one sepiolite and over twenty montmorillonites (the same parent material ion-exchanged with different metal or organic cations) form the basis for the present study.[6] The colloidal solids were obtained by allowing several liters of slurries containing about 10% dry matter to dry by slow evaporation at 353K. The resulting lumps were first fragmented with a hammer and with a ball mill afterwards. For each material, about twenty granulometric fractions were isolated by sieving, with mesh sizes ranging from 20 to 4000 μm.

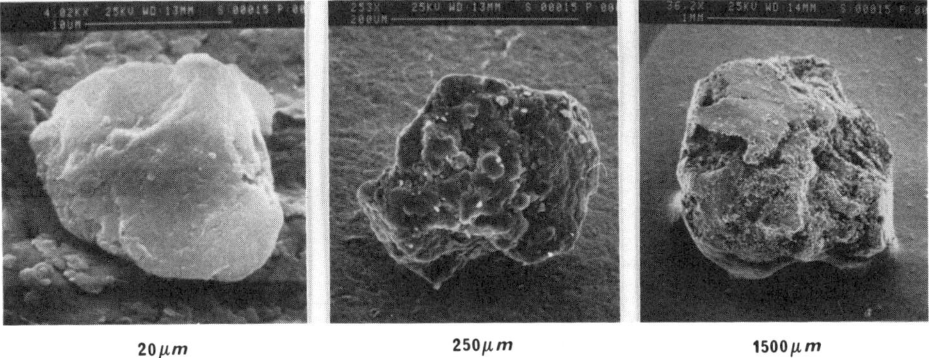

Figure 2 Scanning electron micrographs of three typical fragments obtained by crushing and sieving a sodium- montmorillonite. The sizes indicated are the nominal sizes of the mesh.

The shape of the fragments was examined by scanning electron microscopy and found to be roughly spheroidal in all cases (figure 2). This may seem surprising at first sight since the elementary clay particles are extremely anisotropic. In fact, even the smallest fragments are much larger than the elementary clay particles, so that one expects the milling process, which associates fracture and erosion, to loose the memory of the elementary particle morphology.

The *apparent volumic mass* was calculated by measuring the volume occupied by a given mass of powder in a simple measuring glass tube, after gentle vibration for a fixed time.[7] The *accessible surface area* was measured by physical adsorption of nitrogen at 77K, using the classical B.E.T. method.[8] The *accessible intraparticular porosity* of the fragments was also determined by physical adsorption, by measuring the total volume of liquid nitrogen adsorbed by the fragments at a relative vapor pressure of 0.999. The *accessible microporosity* which, in physical adsorption, is defined as the porosity stemming from voids which are only a few molecular diameters wide (say, 1 nm), was also calculated from the adsorption isotherms, using the t-plot procedure.[8]

4. Scaling behavior of fragments

As an operational approach, we look for simple power law relationships between the characteristic dimension (mesh size) of the fragments, R, and their mass, $w(R)$, their surface area, $s(R)$, and their pore volume $p(R)$:

$$w(R) \sim R^{D_m} \tag{1}$$

$$s(R) \sim R^{D_s} \tag{2}$$

$$p(R) \sim R^{D_p} \tag{3}$$

This allows us to define three exponents (note that the validity of a power law is not at all guaranteed *a priori*).

On the other hand, the number of spheroidal fragments, $N(R)$, which can be packed in a given volume, V_a, large with respect to R, is scaling as

$$N(R) \sim \frac{V_a}{R^3} \tag{4}$$

(this is of course no longer true for grains which are not spheroidal). Thus, if equations 1 to 4 hold, the following relations can easily be derived for the apparent volumic mass, $M_v(R)$, for the specific surface area (surface area per unit weight), $S_w(R)$, and for the internal porosity of the fragments, $P(R)$, respectively:[7,9]

$$M_v(R) = \frac{N(R)w(R)}{V_a} \sim R^{D_m - 3} \tag{5}$$

$$S_w(R) = \frac{N(R)s(R)}{N(R)w(R)} \sim R^{D_s - D_m} \tag{6}$$

$$P_v(R) \sim \frac{N(R)p(R)}{N(R)R^3} \sim R^{D_p - 3} \tag{7}$$

Figure 3 shows typical $\log M_v$ vs $\log R$ plots. Within experimental error, the slope is zero, so that $D_m = 3$. This value is obtained for *all* the materials that we consider here.

It points to the general result that the mass distribution, as well as the total void volume distribution, within the fragments obtained by fragmenting dense colloidal packings is homogeneous, at least as long as the size of the fragments is large compared to the size of the elementary colloidal units. Thus, the fragments are neither mass– nor total pore volume fractals.

The specific surface area accessible to nitrogen molecules has a more subtle behaviour. $D_s = 3$ (*i.e.* the specific surface area is independent of fragment size)

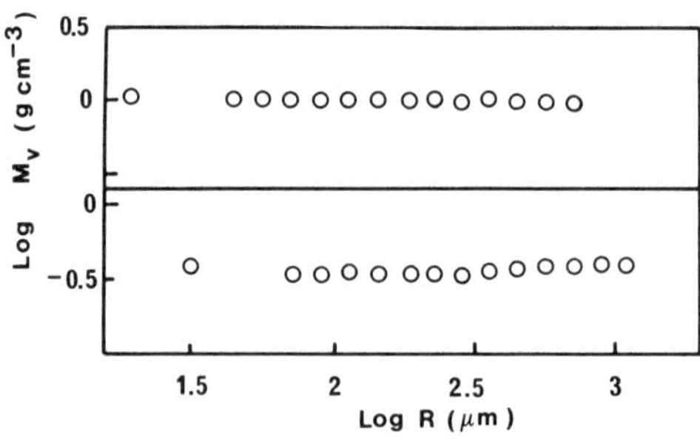

Figure 3 Log-Log plot of the apparent volumic mass versus average fragment size for sodium-montmorillonite (top) and sepiolite (bottom).

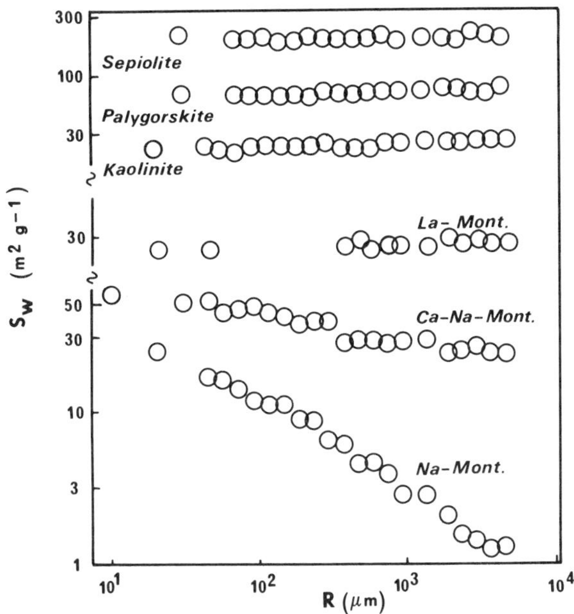

Figure 4 Log-Log plot of the nitrogen BET surface area of sepiolite, palygorskite, kaolinite, and $Na-$, $Ca-$, and $La-$exchanged montmorillonites versus average fragment size.

for all the materials in which the contact areas between elementary units are very limited. This is the case for the fibers (sepiolite), the needles (palygorskite) and the rigid platelets (kaolinite). At first sight, this result may look somewhat "magic" with respect to the usual case of a compact, non porous, body of size R, the specific surface area of which is scaling as R^{-1}. In fact, $D_s = 3$ merely expresses the fact that, at the length scale of the packing fragments, which is much larger than size of the elementary units, the system is an ideal porous body, in which the surface area developed by all the elementary units is space-filling. It corresponds to a porous system in which the accessibility of the elementary units to a probe coming from outside (gas phase nitrogen) is statistically independent of the position of the unit in the packing. It is another trivial situation, which is similar to that of glass beads (the elementary units) in a beaker (the packing fragment). Although the surface of each bead is a two-dimensional space, the total surface of the ensemble of beads looks three-dimensional at the length scale of the beaker.[9]

The most interesting situation occurs with the flexible units, *i.e.* with the fragments of smectitic packings, which, for most of them, have specific surface areas which decrease as R increases. Remarkably, all the cation–exchanged montmorillonites are still characterized by linear $\log S_w$ vs $\log R$ plots, over almost three decades, which justifies *a posteriori* the power law hypothesis (figure 4). The slopes correspond to surface fractal exponents ranging from 2.5 to 3.0, most of them being significantly below 3.0.

Equally remarkable is the fact that the same power laws are obtained for the accessible porosity and microporosity, with non-integer exponents which are in excellent agreement with the surface fractal exponents (figure 5). Thus, smectitic packings submitted to a fragmentation treatment lead to fragments which are, most of the time, surface and pore volume fractals. However, the surface and the void volume which exhibit this fractal character are not the total surface and void space, but the open surface and the open porosity. In other words, what appears to be fractal in the fragments structure is the surface defect geometry.

5. Defect geometry and local order

The relationship between the surface defect geometry and the microtexture of the solids appears in figure 6, where the scaling exponents for the accessible surface and accessible pore volume have been plotted vs the coherence length, L, of the layers stacks within the solid, as calculated from the line broadening of the 001 reflection lines in the X-ray diffraction diagram. L is a direct estimate of the thickness of the ordered aggregates in the solid. Although the data are scattered, the general trend is clear: the (accessible) surface and pore space exponents increase as the thickness of the aggregates increases. In addition (but this is less clear on figure 6), L increases, as expected (see section 2), as the charge and the polarizing strength of the interlayers cations increases.

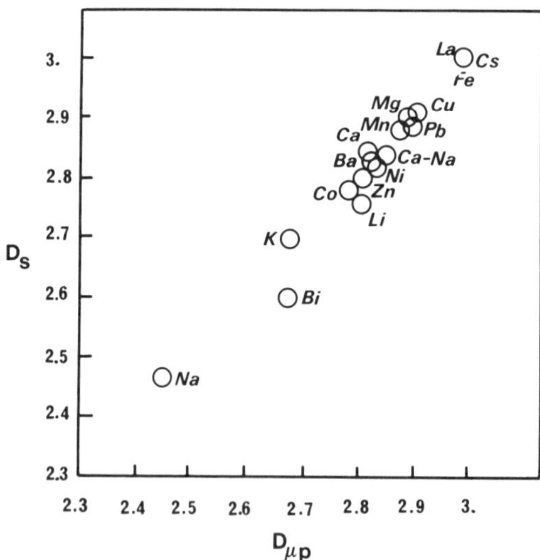

Figure 5 Correlation between the scaling exponents for accessible surface area and those for open microporosity.

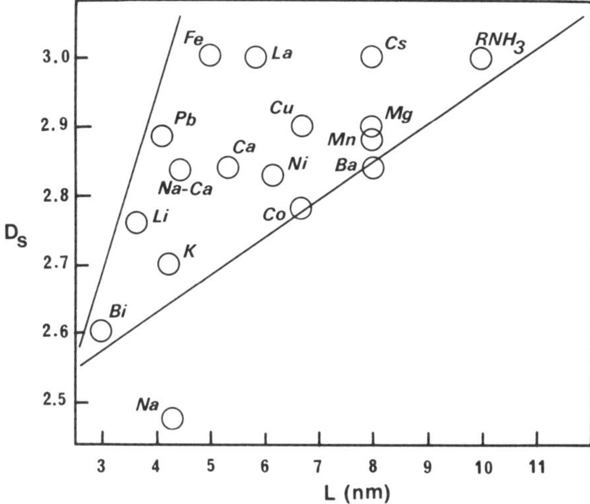

Figure 6 Correlation between the scaling exponent for accessible surface area (or for open porosity) and the coherence length for stacking order within the solid.

The increase of the scaling exponents with aggregate thickness can be rationalized in morphomechanical terms. As pointed out in section 2, the thickness of the aggregates is directly related to their thickness, so that a system of thick layered aggregates is expected to behave like a system of rigid platy crystals, and to yield a card-house-type structure with very local contact areas, a large and totally open porosity and scaling exponents equal to 3. On the other hand, thin aggregates are still flexible and, upon drying, the capillary forces are expected to drive the system into a compact packing structure, owing to the deformability of the particles. This type of arrangement is expected to have a much smaller accessible surface area and a much smaller open porosity. This is sketched in figure 7.

A close look to the fragments of figure 2 shows that the surface defects are not

Figure 7 Two-dimensional sketch of the microtexture of a packing of thick and rigid aggregates of layers (left) and of individual and flexible layers (right)

large voids. Clearly, they ought to be microscopic, with sizes closer to the size of the elementary colloidal particles or their aggregates than to the size of the packing fragments. The equality between the scaling exponents for surface area and those for porosity strongly suggests that the surface defects are thin cracks, with a linear relationship between the length of the cracks and their width. This conclusion is very close to the observations of Skjeltorp and Meakin[10] on the fracture of monolayers.

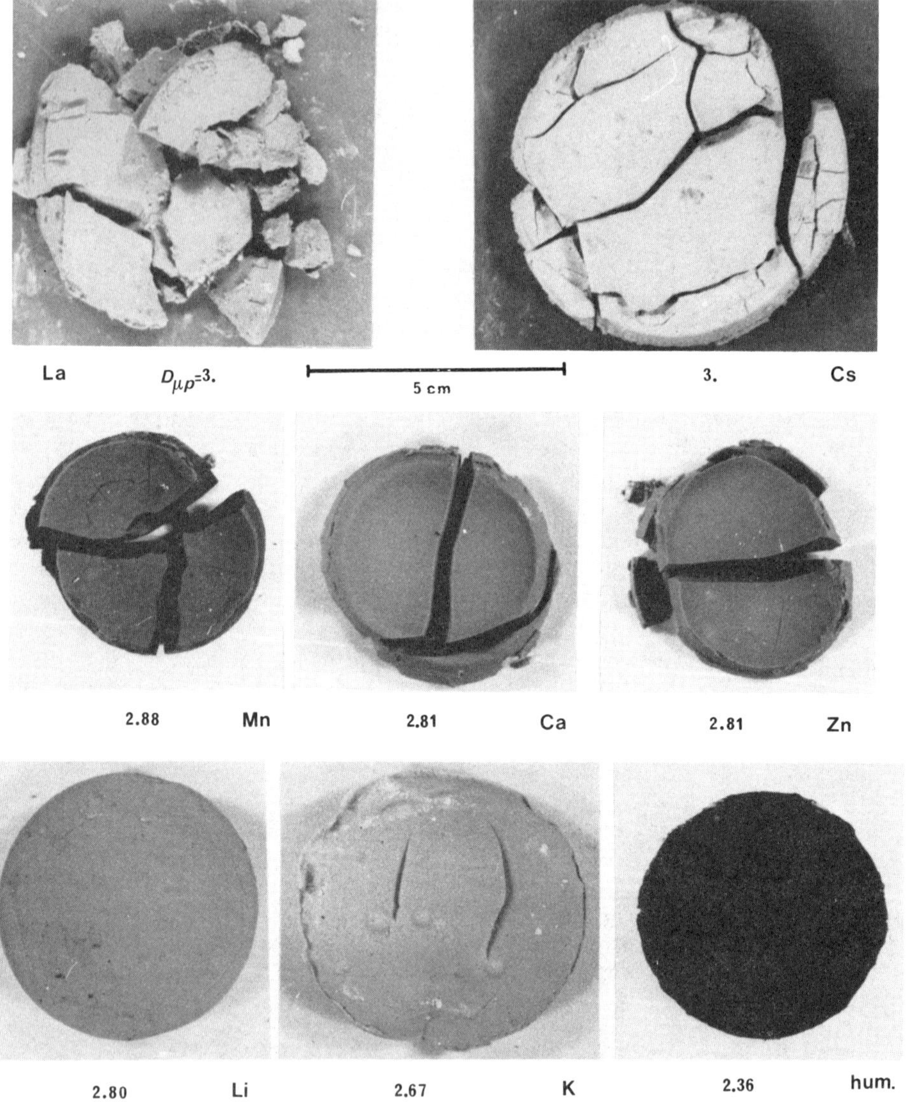

Figure 8 Photographs of dried cakes obtained by filtering several ion-exchanged montmorillonite slurries. The scaling exponent for the open microporosity is indicated.

It founds an indirect support in the fracture patterns of dried cakes obtained by filtering smectite slurries (figure 8). Cakes obtained from slurries containing thick and stiff aggregates develop an extensive fracture network and undergo spontaneous fragmentation down to millimetric sizes. On the other hand, cakes obtained from slurries containing thin and flexible aggregates develop only a very limited visible crack network and remain monolithic.

6. Conclusion

Two main questions have to be raised. The first concerns the nature of the physical process which generates the scaling behaviour revealed by the surface area and open porosity measurements. Although the crushing treatment of the packings is certainly producing cracks, it is likely that some cracks already pre-exist, due to the capillary forces and the stresses developed   during the drying process. Thus, the most reasonable assumption is that crushing acts as a revealer and an amplifier of existing flaws. In that respect, the connection between the geometry of fragmentation and its dynamics[11] has to be done.

The second question concerns the generality of the scaling behaviour of the open porosity. Is it restricted to a particular family of materials (smectites and other layered colloids) or is it more general? In spite of the results presented here which tend to give the first answer, there are arguments for saying that a scaling behaviour might be general. Indeed, a possible reason for the fact that only smectites exhibit scaling surface flaws is that the probe used to evidence them (the nitrogen molecule) is precisely adapted to the flaws developed by these materials, whereas it might be much too small to reveal the flaws in other materials. Unfortunately, the size of molecules which can be used to measure the surface area and open porosity by physisorption is very limited. Impregnation of the fragments by resins, followed by ultramicrotomy, transmission electron microscopy and image analysis is probably the only generalizable technique.

References

[1] J.N. Israelachvili, *Intermolecular and Surface Forces*, Academic Press, London, (1985).

[2] L. Fowden, R.M. Barrer and P.B. Tinker Eds., *Clay Minerals: Their Structure, Behaviour and Use*, The Royal Society, London (1984).

[3] K. Norrish, Disc. Faraday Soc., **18**, 120, (1954).

[4] J.J. Fripiat, J. Cases, M. François and M. Letellier, J. Coll. Interf. Sc., **89**, 378, (1982).

[5] Schultz, H. Van Olphen and F.A. Mumpton Eds., The Clay Minerals Society, Bloomington, 292, (1987).

[6] M. Ben Ohoud, Ph.D. Thesis, Orléans, France (1988).

[7] M. Ben Ohoud, F. Obrecht, L. Gatineau, P. Levitz and H. Van Damme, J. Coll. Interf. Sc., **124**, 156, (1988).

[8] S.J. Gregg and K.S.W. Sing, *Adsorption, Surface Area and Porosity*, Academic Press, New York, (1982).

[9] H. Van Damme, P. Levitz, L. Gatineau, J.F. Alcover and J.J. Fripiat, J. Coll. Interf. Sc., **122**, 1, (1888).

[10] P. Meakin, this volume and references therein.

[11] S. Redner, this volume and references therein.

III Statistical Fracture Models

The section is concerned with some simple models of fracture where disorder is explicitly taken into account. In the first case, Meakin *et al.* introduce many models of fractures where the disorder appears continuously as the time evolves (annealed disorder). Some of these models are direct mechanical analogs of the Dielectric Breakdown models (in the class of DLA type models), making the transition continuous from the previous section. However, the patterns obtained in this model are very different from the ones obtained in usual DLA. The scaling behavior of these fracture patterns remains still an open question since it seems to be sensitive to local details of the model. Other models are presented such as the fragmentation of thin films deposited on a substrate, and where the stresses originate from drying.

The next lecture by Duxbury and Li deals with a "dilution" type disorder (random distribution of voids). Here a basic assumption is that a crack is always unstable and therefore, it suffices to identify the weakest defect in the structure. Using a scaling argument, it is proposed, and checked numerically, that the most sensitive element carries a stress (or a current in dielectric models) which increases with the system size. Therefore, the failure stress decreases to zero with the system size. Extensions of this model to other situations are discussed. Finally the brittle to tough transition with residual bonding is mentioned.

The final lecture by Herrmann and de Arcangelis gives different models of rupture, classified according to the origin of the disorder (annealed — *i.e.* generated in time as the rupture evolves, with no memory — or quenched — *i.e.* present in the material from the beginning, and fixed in time). Deterministic fracture models on beam networks are discussed with a discussion of "memory effects." Fractal patterns obtained using some of these rules are presented. This analysis bridges the gap between the stability analysis where disorder is merely absent, and annealed disorder models which exhibit fractal or self-affine patterns, due to this instability. For one quenched disorder model, some scaling laws on the stress-strain relations are obtained, which allows to take size effects into account. The latter are in this case quite important and result in non trivial exponents in the scaling laws proposed. Then, concerning an annealed model, close to the one studied by Meakin in a previous lecture, the patterns generated in the limit of a deterministic process (vanishing disorder) are studied.

Chapter 7

Simple Stochastic Models for Material Failure

Paul Meakin
Gang Li, Leonard M. Sander, and Hong Yan[†]
Francisco Guinea and Oscar Pla[‡]
Enrique Louis[*]

Central Research and Development Department
E. I. du Pont de Nemours and Company
Wilmington, DE 19880-0356 USA

A variety of simple models for material failure and deformation have been developed in recent years. These models are closely related to and were in part motivated by diffusion-limited aggregation model and other models for non-equilibrium pattern formation. Here several related models for crack growth are described. In a model closely related to DLA the effective fractal dimensionality appears to depend on local boundary conditions as well as on the dependence of the growth probabilities on the local stress and strain. The dependence on local boundary conditions is surprising and may be a consequence of finite size effects. If the growth probabilities are represented by an inhomogeneous power law then crossover effects that depend on the boundary conditions are found. Such effects are also expected for other models such as DLA and dielectric breakdown models. Some recent work concerning models for surface cracking and models with non-central forces are also surveyed. These simple models raise important theoretical questions that may lead to a better understanding of real materials. Improved algorithms for the efficient relaxation of elastic networks are needed for models of this type to continue to complement experimental and theoretical work in a strongly synergistic fashion.

1. Introduction

Simple computer models have been used for almost 20 years[1] to investigate the mechanical failure of materials. In these models the material is represented by

---

[†] Harrison M. Randall Laboratory, University of Michigan, Department of Physics, Ann Arbor, MI 48109-1120 USA

[‡] Instituto de Ciencia de Materiales, Universidad Autonoma de Madrid, E-28049 Madrid, Spain

[*] Departmento de Fisica Aplicada, Universidad de Alicante, Apt 99E-03080 Alicante, Spain

Disorder and Fracture, Edited by J. C. Charmet *et al.*
Plenum Press, New York, 1990

a network of undeformable nodes (of zero or non-zero size) connected by elastic bonds. It is usually assumed that the material fails by a random bond breaking process in which the bond breaking rates or probabilities (P_i) depend only on the forces (f) or strains (δ) associated with the bonds. The assumption that the rate of bond breaking can be expressed as a function of the local force only, results in tractible algorithms that lead to realistic cracking patterns and results that are in surprisingly good agreement with experiments in those cases where comparisons have been made. However, it is difficult to find any fundamental justification for this assumption. This is particularly true in the late stages of failure where the strained bonds are found in a wide range of different environments. There has also been considerable effort devoted to the use of atomic level molecular dynamics simulations (see references [2-10], for example) to explore material failure processes. This is an important approach to the development of a better understanding of material failure processes that is complementary to that described here. Much of this molecular dynamics work has been confined to two dimensional systems and frequently employs simplified pair-potentials.

In recent years there has been a resurgence of interest in simple models for material failure. There seem to be several reasons for this. The success of the Diffusion-Limited Aggregation (DLA) model[11] and in particular the dielectric breakdown version[12] of this model naturally stimulated interest in related problems such as random growth in an elastic medium[13] described by the Navier equation[14,15] instead of random growth processes in which the growth probabilities are determined by a scalar field obeying the Laplace equation (DLA). Interest in using computer simulations to develop a better understanding of complex phenomena has grown rapidly in recent years as the power, availability and ease of use of digital computers has advanced. The ability to display the growth of complex patterns using a variety of graphical devices has also contributed to the development of this area.

In most models the simulation consists of two primary steps that are repeated many times. In the first step, a bond in the network is selected randomly with a probability proportional to its strain dependent bond breaking rate. In the second step, the elastic network is allowed to relax to a new mechanical equilibrium. The simulation proceeds by a sequence of bond breaking and relaxation steps until the material has failed or until a sufficiently large cracking pattern has developed. The relaxation step is very time consuming on the computer. A variety of strategies (over-relaxation,[16] block relaxation,[16] conjugate gradient methods,[17] Fourier acceleration[18] and others) have been developed to improve the rate of convergence. However, at present, there are no really efficient algorithms that are widely applicable to this class of problems. This is a major obstacle to progress in this area.

All materials of practical importance have complex morphologies[19-21] and many processes play an important role in material failure.[3-5,22-25] The development of a comprehensive model that would describe all of these structures and phenomena would be a formidable undertaking that would require many man-years of effort and would be error-prone because of its complexity. In addition, a computer program of this scope and complexity would require enormous amounts of computer time. The results from such a program could also be difficult to interpret (but would certainly be easier to interpret than the results of real experiments providing that an error-free code could be developed). On the other hand, many of the models discussed here include only one fracture mechanism in an (initially) homogeneous elastic network. While such models may contribute to our understanding at a fundamental level, they are unlikely to be of much value in representing the behavior of real materials. Progress is now being made on intermediate level models that include more than one failure mechanism and/or at least a crude representation of defects, grain boundaries, second phases, *etc...*At least in the short term, these models appear to be of the most value in helping us to understand the failure and distortion of real materials.

All of the models described here can be used to explore the kinetics as well as the structural aspects of material failure. In all of the models the bond breaking rates are functions of the local bond strain (δ) so that the bond breaking rate constant for the ith bond in the system is given by

$$k_i = g(\delta_i). \tag{1}$$

In this event the simulations can be made fully time dependent by picking bonds randomly and breaking them if

$$x < \frac{g(\delta_i)}{g_{max}}. \tag{2}$$

Here $g(\delta_i)$ is the bond breaking rate constant for the randomly selected bond. g_{max} is the maximum value of $g(\delta)$ for any bond in the system and x is a random number uniformly distributed over the range $0 < x < 1$. Each time a bond is selected (whether or not it is actually broken) the time is incremented by an amount δt given by

$$\delta t = \frac{1}{N g_{max}} \tag{3}$$

where N is the total number of unbroken bonds in the system. In most cases the bond breaking rate constants k_i are not expressed in absolute units (sec^{-1} *etc...*) so that the time scale is defined only within a constant factor.

2. Crack Growth Models Related to DLA

The use of a diffusion-limited-aggregation[11] (DLA) like model to simulate crack growth was first successfully carried out by Louis and Guinea.[13,26,27] The Navier equation

$$(\lambda + \mu)\partial_i \Big[\sum_j \partial_j u_j \Big] + \mu \sum_j \partial_j^2 u_i = 0 \tag{4}$$

was discretized onto a lattice using a triangular network of bonds and nodes in much the same way that the Laplace equation is discretized onto a lattice in the dielectric breakdown model.[12] In Eq.(3) u_i is the ith component of the displacement field and the quantities λ and μ are the Lame coefficients.[15] A triangular lattice is used rather than a square lattice because a square network of nodes with nearest neighbor central force interactions would have no shear modulus. In these DLA-like models it is assumed that only those bonds on the surface of the crack (which is initiated by removing a bond near the center of the network) are broken as the crack grows. Fig.(1) shows three different ways in which the surface bonds can be defined (other definitions are, of course, possible). The three possibilities will be referred to as model I, model II and model III respectively. In model I, (Fig.(1a), only those bonds joining pairs of nodes that are both on the crack perimeter can be broken. In model II, Fig.(1b), any bond associated with a damaged node (a node with 5 or fewer bonds) can be broken and in model III, Fig.(1c), any of the bonds associated with any of the nodes at the crack perimeter can be broken. Model II was used in the original work of Louis et al.[13,26]

In this model the bond breaking probabilities P_i are given by

$$P_i = \delta_i^\eta / \sum_{j=1}^{N} \delta_j^\eta \tag{5}$$

where $\delta_i = |l_i - l_0|$ is the bond strain (l_i is the bond length and l_0 is the equilibrium bond length) and N is the number of surface bonds. In most simulations the network is dilated isotropically (by a small amount to avoid unwanted non-linearities) and the positions of the nodes at the edges of the network are fixed. In some cases, a constant force is applied to the nodes at the perimeter of the network. Fig.(2)

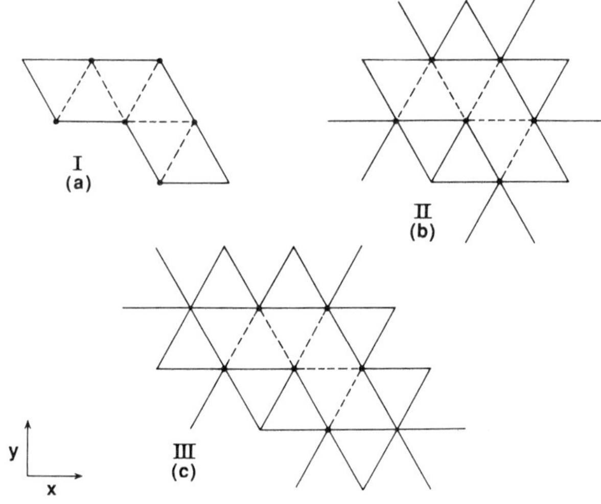

Figure 1 Three alternative definitions of the surface bonds near a crack on a triangular lattice. The crack is defined by the broken (– – –) bonds. In model I, Fig.(1a), the bonds which form the edge of the crack are used. In model II, Fig.(1b), all of the bonds asociated with "damaged" nodes are used. In model III, Fig.(1c), all of the bonds associated with all of the nodes at the edges of the crack (damaged and undamaged) are used.

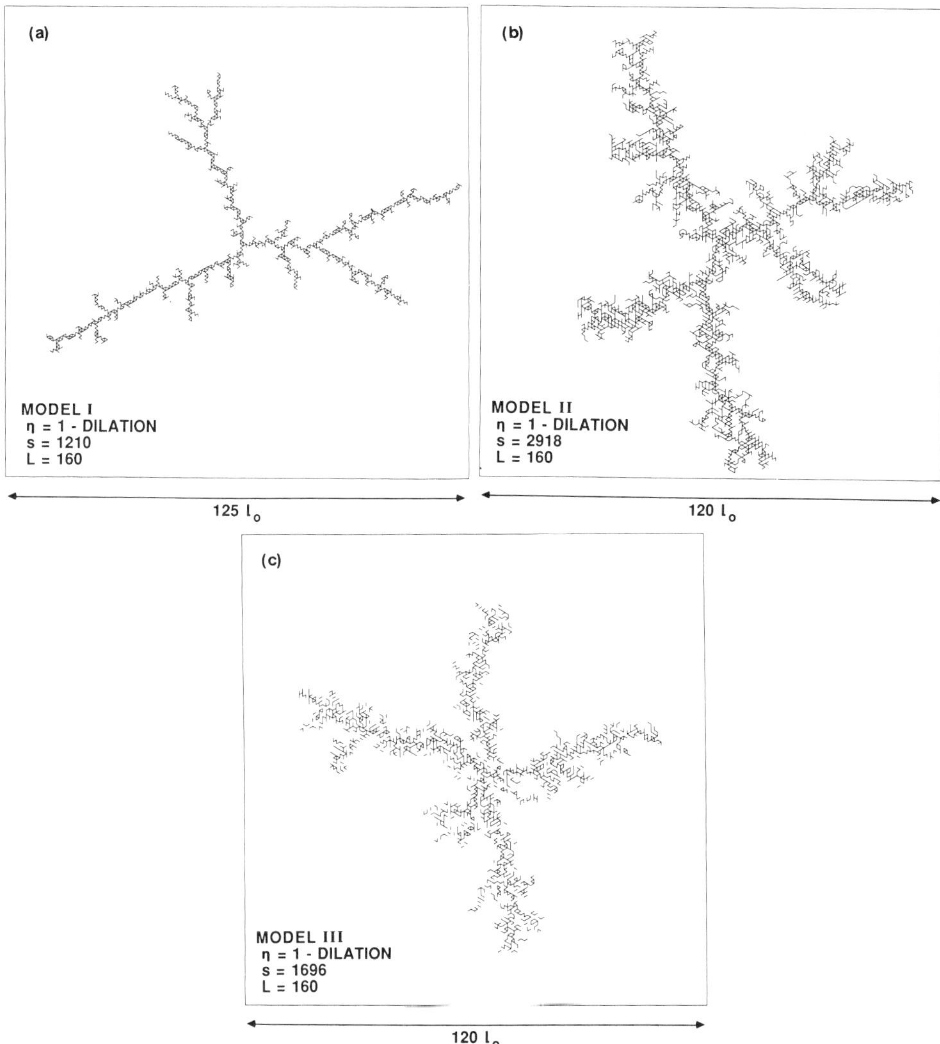

Figure 2 Examples of cracking patterns generated using the DLA-like cracking model with a bond breaking probability exponent η of 1 and Figs.(2a, 2b and 2c) show cracks generated using models in which the bonds capable of breaking correspond to models I, II and III respectively, see Fig.(1).

shows the results of simulations carried out using all three models for the local boundary conditions at the crack surface with a growth probability exponent η of 1. The patterns generated by these models are quite similar to those associated with the DLA model. These patterns are probably self-similar fractals and their fractal scaling properties can be measured using the methods that have been applied to DLA. Unfortunately, the amount of computer time required to grow a cracking pattern consisting of a few thousand broken bonds is comparable to that required to grow a DLA cluster with a few million sites or particles. Fig.(3) shows the dependence of the radius of gyration R_g on the number of broken bonds s obtained from eight simulations with model III boundary conditions and a growth exponent η of 1. The dependence of $\ln(R_g)$ on $\ln(s)$ appears to be quite linear for cracks containing more than about fifty broken bonds, but the accessible linear regime spans less than an order of magnitude in length scales. The slope of about 0.60 corresponds to a fractal dimensionality of $1/0.6$ or $5/3$ (quite close to the value of about 1.7 obtained from the DLA model). The values obtained for the fractal dimensionality D_β from the dependence of R_g on s assuming that

$$R_g \sim s^\beta \tag{6}$$

and $D_\beta = 1/\beta$ are given in Table I.

Simulations have also been carried out using similar models with uniaxial extension[29,31] or shear strain.[13,26,29–31] With shear strain the cracking pattern has an "X" like shape. (If only bonds in tension are allowed to break only one arm of the "X" shape grows.) Fig.(4) shows the results obtained from simulations with $\eta = 1$ and with model I and model III boundary conditions. In this case the shear was in the x direction (starting with an equilibrium triangular lattice in which the nodes have coordinates (x_i, y_i), the network was distorted so that the coordinates became $(x_i + \varepsilon y_i, y_i)$). Very similar results were obtained for shear in the y direction (the x and y directions are not equivalent on the triangular lattice - see Fig.(1)). The shear strain parameter ε was set to a small value (0.01 in most cases). Some additional results obtained from these models are presented in Table I. Fig.(5) shows cracking patterns generated with uniaxial extension. Here the node coordinates are changed from (x_i, y_i) to $(x_i + \varepsilon y_i, y_i)$ at the start of a simulation where (x_i, y_i) are the equilibrium node coordinates.

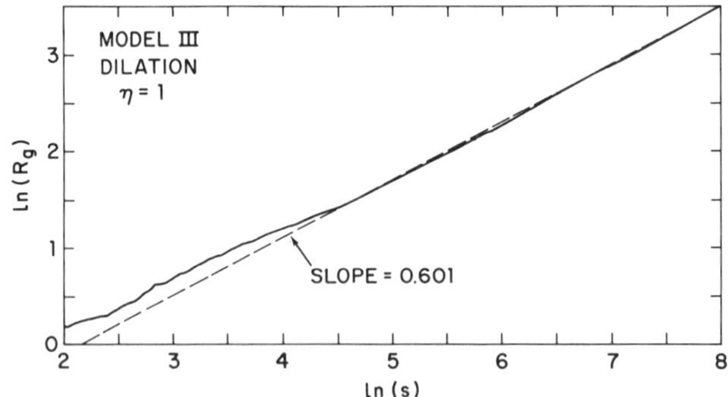

Figure 3 Dependence of the radius of gyration R_g on the crack size (number of broken bonds, s) obtained from 8 simulations with model III surface bonds, a growth exponent, η, of 1 and dilational strain.

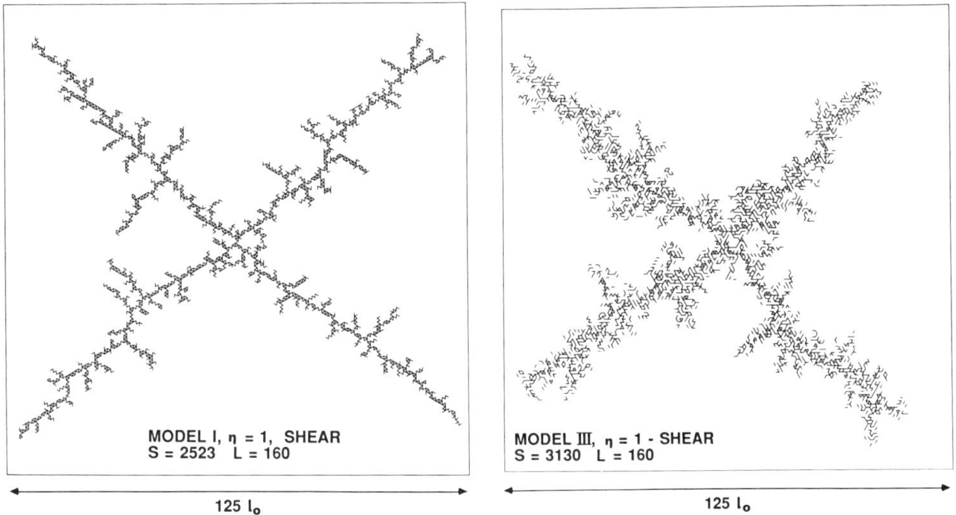

Figure 4 Effect of the selection of surface bonds (bonds capable of being broken) on the cracking patterns obtained with $\eta = 1$ and shear strain.

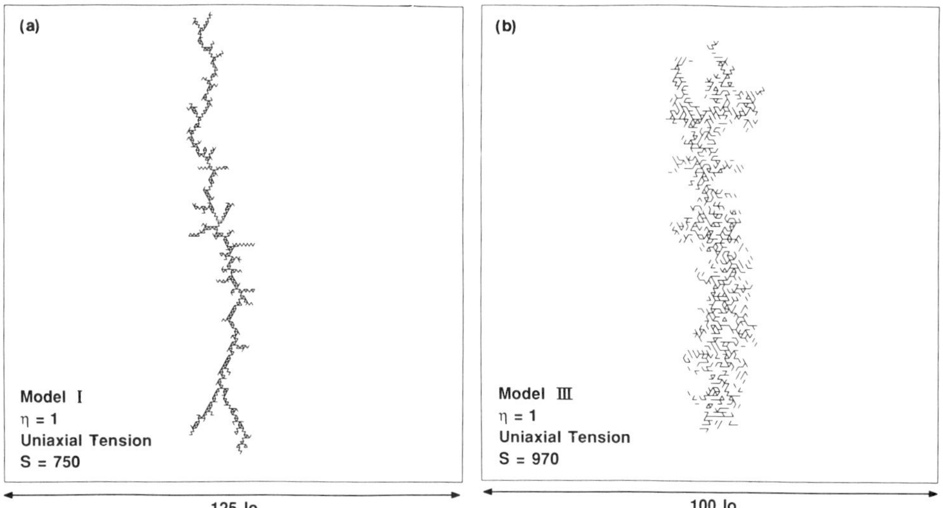

Figure 5 Cracking patterns obtained using uniaxial tension (in the x direction). This figure shows results obtained from simulations carried out using a bond breaking probability exponent η of 1 for model I, Fig.(5a) and model III, Fig.(5b) surface bonds.

Table I. Effective Fractal Dimensionalities.* Obtained from the DLA-like Crack Growth Models

Growth Exponent η	Stress Field	Surface Model	$D_{eff.}$
1	Dilatation	I	1.35
1	Dilatation	II	1.51
1	Dilatation	III	1.66
1	Shear	I	1.42
1	Shear	II	1.62
1	Shear	III	1.65
2	Dilatation	I	1.12
2	Dilatation	II	1.16
2	Dilatation	III	1.45
2	Shear	I	1.17
2	Shear	II	1.49
2	Shear	III	1.40

** The effective fractal dimensionalities were obtained by least squares fitting straight lines to the coordinates $(\log(R_g), \log(s))$ for the larger values of s. Because of both finite size effects and statistical uncertainties the values of D_β given here may not be close to their asymptotic values. Particularly for model III, the simulated clusters exhibit geometric scaling over only a small range of length scales.*

Hinrichsen et al.[29] have investigated a very similar model for the cases of uniaxial tension (or compression) and shear with model I for the surface bonds capable of being broken and a growth probability exponent η of 1. In addition, a few simulations were also carried out using model II for the surface bonds. They used a conjugate gradient method to obtain the distribution of forces in the network after each bond breaking event. This method is substantially more efficient than the overrelaxation method, but cracks containing only about 300 broken bonds on 64x64 node arrays were used in this work. The criterion used for stopping the relaxation process was much more stringent in this work than in other simulations and it is possible that some of the results obtained in other studies may be compromised by incomplete relaxation. However, reasonably accurate results have been obtained from the dielectric breakdown model using quite crude and inefficient algorithms.[12]

For uniaxial compression Hinrichsen et al. found a value of 0.82 for β ($D_\beta \approx$ 1.22) and for shear $\beta \approx 0.74$ ($D_\beta \approx 1.35$). This is in reasonably good agreement with the value of about 1.42 obtained for D_β for model I surface bonds, $\eta = 1$ and shear strain by Meakin et al.[30] (see Table I). However, based on the scaling behavior of the principal values of the inertial tensor of the cracking pattern, Hinrichsen et al. came to the conclusion that the effective fractal dimensionality for the case of shear strain is 1.28 ± 0.06 (from the largest of the two exponents describing the growth of the smaller eigenvalue of the inertial tensor). As a result of this observation, they suggest that the asymptotic ($s \to \infty$) fractal dimensionality might be the same for both shear and uniaxial compresion. They point out that the individual arms of the cracking patterns generated by shear strain resemble the cracking patterns produced by uniaxial compression (compare Figs.(4) and (5)).

Simulations have also been carried out with values for the bond breaking probability exponent η different from 1. For $\eta = 1/2$, the effective fractal dimensionality D_β lies in the range $1.90 \leq D_\beta \leq 1.95$ for models I, II and III boundary conditions with both shear strain and isotropic dilation.[30] The results obtained from several models with $\eta = 2$ are shown in Table I. In view of the earlier results obtained using the dielectric breakdown model, it is not surprising that larger values of the exponent η lead to structures that can be described in terms of a smaller fractal dimensionality.

All of the simulations shown in Figs.(2-5) were obtained using 160x160 arrays of nodes (25600 nodes and 76800 bonds). The simulated cracking process was continued until the crack reached a size s of 3200 broken bonds or approached the edge of the network. The relaxation process is continued until the largest displacement of any node from its equilibrium position is smaller than 0.001-0.01 times the initial bond strain. This would not ordinarily be sufficient to guarantee an accurate convergence, but relaxation of the whole system continues as other bonds are broken and the additional relaxation near the last broken bond ensures almost complete relaxation in this region. It appears that quite good results can be obtained without using a very small convergence threshold. However, there is no simple way of determining what threshold value is needed other than repeating the simulations with a larger threshold and determining if the cracking patterns have changed significantly.

In order to draw more definitive conclusions, much more extensive, larger scale simulations would be needed. This is not possible with present algorithms and computer resources. It is clear from this work that much more efficient algorithms will be needed to make further progress.

3. Inhomogeneous DLA-like Cracking Models

In the most simple models for crack growth the dependence of the bond breaking rates on the bond strain is a simple homogeneous power law. Under these conditions, it is reasonable to expect that the cracking patterns will also be homogeneous and their scaling properties may be represented by a single exponent or fractal dimensionality. In view of the non-linear behavior of most real materials, it seems unlikely that a homogeneous power law will provide a realistic description of the crack growth process. A step towards the development of a more realistic model is to represent the dependence of the bond breaking rates or probabilities by an inhomogeneous power law

$$P_i \propto a\delta_i^{\eta_1} + b\delta_i^{\eta_2} \qquad (\eta_1 < \eta_2). \tag{7}$$

In contrast to the models with a homogeneous power law, the cracking patterns generated using an inhomogeneous power law depend on the macroscopic stress or strain applied to the systems. For low strain levels the first term in Eq.(7) will dominate and for high strain levels the second term will dominate. Consequently, the evolution of the cracking pattern depends on the boundary conditions used in the simulations.

Measurements carried out for simulations with growth probabilities described by a homogeneous power law ($\eta = 1$, 2 and 3 in Eq.(5)) demonstrate that the maximum stress or strain increases with increasing crack size for simulations carried out with a "clamped" outer boundary. Consequently, we expect to see a crossover from a relatively compact pattern at early stages to a more stringy pattern at late stages. Ideally, we might expect to see a crossover from a pattern characteristic of $\eta = \eta_1$ to a pattern characteristic of $\eta = \eta_2$. In practice, the entire crossover cannot easily be seen in a single simulation.

Fig.(6) shows the results of simulations carried out with bond breaking rate constants given by

$$k_i = \delta_i + b\delta_i^4 \tag{8}$$

The initial strain δ was 0.1. Cracks were generated with four different values of b (125, 64, 27 and 8 respectively in Figs.6a, 6b, 6c and 6d). Model II, see Fig.(1), was used for the local boundary conditions. In Fig.(6d) the parameter b is quite small and the cracking pattern resembles, quite closely, that shown in Fig.(2b). The crossover can be most clearly seen in Fig.(6b) ($b = 64$). For significantly larger values of b ($b = 125$, Fig.(6a)) the entire pattern resembles the cracking pattern obtained with an exponent η larger than 1. The crossover (for $b = 64$) can be more easily seen by displaying the cracking pattern at several stages during its growth.

Similar simulations have been carried out for other values of the exponent η_2. As η_2 becomes larger, the range of strains or stresses over which the crossover takes place becomes smaller and the crossover can be more easily seen. Fig.(7) shows four stages in the growth of a crack with the parameters $a = 1$, $b = 65536$, $\eta_2 = 1$ and $\eta_2 = 9$.

We have also examined cracking patterns using models in which the bond breaking probability becomes independent of the local stress or strain for high stresses or strains. In these models the bond breaking probabilities are given by

$$Pi \propto \exp(-\delta_0/\delta_i) \tag{9}$$

or

$$Pi \propto \tanh(\delta_0/\delta_i). \tag{10}$$

In the first case, Eq.(9), the bond breaking probability decreases very rapidly with decreasing strain for small strains. The overall cluster shape is quite similar for a quite wide range of values for the parameter δ_0 but the local structure becomes less dense as δ_0 is increased. A similar behavior is observed in the second case, Eq.(10), but the local structure is more compact for clusters grown using Eq.(10) than those grown using Eq.(9) for the same value of δ_0.

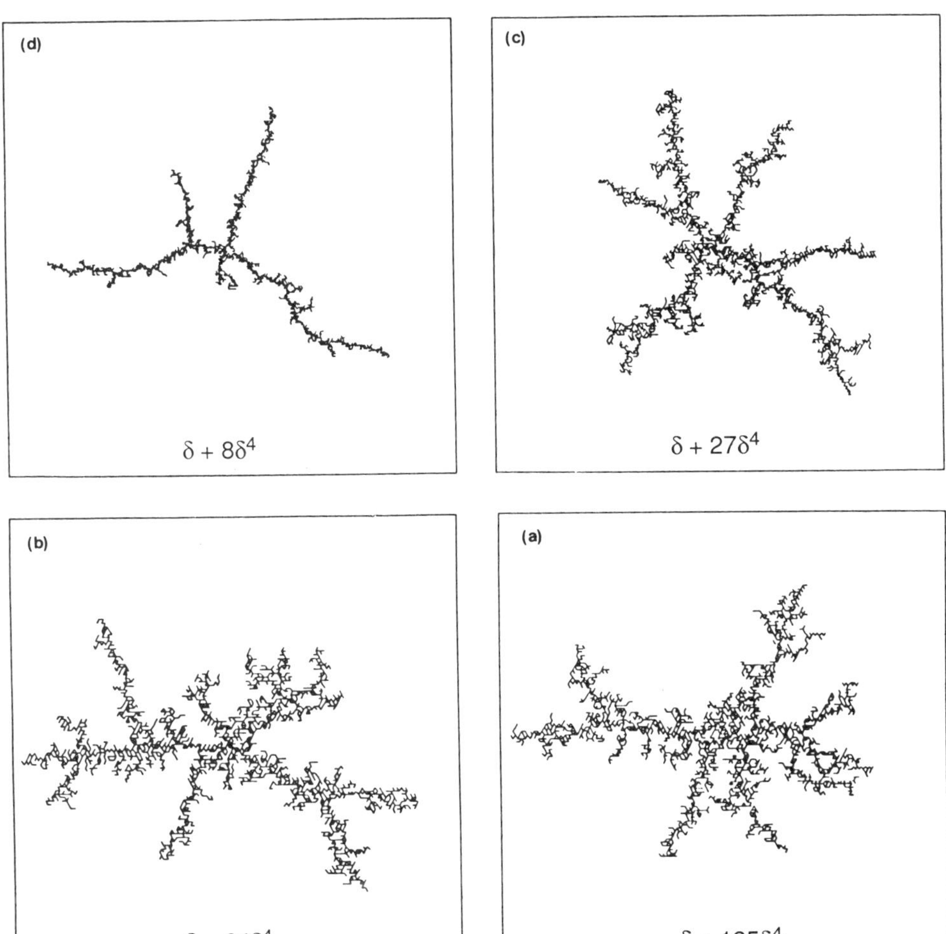

Figure 6 Cracking patterns generated using bond breaking probabilities given by $P_i \propto \delta_i + b\delta_i^4$ with four different values of the parameter b (125, 64, 27 and 8 in Figs.6a, 6b, 6c and 6d, respectively).

Finally, simulations have been carried out using the growth law

$$P_i = 0 \qquad (\delta < \delta_0) \tag{11a}$$
$$P_i \propto \delta - \delta_0 \qquad (\delta > \delta_0) \tag{11b}$$

For simulations carried out with an initial strain of 0.1 and $\delta_0 = 0.09$, the cracking pattern is very stringy on short length scales (this is a consequence of the complete cessation of growth in the screened regions). On longer length scales the pattern is more highly branched than a crack generated by homogeneous growth processes with $\eta = 1$. A typical pattern is shown in Fig.(8).

4. Models for Surface Cracking

Systems in which a thin surface layer has quite different properties than an underlying substrate have many important applications. A few examples include paint films on metal or wood, ceramic coated metals, optical coatings on glass and metal coated polymers. A phenomenon that frequently limits the performance and reliability of these systems is cracking of the surface layers. Since the substrate and surface layer have quite different properties, large stresses frequently occur in the

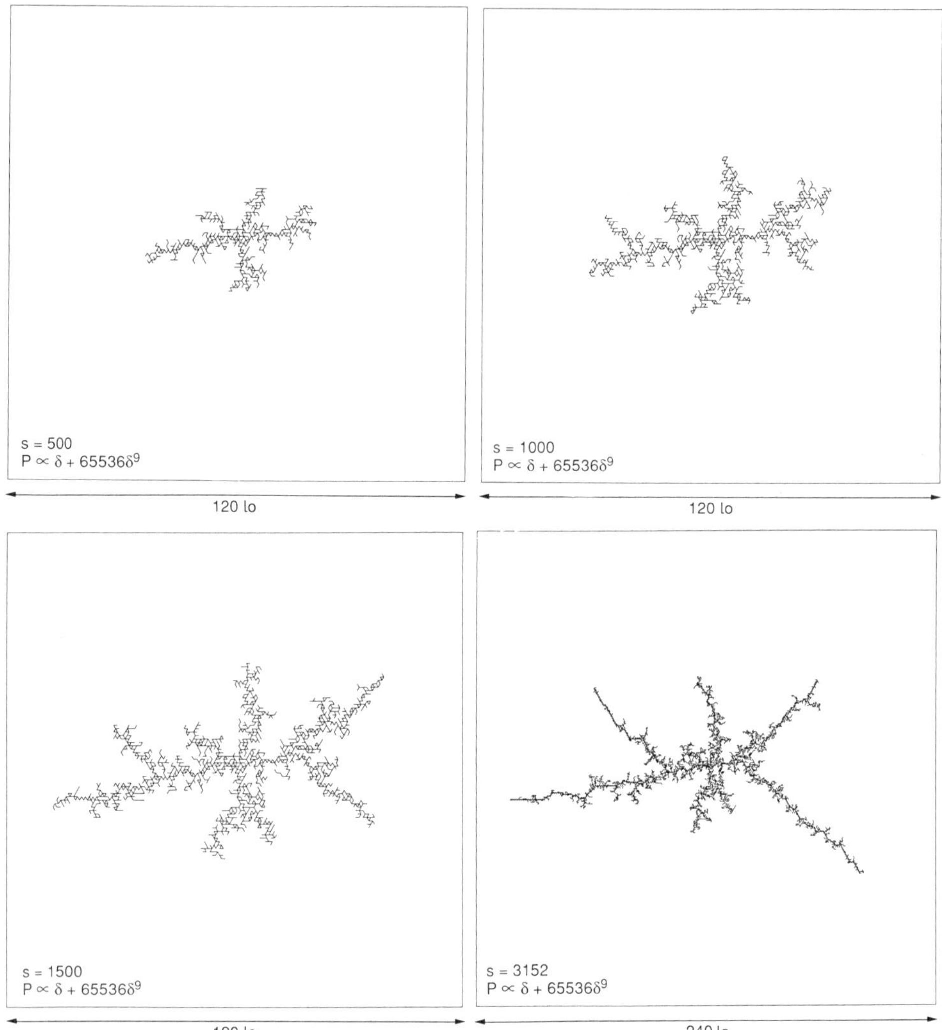

Figure 7 Four stages in a crack grown simulation carried out using bond breaking probabilities given by $P_i \propto \delta_i + 65536\delta_i^9$.

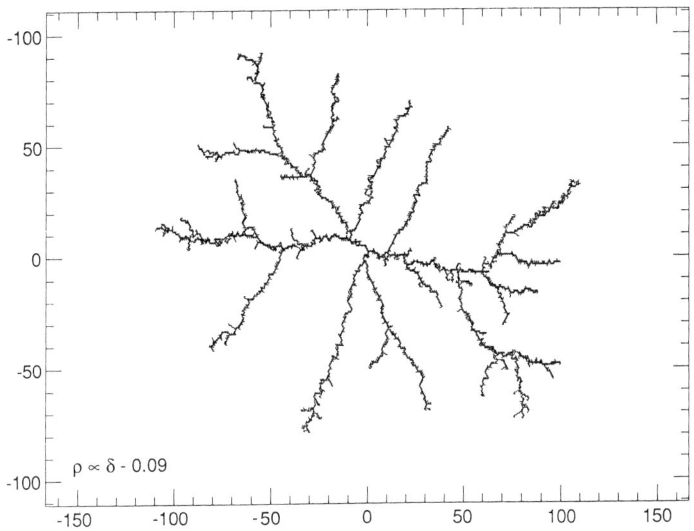

Figure 8 A cracking pattern generated using the growth law $P_i \sim (\delta_i - 0.09)$ using model II local boundary conditions with an initial strain δ of 0.1.

surface layers. The resulting cracking pattern frequently resembles that associated with mud cracking in dried up lake beds.

Fig.(9) shows a model used to simulate this type of cracking process. The surface film is represented by a network of bonds and nodes described in the previous section. In the harmonic approximation, the elastic energy associated with this layer is given by

$$H = \frac{1}{2} k \sum_i \delta_i^2 \qquad (12)$$

where the summation is over all of the bonds in the network and it is assumed that the force constant k is the same for all bonds. In addition, each node in the network is attached to the substrate by a weak bond (represented by the dashed lines in Fig.(9)). The force exerted by this weak bond on the ith node is given by

$$\mathbf{f} = k_2(\mathbf{r}_i^0 - \mathbf{r}_i) \qquad (13)$$

where r_i is the position in the ith node and r_i^0 is the position of attachment to the substrate (its position at the start of the simulation). In this model, only the "strong" bonds in the surface layer are allowed to break (all of the nodes remain attached to the substrate). The bond breaking probabilities are given by

$$Pi \propto e^{\frac{h(\delta_i)}{k_B T}} \qquad (14)$$

where k is the force constant for bonds in the surface layer and k_B is the Boltzmann constant. In reduced units ($k_B T = 1$) the bond breaking probability for the ith bond is given by

$$P_i \propto e^{h(\delta_i)} \qquad (15)$$

It is usually assumed that the function $h(\delta)$ has a simple form such as

$$h(\delta) = \frac{k\delta^2}{2} \tag{16}$$

or

$$h(\delta) = v\delta \tag{17}$$

In a typical simulation[32] the initial bond extension is about 10% ($l = 1.0$ at the start of the simulation and $l_0 = 0.90909.$ or $l/l_0 = 1.1$). Fig.(10) shows the results of a small scale simulation carried out with a triangular network consisting of a 50×50 array of nodes with 7500 connecting bonds. The function $h(\delta)$ was given by Eq.(15) with values of 1600 and 16 for k and k_2 respectively. The initial bond extension ($l - l_0$) was 0.0909. Fig.(10b) shows features (labeled "S") that resemble shear bands. Fig.(11) shows results obtained from much larger scale simulations using a network of 200×200 nodes joined by 120,000 bonds at the start of the simulations. In this figure only the locations of the broken bonds in the original (undistorted) network are shown. All four parts of Fig.(11) were obtained using a value of 100 for k/k_2. The effects of increasing k from 600 to 2400 are shown. For large values of k, the first cracks to be formed are quite linear; but as the cracking process continues, the cracks become more irregular.

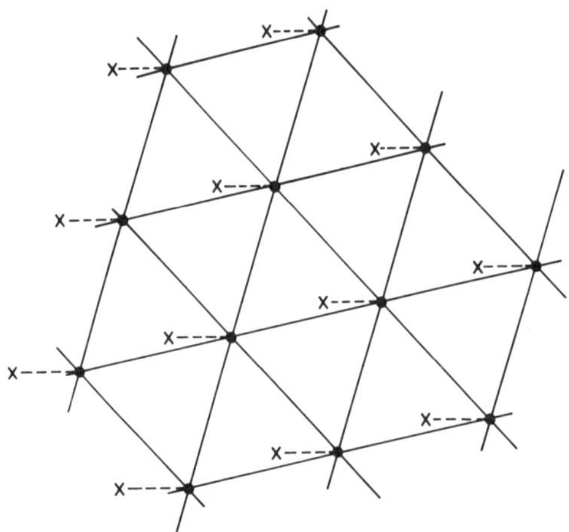

Figure 9 A schematic representation of a model for elastic fracture in thin films. The nodes (large dots) are connected by strong bonds to form a triangular lattice. Each node is joined to the underlying structure by a weak bond (– – –) at its original position at the start of the simulation. Throughout the simulation the distance from the nodes to the underlying substrate is constant (only horizontal motion is allowed). This figure shows the original configuration in which each node is associated with six bonds in the surface network.

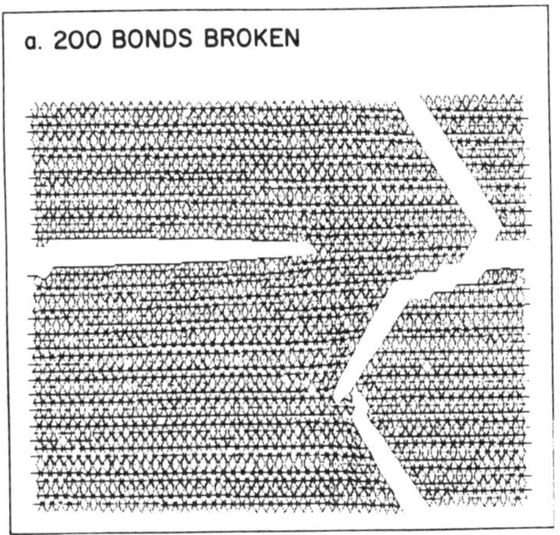

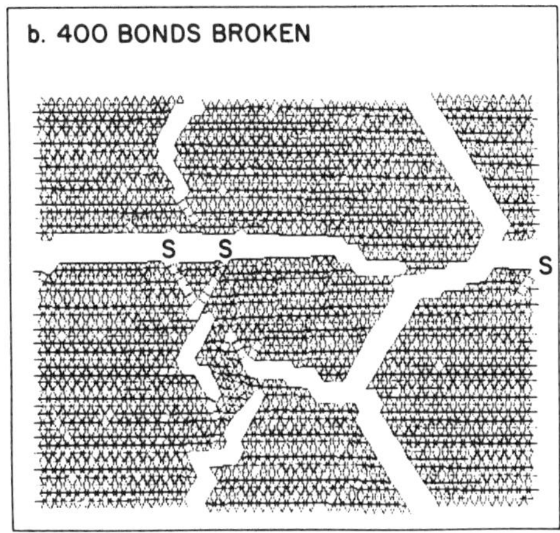

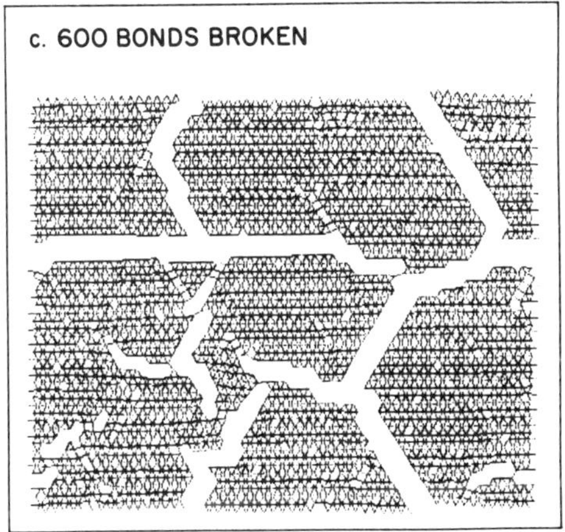

Figure 10 Results obtained from a small-scale simulation (2500 nodes) with $k = 1600$ and $k_2 = 16$. The system is shown after 200 Fig.(10a), 400 Fig.(10b) and 600 Fig.(10c) bonds broken. In Fig.(10b), three of the features that resemble shear bands are indicated with the letter S.

Fig.(12) shows the results of a similar simulation carried out using a square network of bonds and nodes with the parameters $k = 1600$, $k_2 = 16$, $l_0 = 0.9090909$ and $l = 1.0$ (at the start of the simulation). The cracking patterns shown in Figs.(11) and (12) resemble quite closely those shown in thin film deposits (Refs. [33-36], for example) and in other systems such as drying concrete.[37] An important difference between this cracking model and DLA-like models is the range of the effect of a growth event on the stress/strain field. In DLA and the DLA-like cracking models, growth of the pattern has a long range non-local effect on the stress/strain field. In the surface cracking model the weak interaction between the surface layer and the substrate is sufficient to strongly localize the effect of both local and extended defects (cracks) on the stress/strain field. Without this localization (and the use of over-relaxation with extra relaxation cycles in the vicinity of the last broken bond) these simulations would not be practical. Fig.(13) shows some results from time-dependent simulations. These and other simulations indicate the following scenario for large values of k and $k/k_2 \gg 1$. Initially isolated defects are formed and the rate of bond breaking is relatively slow. However, the stresses associated with bonds near these defects are larger than those further from defects and they have a relatively high probability of breaking. As the defects grow larger, the stress concentration effects become more important and once a crack has developed it grows very quickly. The fast growth of the first few cracks reduces the stress in the

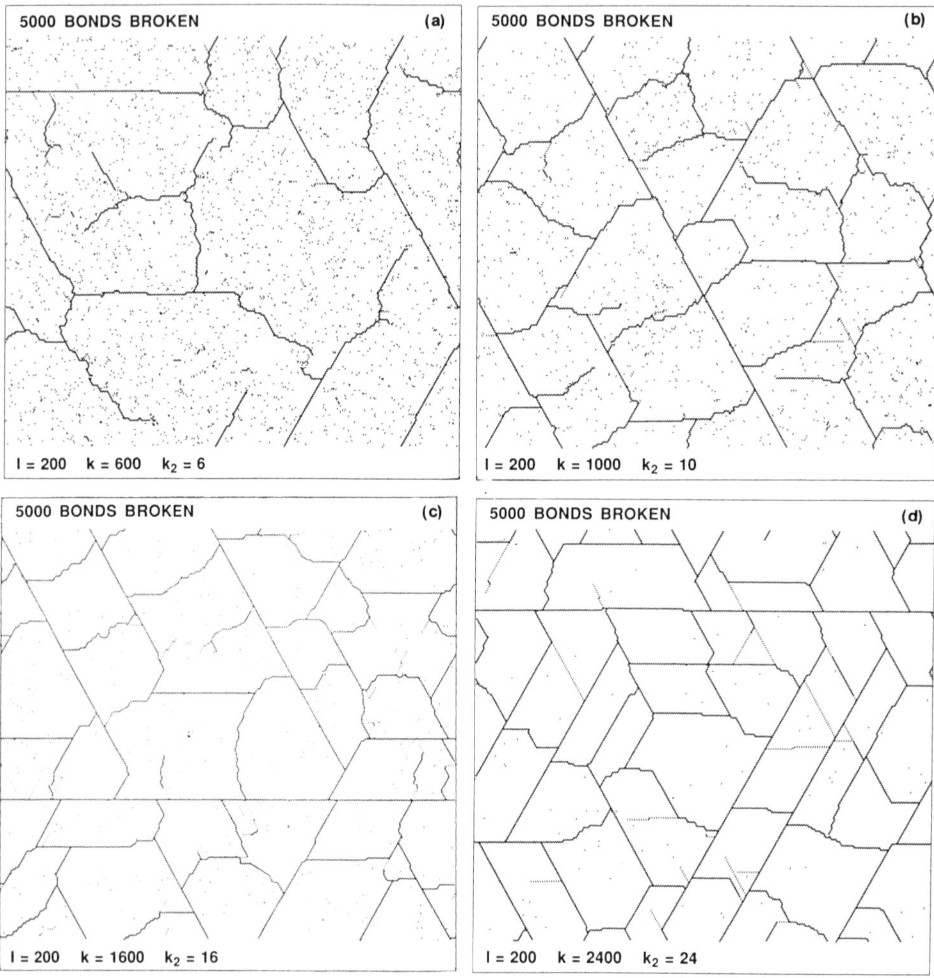

Figure 11 Simulations carried out using the surface cracking model with $k/k_2 = 100$ and four different values for k (600, 1000, 1600 and 2400 for Figs.11a, 11b, 11c and 11d, respectively). These simulations were carried out using a 200x200 node triangular lattice with $l_0 = 0.90909$ and an initial bond length of 1.0.

L = 300, k = 1600, k₂ = 16, 7500 BONDS BROKEN

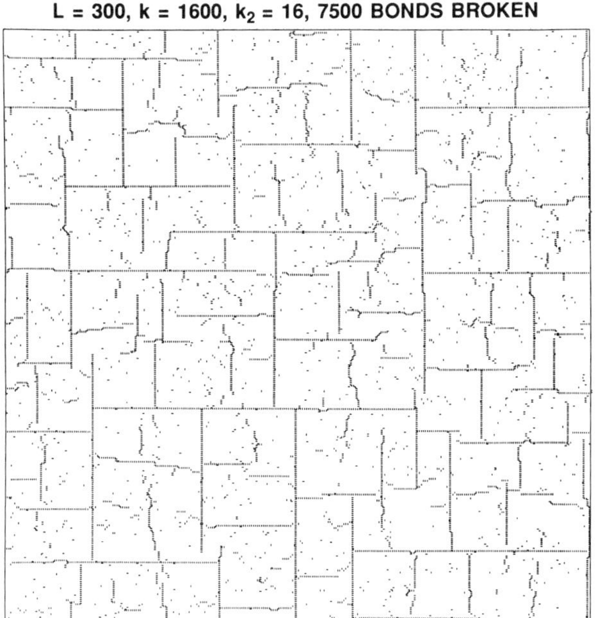

Figure 12 The results of a simulation of elastic fracture in surface film attached to a rigid substrate using a square network instead of a triangular network.

surface film and the rate of bond breaking becomes slower and slower as more of the elastic energy in the film is released. As the stress is progressively released, the cracks become less linear because of the reduced effects of stress concentration at the growing crack tips.

This model has been used to investigate several aspects of surface film fracture such as the "strength" of attachment of the underlying substrate, viscoelastic effects and the effects of pre-existing defects and heterogeneities. For polymers, a large body of evidence suggests that the reduction in the activation energy is linearly dependent on the stress (see reference [22], for example). Consequently, simulations have also been carried out using models in which Eq.(16) is replaced by Eq.(17). For large values of v ($v \gg \delta_0^{-1}$, where δ_0 is the initial bond strain) this model gives results that are qualitatively similar to those described above.

5. The Effects of Non-Central Forces

For all of the models described earlier in this survey, central force interactions were assumed and the network can be described in (in the low strain linear regime) terms of an elasticity Hamiltonian of the form

$$H = \frac{1}{2} \sum_{<ij>} k_{ij} [\hat{r}_{ij} \cdot (\mathbf{u}_i - \mathbf{u}_j)]^2 \tag{18}$$

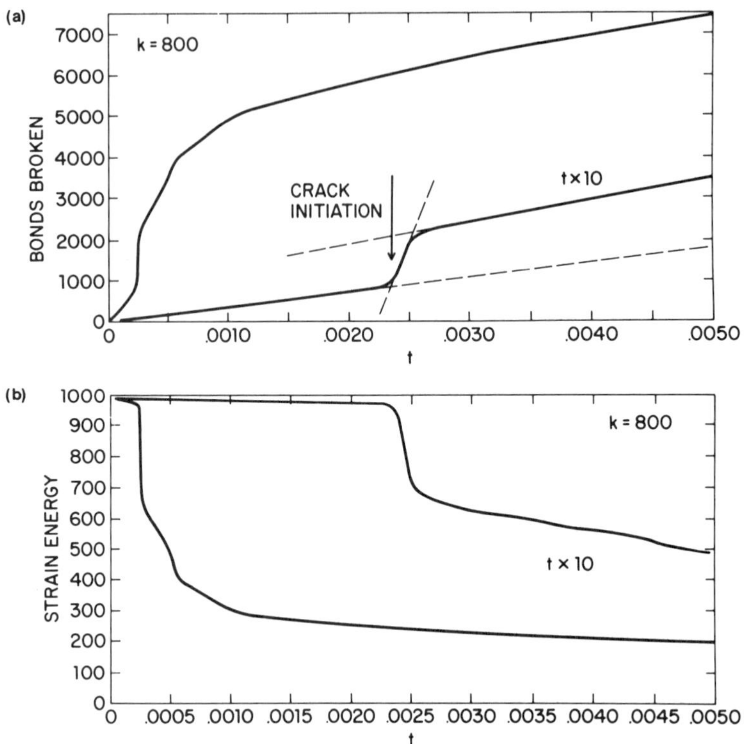

Figure 13 Fig.(13a) Time dependence of the number of broken bonds in a surface cracking model simulation and Fig.(13b) the total strain energy in the surface layer obtained from simulations similar to those illustrated in Fig.(10) with a value of 800 for k. These figures indicate a slow initiation period in which little strain energy is released and only isolated defects are formed, followed by a period of rapid crack propagation in which about 30% of the strain energy is released. This rapid release of the stress in the surface layer slows down subsequent crack initiation and propagation events. As more cracking occurs, the stress becomes still smaller and further slowing down is observed.

where k_{ij} is the force constant for the bond joining the ith and jth nodes, $\mathbf{u}_i$ and $\mathbf{u}_j$ are the displacement vectors for these nodes and $\hat{r}_{ij}$ is a unit vector in the direction of the bond. Recently, Herrmann et al.[37] have developed a probabilistic fracture model based on a square network of elastic beams. In molecular networks the elastic energy of the network can be described in terms of both bond stretching and bond bending contributions. This approximation has enjoyed considerable success in the area of molecular spectroscopy.[38] An alternative (and more simple) approach is to use the Born model for the elastic potential energy. The Hamiltonian can be written as

$$H = \frac{1}{2} \sum_{<ij>} V_{ij} = \frac{1}{2}(\alpha - \beta) \sum_{i,j} \left[(\mathbf{u}_i - \mathbf{u}_j) \cdot \hat{r}_{ij} \right]^2 + \frac{1}{2}\beta \sum_{i,j} (\mathbf{u}_i - \mathbf{u}_j)^2, \qquad (19)$$

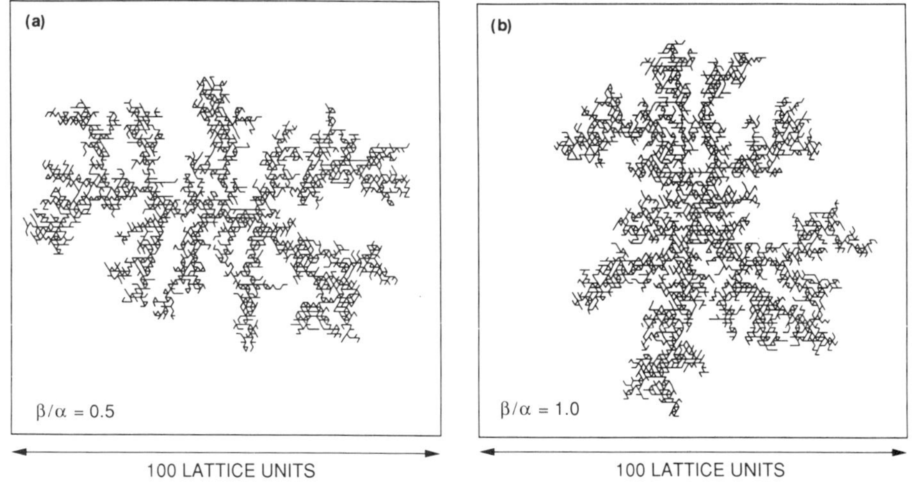

Figure 14 Cracking patterns generated using a DLA-like cracking model with a Born potential elastic network. Fig.(14a) shows a crack generated with $b/a = 0.5$ and Fig.(14b) shows a pattern generated with $b/a = 1.0$.

Here the first term is the central force contribution. By varying the ratio β/α the lattice model can be tuned from the scalar ($\beta = \alpha$) to vector ($\beta = 0$) behavior. In many instances[39] the Born model provides a qualitatively correct description of the lattice properties. However, the Born model is not rotationally invariant.[40]

Simulations were carried out using procedures very similar to those described in the section of central force DLA models using model II boundary conditions at the crack surface, Fig.(1), and a growth probability exponent η of 1. The bond breaking probability for the bond joining the ith and jth nodes was taken to be proportional to $V_{ij}^{1/2}$, Eq.(19). The results of simulations carried out using uniform dilation with a triangular lattice are shown in Fig.(14) for $\beta/\alpha = 0.5$ and $\beta/\alpha = 1.0$ ($\beta/\alpha = 0$ corresponds to the central force models and gives results equivalent to those described earlier). The cracking patterns in the presence of non-central forces are much more compact than those generated with the central force model. Table II shows the effective fractal dimensionality obtained from the dependence of R_g on s, Eq.(6), for several values of β/α. For each value of β/α 8-10 patterns were generated using lattices of size 100×100. The effective fractal dimensionality increases rapidly with β/α for small values of β/α and then becomes quite insensitive to this ratio until β/α approaches 1. The increase in the fractal dimensionality is not unexpected since the additional constraints associated with the Born model may provide an effective mechanism for strain propagation onto the otherwise well screened interior regions of the cracking patterns. A more precise evaluation of the dependence of D_β on β/α would require much larger scale simulations (beyond our current resources) or more detailed theoretical understanding.

Table II. Dependence of the effective fractal dimensionality D on the ratio α/β obtained from simulations carried out using DLA-like cracking models with Born model elastic networks. Model II boundary conditions were used to obtain these results.

(triangular Lat., isotropic dilation)

β/α	0.	.01	.05	.1	.2	.5	.8	1.
D	1.61	1.68	1.73	1.72	1.81	1.83	1.81	1.85
$\pm$	0.01	0.02	0.04	0.02	0.02	0.02	0.02	0.02

(square Lat., isotropic dilation)

β/α	.01	.2	.5	1.
D	1.49 ± .03	1.72 ± .03	1.74 ± .02	1.72 ± .03

De Gennes[39] has pointed out that the case $\beta/\alpha = 1$ corresponds to a scalar elasticity problem analogous to dielectric breakdown. But because of the coupling between the x and y components of the displacements in both the relaxation and the bond breaking probabilities, we do not necessarily expect to recover the fractal dimensionality associated with the dielectric breakdown model[12] in our simulations.

Similar simulations have also been carried out using a square lattice version of the Born model with local boundary conditions corresponding to model II for the triangular lattice (all bonds associated with "damaged" nodes can be broken). Some of the results from this model are also shown in Table II. Qualitatively, the effects of increasing β/α are similar to those observed for the triangular lattice model. In this case, however, the effective fractal dimensionality found for $\beta/\alpha = 1$ (1.72 ± 0.03) is quite similar to that obtained from square lattice DLA or dielectric breakdown models ($D \approx 1.71$) .

Summary

In recent years a variety of simple probabilistic models for crack growth have been explored. These models lead to the formation of cracking patterns that are quite sensitive to model details. In some cases these patterns resemble those observed in real systems.[42,43] However, the phenomenology of crack growth is extremely diverse and these encouraging observations may be mere coincidence. It is apparent that a much more serious and detailed comparison between cracking models and carefully controlled experiments are needed if these simple models are to be taken seriously by those concerned with materials failure processes.

The experimental study of simple model systems may provide an important bridge between computer simulations and the failure of commercially important materials such as steel, concrete and polymers. An encouraging start has been made in this direction,[44,45] but much more work is needed.

There is also a clear need for a better understanding of the simple models discussed here. However, we are finding that even the most simple non-equilibrium growth models (such as DLA) may represent a formidable theoretical challenge.

Finally, it should be clear from this survey that the development of much improved algorithms for the relaxation of elastic networks is urgently needed.

Acknowledgment

The Michigan group is supported by NSF grant DMR 88-15908. We all acknowledge support by NSF grant CCB-8604032.

References

[1] A. V. Dobrodumov and A.M. El'yashevich, Sov. Phys. Solid State, **15**, 1259, (1973) (Fiz. Tverd Tela, **15**, 1891, (1973)).

[2] A. Paskin, A. Gohar and G.J. Dienes, Phys. Rev. Lett., **44**, 940, (1980).

[3] *Modeling of Environmental Effects on Crack Growth Processes*, E.H. Jones and W.W. Gerberich eds., Metallurgical Society, Warrendale, PA (1986).

[4] *Chemistry and Physics of Fracture*, R.M. Latanision and R.H. Jones eds., NATO ASI Series E130, (1987).

[5] *Atomistics of Fracture*, R.M. Latanision and J.R.Pickens eds., Plenum Press, New York, (1983).

[6] D. Shchukin and V.S. Yuschenko, J. Material Sci., **16**, 313, (1981).

[7] B.K. Chakrabarti, D. Chowdhury and D. Sauffer, Z. Phys., **B62**, 343, (1986).

[8] *Computer Simulations in Materials Science*, R.J. Arsenault, J. R. Beeler, Jr. and D.M. Esterling eds., American Society of Metals, Metals Park (Ohio), (1986).

[9] *Computer Simulations for Materials Applications*, R. J. Arsenault, J.R. Beeler, Jr. and J.A. Simmons eds., National Bureau of Standards, Gaithersburg, MD, (1976).

[10] *Interatomic Potentials and Simulation of Lattice Defects*, R. C. Gehlen, J.R. Beeler and R.I. Jaffee eds., Plenum Press, New York, (1972).

[11] T.A. Witten and L.M. Sander, Phys. Rev. Lett., **47**, 1400, (1981).

[12] L. Niemeyer, L. Pietronero and H.J. Wiesmann, Phys. Rev. Lett., **52**, 1033, (1984).

[13] E. Louis, F. Guinea and F. Flores in *"Fractals in Physics"*, *Proceedings of the Sixth International Symposium on Fractals in Physics ICTP*, L. Pietronero and E. Tosatti eds., North Holland, Amsterdam, (1985).

[14] A. H. England, *Complex Variable Methods in Elasticity*, William Clowes and Sons, London, (1971).

[15] L.D. Landau and E.M. Lifshitz, *Theory of Elasticity*, Pergamon, Oxford, (1975).

[16] D.N. de G. Allen, *Relaxation Methods*, McGraw-Hill, New York, (1954).

[17] L.M. Dudkin, I. Rabinovich and I. Vakhatinsky, *Iterative Aggregation Theory*, Marcel Dekker, New York, (1987).

[18] G.G. Batrouni, A. Hansen and M. Nelkin, Phys. Rev. Lett., **57**, 1336, (1986).

[19] B. Wunderlich, *Macromoleular Physics*, Academic, New York, (1973).

[20] W.O. Kinger, H. Bowen and D.R. Uhlmann, *Introduction to Ceramics* (second edition), Wiley, New York, (1976).

[21] R.W. Cahn, *Physical Metallurgy*, North Holland, Amsterdam, (1970).

[22] H.H. Kaush, *Polymer Fracture*, Springer-Verlag, Berlin, (1987).

[23] *Fracture: An Advanced Treatise*, H. Liebowitz ed., Academic Press, (1968).

[24] S.T. Rolfe and J.M. Barsom, *Fracture and Fatique Control in Structures*, Prentice Hall, Englewood Cliffs, N J, (1977).

[25] R.O. Ritchie, W.W. Gerberich and J. H. Underwood in *Encyclopedia of Physical Science and Technology*, Vol. 5, 594

[26] E. Louis and F. Guinea, Europhys. Lett., **3**, 871, (1987).

[27] L. Fernandez, F. Guinea and E. Louis, J. Phys., **A21**, L301, (1988).

[28] B.B. Mandelbrot, *The Fractal Geometry of Nature*, W.H. Freeman and Company, New York, (1982).

[29] E. Hinrichsen, A. Hansen and S. Roux, Europhys. Lett., **8**, 1, (1989).

[30] P. Meakin, G. Li, L.M. Sander, E. Louis and F. Guinea, J. Phys. A, to appear.

[31] P. Meakin, Crystal Properties and Preparation, **17-18**, 1, (1988).

[32] P. Meakin, Thin Solid Films, **151**, 165, (1987).

[33] S.F. Pellicori, Thin Solid Films, **113**, 287, (1984).

[34] R. Jarvinen, T. Mantyla, and P. Kettunen, Thin Solid Films, **114**, 311, (1984).

[35] J.M. Phillips, L.C. Feldman, J.M. Gibson and M.L. McDonald, Thin Solid Films, **107**, 217, (1983).

[36] E. Namgoong and J.S. Chun, Thin Solid Films, **120**, 153, (1984).

[37] H.J. Herrmann, A. Hansen and S. Roux, Phys. Rev., **B39**, 637, (1989).

[38] E.B. Wilson, J.C. Decius and P.C. Cross, *Molecular Vibrations*, McGraw Hill, New York, (1955).

[39] P.N. Sen and M.F. Thorpe, Phys. Rev., **B15**, 4043, (1977).

[40] S.A. Alexander, J. Phys. (Paris) **45**, 1939, (1984).

[41] P.G. de Gennes, J. Phys. L, **1**, 37, (1976).

[42] For the surface cracking model, the cracking patterns are quite similar to those found in Refs.[33-36].

[43] Cracking patterns similar to those generated by the DLA models have been observed in the stress corrosion of Kovar ($Fe_{29}Ni_{17}Co$) (W. C. Cieslak, Corrosion Science, **44**, 517, (1988)).

[44] A.T. Skjeltorp and P. Meakin, Nature **335**, 6189, (1988).

[45] A.T. Skjeltorp in *Time-Dependent Effects in Disordered Materials*, NATO ASI Series B167, P. Pynn and T. Riste eds., Plenum Press, New York, (1987).

Chapter 8

Scaling Theory of the Strength
of Percolation Networks

Phillip M. Duxbury and Yongsheng Li

CFMR and Physics/Astronomy, MSU
East Lansing, MI 48824-1116 USA

A brief review of percolation models of breakdown is given. The most surprising result is that brittle networks exhibit a dilute limit singularity, in which any finite amount of disorder drastically reduces the network strength. A logarithmic size effect also occurs in this limit. Since brittle strength is an extreme variable, extreme statistics apply and sample to sample fluctuations in the strength of diluted networks are found to be logarithmic. These brittle networks may be toughened, and the brittle to tough transition may be quantified by studies of the network damage as a function of the toughening parameters. In the case of toughening by residual bonding, a sharp brittle to tough transition appears to occur.

1. Introduction

There has been intense study of composites[1] , in which the composite properties are measured or calculated as a function of the volume fractions of the composite components. The theory of low moment properties such as transport and elastic moduli are fairly well understood, but similar predictions for extreme properties such as brittle failure, impact strength, yield strength and creep are less well understood. Randomness plays a dominant role in extreme properties, and there has been recent progress in the use of statistical mechanics models and methods to study these problems. In this paper, we introduce and briefly review progress on one statistical mechanics model for breakdown processes, the percolation model,[2–4] that plays a central role in the new ideas developing in this area. A more comprehensive review is available elsewhere.[5]

The model is most easily defined on a randomly diluted lattice, where we define f to be the volume fraction of removed bonds.[6] To model breakdown, each bond of the lattice is given a failure characteristic, and we illustrate this in Figure 1 below for the case of local brittle response;

Upon application of an external current or stress, one or more of the bonds of

Disorder and Fracture, Edited by J. C. Charmet *et al.*
Plenum Press, New York, 1990

the network may fail, and in this way the random network may become progressively more damaged and, for sufficiently large external load, will eventually fail. Both analytic and numerical studies have been carried out using this model.

2. Analytic Studies

The numerical studies to be discussed later indicate that the diluted networks with bonds obeying local brittle response are themselves brittle, so that once damage begins locally, the same or a slightly higher external load will lead to catastrophic failure. Given this, it is possible to do an analytic analysis that gives the strength of these networks as a function of the initial void volume fraction,[5,7,8] and the results are;

$$j_b(f) \approx \frac{j_0 \xi^{-(d-1)}}{1 + k_e (d \ln(L/\xi)/\xi_1)^\alpha}. \tag{1}$$

Where $j_b(f)$ is the current density necessary to cause failure in a network of size L with void volume fraction f, and $j_0 = j_b(0)$. ξ is the percolation correlation length, while $\xi_1 = -ln(f)$. k_e is an undetermined constant of order 1, while α lies in the range $1/(2(d-1)) < \alpha < 1$. This formula interpolates smoothly between the different singular behaviors in the limits $f \to 0$ and $f \to f_c$, and provides a good approximation to strength behavior over the whole dilution range. A similar ex-

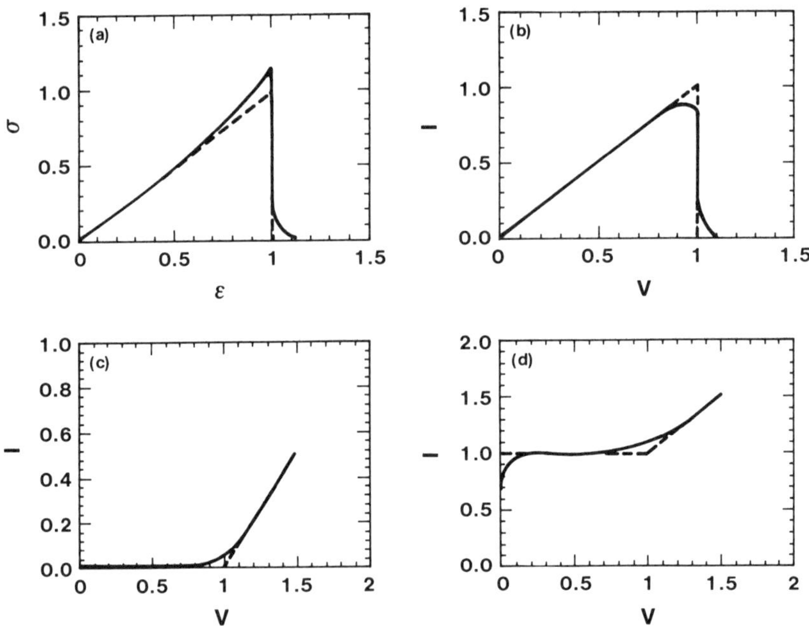

Figure 1 Examples of breakdown phenomena: a) The failure of a spring; b) The failure of a fuse; c) The failure of a dielectric and; d) A superconductor characteristic. In all of these figures, the dotted line is a discontinuous idealization of the more realistic (solid line) characteristics.

pression may be derived for the dielectric strength of metal-loaded insulators[5,7−9];

$$E_b(f) \approx \frac{E_0 \xi^{-1}}{1 + k_d (d \ln(L/\xi)/\xi_1)^\beta} \tag{2}$$

Where $E_b(f)$ is the dielectric strength of an insulator containing volume fraction, f, of random metal inclusions and $E_0 = E_b(0)$. k_d is an undetermined constant and $1/2 < \beta < 1$. For tensile fracture of brittle porous material, the analogous result is [5,8−10];

$$\sigma_b(f) \approx \frac{\sigma_0 \xi^{-x/\nu}}{1 + k_m (d \ln(L/\xi)/\xi_1)^\alpha} \tag{3}$$

where $\sigma_b(f)$ is the strength of a brittle material containing volume fraction f of random pores, and $\sigma_0 = \sigma_b(0)$. k_m is an undetermined constant and x lies in the range $(d-1)\nu < x < d\nu$. The results (1)–(3) are derived by combining standard ideas about scaling in the critical regime, with new "dilute limit scaling" ideas valid near the pure limit. The dilute limit scaling ideas only apply to extreme properties, and lead to a rather surprising singular behavior near the pure limit. This singular behavior is clearly evident in the numerical simulations described in the next section, and for example, is illustrated in Figure 2 for the electrical case. Controlled experiments to test the scaling behaviors of equations (1)–(3) would be very useful. Some interesting limiting cases have been tested. Critical scaling in the mechanical case has been studied using perforated sheets, and qualitative agreement with predicted critical exponents was found.[11,12] Dilute limit scaling has been tested in the dielectric breakdown case, and good agreement with dilute limit scaling was found.[13]

The novel features of extreme scaling in composites are also evident in sample to sample fluctuations in network strength.[5,14] In contrast to transport and elastic moduli which self average so that sample to sample fluctuations are of order $(L/\xi)^{-d/2}$, strength statistics conform to extreme distributions, and in percolation models near the pure limit, this leads to sample to sample fluctuations of order $1/\ln L$.[5] This is the origin of the large scatter characteristic of strength measurements. The extreme distribution corresponding to the three cases for which the average behavior is given in equations (1)–(3) are available elsewhere.[5,7−11]

3. Numerical Simulations

The key ingredients of numerical studies of percolation models of breakdown (and any other damage model), is the criterion for local failure, and the manner in which the load previously carried by the failed member or region is redistributed after local failure - i.e. the *damage evolution algorithm*. Most of the numerical modeling on percolation networks and other types of disordered networks has been carried out using the "Hottest Bond Algorithm" defined as follows[3]:

Hottest Bond Algorithm:

1. Solve for the loads carried by each bond in the network using:

   (a) Kirchhoff's Laws (Electrical Failure)
   (b) Force and moment equilibrium (Mechanical Failure)
   (c) Laplace's Equation (Dielectric Breakdown)
   (d) London or Landau-Ginsburg Equations (Critical Current)[15]

2. Remove the bond carrying the largest load.

3. If the network has failed, exit, otherwise return to 1.

There are clearly several approximations involved in this algorithm, the most obvious one being that the true dynamics of the problem have been replaced by a procedure which assumes that equilibrium occurs after each local failure. A second feature of the algorithm is that it corresponds to a peculiar loading condition, that continually adjusts the external load so that the bond carrying the largest load is just at its threshold value. In many cases, this is a serious flaw in the algorithm. For the calculation of brittle tensile strength, however, the precise form of the algorithm is irrelevant, but the structure of the crack pattern does depend strongly on the loading conditions, and hence on the assumptions inherent in the algorithm. Calculations of the strength of brittle electrical networks as a function of initial void volume fraction are shown in Figure 2.

The dilute limit singularity present in strength scaling is clearly evident in this Figure. It is possible to interpolate between the case of extreme scaling (breakdown) and low moment scaling (transport and elastic moduli) by studying the

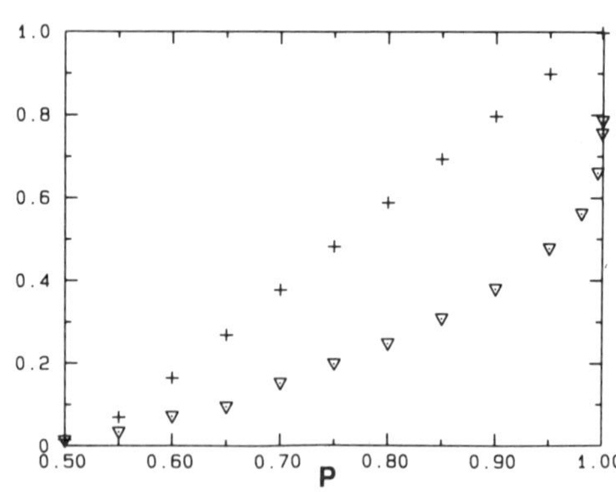

Figure 2 Plots of the conductance (+) and the breakdown current (∇) as a function of dilution for 100×100 random resistor networks. Each point is an average over 50 realizations at fixed p.

moment spectrum of the local load distribution.[16] A second important and at first surprising effect occuring in percolation models of breakdown is a logarithmic size effect in strength (consider equations (1) to (3) with f fixed). A good deal of the statistical mechanics analysis carried out so far has focused on this and other size effects in the strength of random systems. In particular, in the models of Herrmann and collaborators[17] (see also these proceedings), they find numerically that the strength of random networks decays as $L^{-1/4}$, for a broad class of bond disorder. It should be noted however, that to the accuracy of current numerical calculations it is not possible to convincingly distinguish between the logarithmic behavior found in percolation models (see equations (1) – (3), and the scaling form used by Herrmann et al[17].

4. The Brittle to Tough Transition

Percolation models may show a transition from brittle behavior to tough (also called ductile by some workers) behavior.[14,18–20] Three microscopic mechanisms have been studied:

(i) Crack Deflection due to second phase inclusions;

(ii) Toughening due to broad distributions of local failure thresholds and;

(iii) Crack Arrest by Residual Bonding.

In the numerical calculations, the brittle to tough transition is identified as the point at which the damage changes from being localized (brittle) to distributed (tough). This is quantified by the number of bonds that break during the failure process. The important scaling behaviors are;

$$N_b \approx L^{(d-1)} \quad for\ brittle\ failure \tag{4}$$

and

$$N_b \approx L^x \quad with\ x > d - 1\ for\ tough\ materials. \tag{5}$$

In the residual bonding case, a rather sharp brittle to tough transition occurs, as is seen from Figure 3.

Although the brittle to tough transitions such as those occuring in the residual bonding model illustrated in Figure 3 have been qualitatively observed in several contexts, the size dependences and a reasonable analytic theory of the transition is still lacking. Some experimental evidence for this sort of a brittle to tough transition in porous gold systems is now available.[21]

Acknowledgements

We thank the Composite Materials and Structures Center at Michigan State

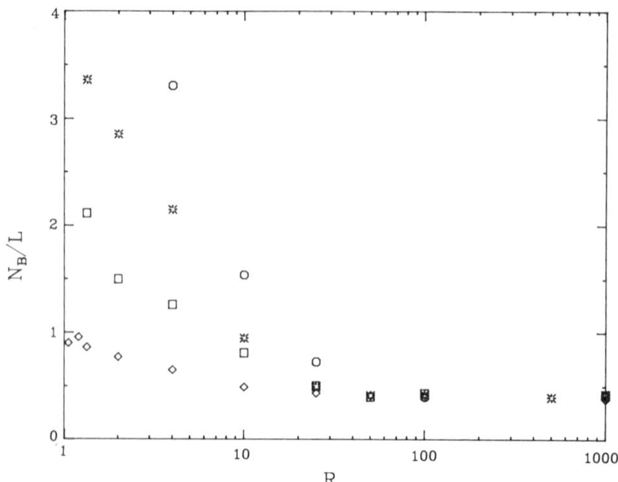

Figure 3 The number of bonds broken during failure of the network as a function of the residual bonding parameter R. The calculations are for $p = 0.75$ and: $L = 10$ ($\Diamond$); $L = 20$ ($\square$); $L = 30$ ($*$) and; $L = 40$ (o).

University, and the Petroleum Research Fund administered by the American Chemical Society for Support. Thanks too to Paul Leath and Paul Beale for enjoyable collaborations, and to Sid. Redner and Hans Herrmann for useful discussion and for keeping us informed about their studies in this area.

References

[1] G.K. Batchelor, Ann. Rev. Fluid. Mech. **6**, 227 (1974);
D. Hull, *An Introduction to Composite Materials*, Cambridge University Press (1981);
S.K. Battacharya, *Metal Filled Polymers*, Marcel Dekker Inc., (1986)

[2] H. Takayasu, Phys. Rev. Letts. **54**, 1099, (1985)

[3] L. de Arcangelis, S. Redner and H.J. Herrmann, J. Physique Lett. **46**, L585, (1985)

[4] M. Sahimi and J.D. Goddard, Phys. Rev. **B 33**, 7848, (1986)

[5] P.M. Duxbury and Y.S. Li, *Scaling Theory and Fluctuations of the Strength of Composites*, in Proceedings of the SIAM conference on *Random Media and Composites*, Leesburg Virginnia (1988)

[6] see e.g. D. Stauffer, *Introduction to Percolation Theory*, Taylor and Francis (London, 1985)

[7] P.M.Duxbury, P.D. Beale and P.L. Leath, Phys. Rev. Lett. **57**, 1052, (1986)

[8] P.M. Duxbury, P.L. Leath and P.D. Beale, Phys. Rev. **B 36**, 367, (1987);
Y.S. Li and P.M. Duxbury, Phys. Rev. **B 36**, 5411, (1987)

[9] P.D. Beale and P.M. Duxbury, Phys. Rev. **B 37**, 2785, (1988)

[10] P.D. Beale and D.J. Srolovitz, Phys. Rev. **B 37**, 5500, (1988)

[11] K. Sieradski and R. Li, Phys. Rev. Lett. **56**, 2509, (1986)

[12] L Benguigui, P. Ron and D.J. Bergman, J. Physique **48**, 1547, (1987)

[13] R.W. Coppard, L.A. Dissado, S.M. Rowland and R. Rakowski, J. Phys. Con. Mat. **1**, 3041, (1989)

[14] P.M. Duxbury and P.L. Leath, J. Phys. **A 20**, L411, (1987)

[15] P.L. Leath and W. Tang, Phys. Rev. **B 39**, 6485, (1989)

[16] Y.S. Li and P.M. Duxbury, Phys. Rev. **B 40**, 4889, (1989)

[17] H.J.Herrmann, A. Hansen and S. Roux, Phys. Rev. **B39**, 637, (1989)

[18] A. Margolina and S. Wu, J. Mat. Sci. **21**, 1819, (1986)

[19] B. Kahng, G.G. Bartrouni, S. Redner, L. De Arcangelis and H.J. Herrmann, Phys. Rev. **B37**, 7625, (1988)

[20] Y.S. Li and P.M. Duxbury, Phys. Rev. **B38**, 9257, (1988)

[21] K. Sieradski, *Private Communication*, (1989)

Chapter 9

Scaling in Fracture

H.J. Herrmann and L. de Arcangelis

Service de Physique Théorique, CEN Saclay
F-91191 Gif sur Yvette, France

Cracks can be grown on a lattice by deterministic rules and their patterns can be fractal due only to the interplay of anisotropy and memory. We show cracks of various shapes generated by different breaking criteria. A quantity of experimental interest is the breaking characteristics of the system (force vs. displacement): we discuss its universal scaling behaviour. Moreover, the distribution of local strain has multifractal scaling properties just before the system breaks fully apart. All the presented results appear to be universal with respect to different distributions of quenched disorder in the thresholds and other models for fracture.

1. Introduction

The laws that govern the fracture[1] of disordered media are technologically of outstanding interest since they cover phenomena ranging from the cracking of glass to the tearing of cloth, from the aging of concrete to stress corrosion. A large effort has been recently undertaken by statistical physicists to find, within the wide variety of experimental situations, generic features due to the underlying instabilities and their interplay with noise, anisotropy or memory effects. Recently two new approaches in this direction have been proposed[2,3] one inspired by random resistor networks[4] and another using DLA[5] as a guideline.[6,7]

In this paper we first discuss how to grow a single crack on a lattice by means of deterministic rules. Three different breaking criteria are presented, in which a memory effect is taken into account together with material dependent parameters. Depending on the breaking rule, the generated crack can be fractal. In section 2 we present a model for fracture derived by the discretization of the Lamé equation and we discuss in section 3 the different deterministic criteria and the corresponding crack patterns obtained.

If one attemps to describe the breaking of a real material the introduction of disorder is a crucial feature of the model. We next study the failure of a two dimensional lattice with quenched disorder in the breaking threshold of each element. In section 4 we discuss some scaling behaviour found for the breaking characteristics

Disorder and Fracture, Edited by J. C. Charmet *et al.*
Plenum Press, New York, 1990

of the whole system and for the distribution of local strain in the system. This distribution has multifractal properties just before the system breaks fully apart, whereas it exhibits constant gap scaling at the maximum of the breaking characteristics, that is at the end of the disorder controlled regime when the catastrophic regime sets in.

2. A Vectorial Model for Fracture

We now focus on the growth of a single, connected crack in an elastic medium. For the case of symmetric elasticity the medium is usually described by the Lamé equation[8]

$$(\lambda + \mu)\nabla(\nabla\mathbf{u}) + \mu\nabla^2\mathbf{u} = 0 \tag{1}$$

where $\mathbf{u}$ is the displacement vector and λ and μ are the Lamé coefficients. On the external boundary an imposed displacement is fixed. We start with the situation of a single microscopic crack already present in the system and study how this small crack grows. On the surface of the crack the stress normal to the surface of the crack is zero and the crack will grow in the direction perpendicular to the surface at the point where the strain parallel to the surface is largest. The detailed growth law depends on the microscopic mechanism like how the elastic energy is transported away from the growing tip. It is reasonable and quite general to assume a normal growth velocity $\mathbf{v}_n$ of the form

$$\mathbf{v}_n \propto ((\partial_\parallel \mathbf{u}_\parallel)^2 + r \cdot \partial_\parallel^2 \mathbf{u}_\perp)^\eta \tag{2}$$

where r and η are material dependent parameters. For $\eta = 1$ this growth law is inspired by the von Mises yielding criterion,[1] but it is not possible to derive it from first principles. Physically r is the affinity of the breaking process to the bending mode (second term on rhs of Eq.(2)) as compared to cleavage (first term on rhs of Eq.(2)). Time derivatives in Eq.(1), like inertia terms, are neglected. We do not consider plasticity or non-linear elasticity.

In the present approach we also do not consider viscoelastic effects, that is we assume that the relaxation of the local strain due to the growth of the crack is much faster than the velocity of the crack itself propagating through the system. One can, however, easily introduce a finite relaxation time of the elastic system with respect to the typical time to grow the crack by a given length if one uses numerical relaxation techniques and in this way viscoelasticity can be effectively implemented.

The elastic equations can be discretized on a square lattice by using the beam model.[9] In fact the beam model is even richer than Eq.(1) because it allows for local rotations in the medium, *i.e.* discretizes the equations of motion of asymmetric elasticity.[10] The discretization on the lattice introduces anisotropy and a cutoff at small length scales, two physical effects that are very often present in real materials.

A detailed description of the beam model is given in Ref[2]. We implement the model as follows: For each beam that is eligible for being broken, that is each of the beams neighboring the existing crack, one calculates the quantity p defined by

$$p = (f^2 + r \cdot max(|m_1|, |m_2|))^\eta \qquad (3)$$

where f is the traction (and/or compression) force applied on the beam and m_1 and m_2 are the moments that are acting at the two ends of the beam; this p determines according to Eq.(2) if the beam will be broken. Each time a beam is broken the shape of the crack and consequently the boundary conditions of the equation of motion change and one has to solve the discretized equation again if one wants to know which beam to break next.

3. Deterministic Growth of a Fractal Crack

Recent numerical simulations of central-force media with a breaking probability proportional to the elongation of the springs have shown evidence for the cracks formed to be fractal.[6,7] The fractal dimension of these cracks seems to depend strongly on the type of external force that is applied (uniaxial tension, shear, uniform dilation) but since only very small cracks can be grown, precise statements are difficult to make. Here, we will investigate some of the origins of this fractal behaviour and try to obtain a much better accuracy by considering deterministic models.

We consider a finite square lattice of linear size L, with periodic boundary conditions in the horizontal direction. On top and bottom we impose an external shear. We remove one beam in the center of the lattice which represents the initial microcrack. Next we consider the six nearest-neighbor beams of this broken beam. These include the two beams that are parallel to the broken beam and the four perpendicular beams that have a common site with the broken beam. This choice of nearest-neighbors comes from the fact that the actual crack consists of the bonds that are dual to the set of broken beams.[2] Other connectivity conditions have also been used.[7] The Lamé equation is solved by the conjugate gradient method[11] to very high precision (10^{-20}) and the p's of Eq.(3) are calculated for each of the nearest-neighbor beams. We set $p = 0$ for a beam that is not a nearest-neighbor to the crack. Now various criteria for breaking are possible:
I. One breaks the beam with the largest value of p;
II. One breaks the beam for which $q_0 = p + f_0 \cdot p_{-1}$ is largest, where p_{-1} is the value of p that this beam had before the previous beam was broken; f_0 is a memory factor;
III. On each beam of the lattice we put a counter c which is set to zero at the very beginning. Each time one has obtained the p's one calculates $\alpha = (1 - c)/p$ and breaks the beam which has the smallest α, namely α_{min}. After the beam has been broken each counter c is set to $c = \alpha_{min} \cdot p + f \cdot c_{-1}$, where c_{-1} is the value the counter had before the breaking and f is another memory factor.

All three breaking criteria described above are deterministic. Criterion III corresponds for $f = 1$ to the limit of infinite noise reduction.[12] Noise reduction was invented to reduce statistical noise in DLA simulations and has also been applied recently to central-force breaking.[13] What noise reduction certainly does is to introduce a memory effect with long range time correlations. Physically the three breaking criteria defined above correspond to three different situations. Criterion I describes ideally brittle and fast rupture. Criterion II contains a short time memory one would expect in cracks that produce strong local deformations at the tip of the crack, as it happens in most realistic situations and where this local damage does not heal much faster than the speed of the crack. Criterion III could be applied to situations of stress corrosion or static fatigue. The memory factors f_0 and f measure the strength of these time correlations. In criterion III the limit $f \to 0$ gives criterion II with $f_0 = 1$, and in criterion II the limit $f_0 \to 0$ gives criterion I.

Let us next discuss the results that one finds[14] for the above model. If one breaks according to criterion I, as shown in Fig. 1 for $L = 50$ and $r = 0.28$, cleavage tends to have the crack grow in the diagonal direction while the bending mode favours a horizontal rupture. The competition between these two effects can lead to complex branched structures. The exact shape of these cracks strongly depends on r and the system size. For any finite r the horizontal rupture will eventually win if the system is large enough while for $r = 0$ one obtains diagonal cracks with eventual kinks. For this reason the cracks will not be fractal. In the analogous scalar model (*i.e.* DLA), however, only straight lines will be formed in criterion I; the different behaviour here is due to the fact that competing directions are possible in a vectorial model.

Let us now consider cracks grown using criterion II. We see in Fig. 2 and Fig. 3 cracks with $\eta = 1.0$ and $f_0 = 1.0$ obtained by breaking each beam only through tension. In Fig. 2 we show a crack grown with $r = 0$ in a system of size $L = 118$. Over four hours on one CrayXMP processor were needed to generate this

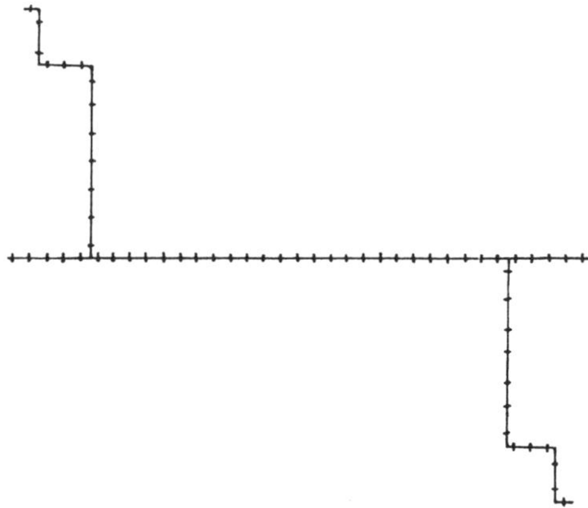

Figure 1 Crack grown in a 50×50 system if an external shear is applied. Beams break under traction or with an affinity of $r = 0.28$ in the bending mode. Only the bond with the largest value of q breaks (criterion I). The first broken beam was vertical (taken from Ref.[14]).

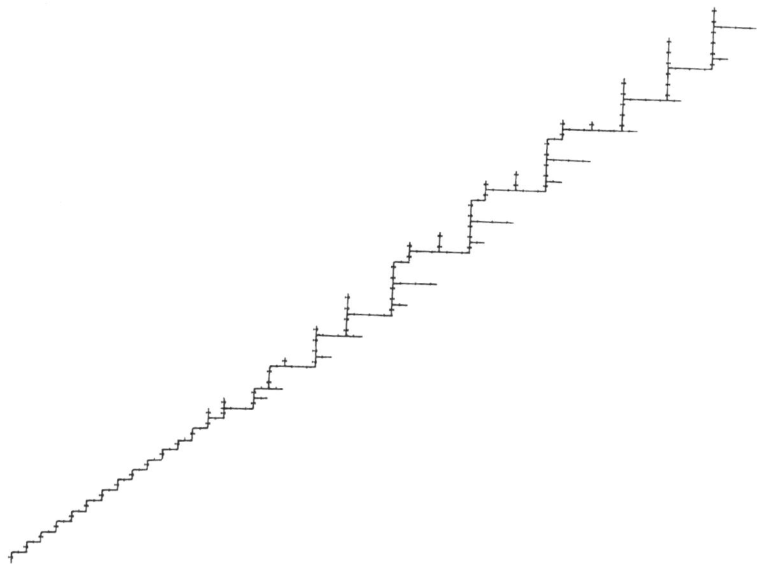

Figure 2 Upper half of a crack grown in a 118×118 system under external shear using criterion II with $f_0 = 1$, $\eta = 1.0$, $r = 0$ (breaking through traction only) and vertical first broken beam.

structure, of which we only show the upper part of the crack the lower part being reflexion symmetric. The very slight curvature of the crack is a finite size effect of the lattice which is, however, such a weak effect that its influence cannot be noticed quantitatively for instance in the value of the fractal dimension that we will discuss next. In Fig. 3 we show instead the whole crack grown in a system $L = 60$ for $r = 0.1$ under external shear. Again we see that the bending mode enhances the horizontal rupture.

If we count the number of broken beams inside a box of length l around the first broken beam and plot it as a function of l in a log-log plot ("sand box method") we find lines with slopes larger than unity which means that the cracks are fractals. For the system size $L = 118$ we find (Fig. 4) for the fractal dimensions d_f values that depend on η: $d_f = 1.3$ for $\eta = 1.0$, $d_f = 1.25$ for $\eta = 0.7$, $d_f = 1.15$ for $\eta = 0.5$ and $d_f = 1.1$ for $\eta = 0.2$. The structures are self-similar around the origin and probably directed fractals.[15] Changing the elastic constants (*i.e.* the Lamé coefficients) just changes the opening angle of the crack. If in Eq.(3) one uses an exponential instead of a power law, the structures seem to be dense.[15]

The effect that using criterion II gives fractal structures is novel and very distinct from what is seen in the scalar case of DLA. It shows that neither noise nor long range time correlations are necessary to obtain fractal breakdown. The origin of fractality is the competition between a global stress perpendicular to the

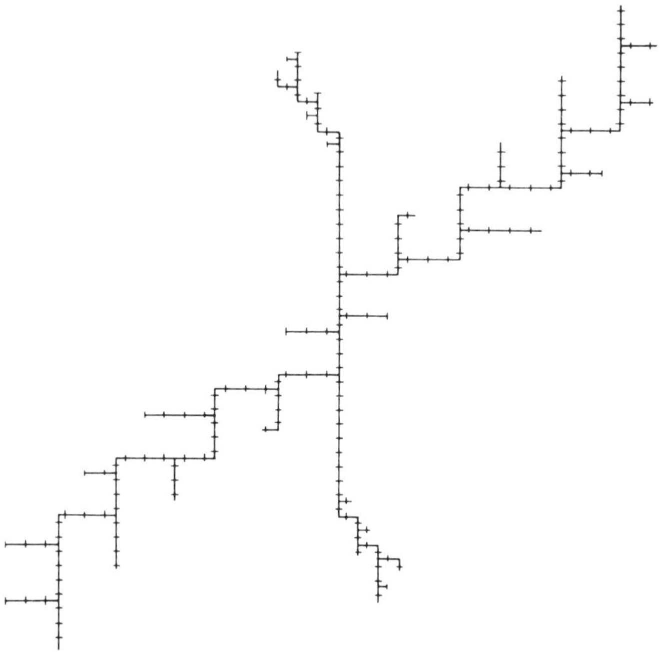

Figure 3 Crack grown in a 60×60 system under external shear using criterion II with $f_0 = 1$, $\eta = 1.0$, $r = 0.1$ and vertical first broken beam. Here breaking occurs only through traction.

diagonal and a local stress that tends to continue a given straight crack due to tip instability. Again we see the important role of the interplay of different directions which is only possible in a truly vectorial model. The relevance of a short memory in criterion II indicates that there might be a relation between this case and the models that have been put forward for snow flakes.[16]

In Fig. 5 we show a crack grown using criterion III for $r = 0$, $f = 1$, $\eta = 1$ and $L = 118$. The physical situation is similar to that seen in criterion II, only the fractal dimension is higher. This case can be directly compared to results obtained for DLA in the limit of infinite noise reduction[13,17] where needles, not fractals, are predicted.

In Fig. 6 we compare a deterministic crack with an experimental example of stress corrosion cracking in an alloy.[18] Due to the heuristic nature and simplicity of our model it makes no sense to compare numerical values of fractal dimensions. It seems also clear that the inhomogeneities of the medium in the experimental crack are important. The vague similarity that one can see between the two patterns in Fig. 6 seems to indicate that the effects found in our model may explain to a certain degree the branching behaviour of experimental cracks.

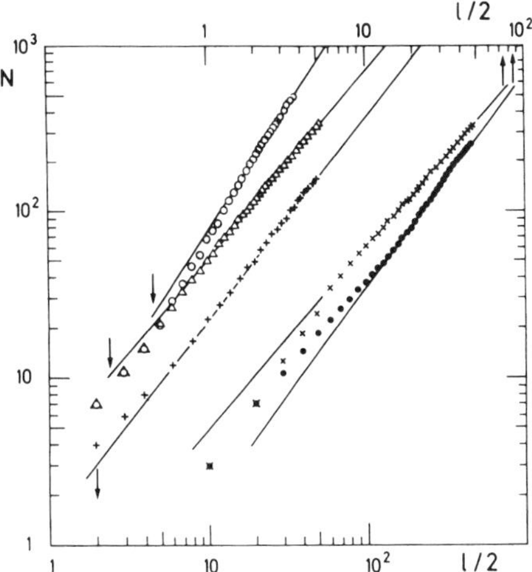

Figure 4 Log-log plot of N, the number of broken bonds in a box of side l, versus $l/2$ for different cracks grown in a system size $L = 118$ under external shear and with breaking occurring only by traction. Cracks obtained with criterion II, $f_0 = 1$ and $r = 0$: ($\times$) for $\eta = 0.2$ $d_f = 1.10$, ($\triangle$) for $\eta = 0.5$ $d_f = 1.15$, ($+$) for $\eta = 0.7$ $d_f = 1.25$ and ($\bullet$) for $\eta = 1.0$ $d_f = 1.30$; and one crack grown with criterion III, $f = 1$, $r = 0$ and $\eta = 0.5$ with $d_f = 1.47$ ($\circ$).

Figure 5 Upper half of a crack grown in a 118×118 system using criterion III with $f = 1$, $\eta = 1$, $r = 0$ and vertical first broken beam when the beams break under traction only.

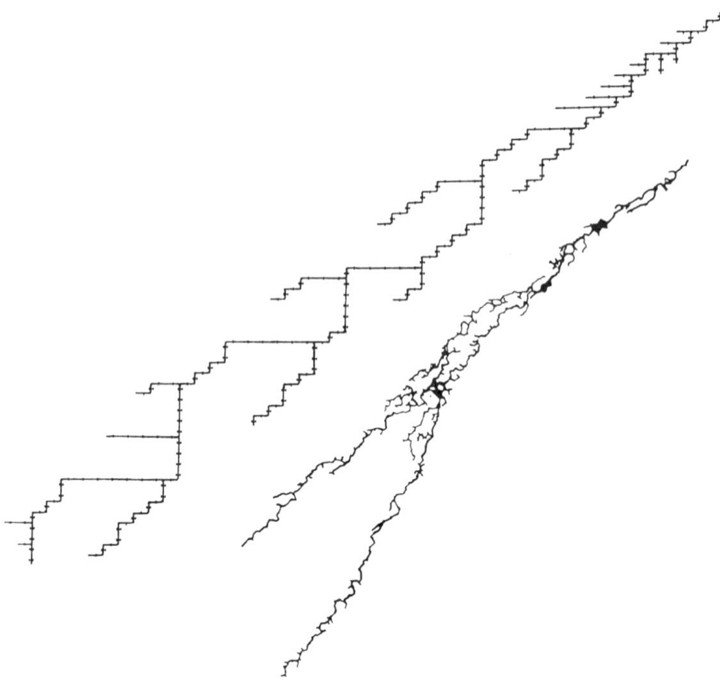

Figure 6 Numerical and experimental cracks. The upper shape is the upper half of a crack grown in a 118×118 system under external shear with $r = 0$ and a vertical initially broken beam using criterion II with $f_0 = 1$ and $\eta = 0.2$. The lower picture shows the morphology of cracking in Ti - 11.5 Mo - 6 Zr - 4.5 Sn aged 100 hrs at $750^\circ K$ and tested in 0.6M LiCl in methanol at -500mV under increasing stress intensity (taken from Ref.[18]).

4. Scaling laws for the fracture of disordered media

The model discussed in the last section contains no disorder, which is a rather unrealistic simplification. In fact, the role of disorder is crucial in fracture if one is interested in a quantitative comparison with real forces or displacements. In the following we will consider a finite two-dimensional lattice $L \times L$ with periodic boundary conditions in the horizontal direction and fixed bus bars on top and bottom on which the external strain (elongation or shear) will be applied. Each bond is supposed to be ideally fragile; *i.e.* to have a linear elastic dependence between force f and displacement δ with unit elastic constant up to a certain threshold force f_c where it breaks (see insert of Fig. 7). We will discuss a model[19] in which the thresholds are randomly distributed according to some probability distribution $P(f_c)$. Once a force beyond f_c is applied to a bond, this is irreversibly removed from the system. As the external strain is increased one can watch bonds breaking one by one until the system falls apart altogether.

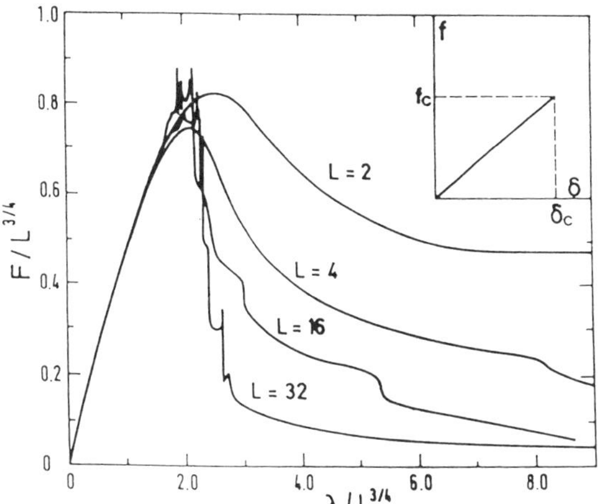

Figure 7 Breaking characteristics of the beam model with $x = 0.5$, $r = 1$ with both axis scaled by $L^{-3/4}$ for different sizes L. The data have been smoothened to reduce statistical fluctuations.[6] The insert shows the characteristics of a single beam (taken from Ref.[24]).

For the beam model one has to introduce actually two random thresholds f_c and m_c and break a beam if[20]

$$(f/f_c)^2 + max(|m_1|, |m_2|)/m_c \geq 1 \qquad (4)$$

which is the equivalent to Eq.(3). We distribute the thresholds according to $P(f_c) = (1 - x)f_c^x$ with $0 < f_c \leq 1$ and $P(m_c) = (1 - x)r^{x-1}m_c^x$ with $0 < m_c \leq r$ where x and r are parameters.

It is also possible to consider models in which the bonds are not beams but simpler objects, like electrical resistors[21] or springs which can only be subjected to a central force.[22] In these cases the breaking criterion is $\lambda = |i|/i_c \geq 1$ where i is the current or the central force respectively. The threshold is again randomly distributed and for the electrical case we will also consider a Weibull distribution, namely $P(i_c) \propto i_c^{m-1}e^{-(i_c/i_0)^m}$ where m and i_0 are parameters, and the power law distribution, $P(i_c) = (1 - x)i_c^{-x}$.

A quantity of technological interest is the breaking characteristics[20−23] which monitors the value of the external force and the external displacement applied during the fracture process. For the random fuse network model this characteristics is given by the relation current-voltage for the whole system. As we will see, even if the microscopic characteristics of each element are ideally fragile, the macroscopic one for the entire network is much richer of information. Each time a bond is broken we then record its value of λ, which exactly equals the external displacement to be applied to break that single element, and the external force.

The relation $F(\lambda)$ is the breaking characteristics for the system and it is shown in Fig. 7. We can distinguish two different regimes: an initial regime, controlled by the disorder, in which small and isolated cracks are created in the system, removing the very weak bonds in an uncorrelated fashion. In this regime the external force increases monotonically up to a certain maximum value and there are weak statistical fluctuations. Moreover, this regime progressively disappears if the quenched disorder in the threshold distribution is made smaller and smaller (Weibull distribution with higher values of m).

In the second regime, that we can call the "catastrophic" regime, the correlations between cracks start to dominate the process and one macroscopic crack forms to attain the final breaking of the system. In this regime statistical fluctuations are rather strong and it is observed that for decreasing quenched disorder the characteristics bends back toward smaller values of the force, corresponding to the situation of self-sustained breaking of the system, where the macroscopic crack propagates even if the external force is actually lowered.

To determine the scaling law dependence of the force and the displacement, in Fig. 7 are plotted the results of the beam model for various system sizes L rescaled with a power law of L in order to find the best collapse of the data of all sizes on the single curve. This collapse is obtained very well in the disordered regime, $i.e.$ in the regime before the maximum is reached. Here there is numerical evidence for the scaling law

$$F = L^{\alpha} f(\lambda L^{-\beta}) \tag{5}$$

with $\alpha \approx \beta \approx 0.75$. This law can be checked for all three models,[23] all distributions ($0.8 \leq x \leq -1, 2 \leq m \leq 5$) and for both, external extension and shear (in the elastic case) with exponents that agree with $3/4$ within 5 to 10%. In the catastrophic regime it is not possible to scale the data and the curve seems to become more vertical as the system size increases.

In the same way, it is possible to find a scaling relation for the external displacement as function of n, the number of broken bonds

$$\lambda = L^{\alpha} g(nL^{-\gamma}) \tag{6}$$

with $\gamma \approx 1.7$ and the same universal range of validity as for Eq.(5). The exponents α, β, γ in Eq.(5) and (6) are independently fitted exponents and the value of α should be the same if found from the scaling law (5) or (6). The interesting result is that the value of the exponents α, β, γ are universal with respect the distribution of quenched disorder, provided that it is wide enough so to generate the first regime in the process, and the nature of the model considered.

The study of the breaking characteristics can be completed by the finite size scaling analysis of same quantities typical of the process, like the number of bonds cut, the largest crack and so on. We find, for instance, that the number of bonds cut in the first regime, the disorder controlled one, scales with the size of the system as $L^{1.7}$, as seen in Fig. 8a, in good agreement with the scaling law (6). As the quenched disorder in the distribution decreases, the exponent crosses over towards a zero value as the first regime disappears. Moreover, the total number of bonds

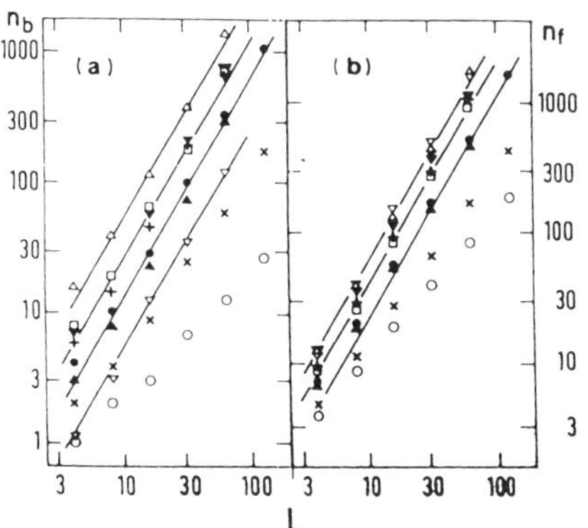

Figure 8 Number of bonds cut when (a) the maximum force is reached and (b) when the system falls apart in a log log plot against L for the scalar model with $x = 0$ (□), $x = 0.5(\triangle), m = 2(\bullet), m = 5(\times)$ and $m = 10(\bigcirc)$ and the beam model $(r = 1)$ for an external elongation with $x = 0(+)$, $x = 0.5(\nabla)$, $x = -1(\blacktriangle)$ and for an external shear with $x = 0(\blacktriangledown)$. The full lines are guides to the eye of slope 1.7 (taken from Ref.[24]).

cut during the whole process is also found to scale (Fig. 8*b*) as $L^{1.7}$, crossing over towards a linear behaviour in L for decreasing disorder. As the majority of bonds is burned in the first regime, the scaling behaviour of the number of bonds cut in the disorder regime and in the whole process should be the same, evidence found within our error bars. This result is also in agreement with the value found for the fractal dimension of the macroscopic crack that causes the failure of the system, $d_f = 1.1 \pm 0.1$. Finally, we would like to stress the striking result of the total number of bonds cut scaling with system size with an exponent ~ 1.7. Ref.[4] predicts in fact $n \sim L$ for the case $x = 0$. Moreover, this finding is universal with respect to the three models,[23] the distribution of quenched disorder and the external boundary conditions within our error bars.

In order to investigate the local properties of the system, we study the distribution of local currents (local forces, shears and moments in the elastic case) across each fuse of the network, $n(i)$, and its moments[20−23] $M(q) = \sum_i n(i)i^q \sim L^{-p(q)}$ In the present context, we consider the distribution of local currents just before the last fuse is burned and the system breaks apart. For the of very weak disorder (*e.g.* Weibull distribution with high values of m), the distribution as function of the logarithm of the current is strongly peaked around its most probable value and it spreads towards smaller values of the current as the quenched disorder increases.

It is possible to describe this distribution by studying its moments and their scaling behaviour with system size. In particular, we want to find out if the moments exhibit a constant-gap scaling with L, as the moments of the cluster size distribution in percolation, or instead they scale with an infinite number of independent exponents as the voltage moments in percolation at the threshold.[24] In the latter case, this hierarchy of exponents would be the expression of the multifractal[24] properties of the system, it would mean that there is not one single value of the current leading the critical behaviour but each moment $M(q)$ is dominated by its typical current that determines its critical exponent.

To answer this question, we analyse the moments $M(q)$ renormalized as $m_q = [M(q)/M(0)]^{1/q}$, averaged over all configurations just before their respective last bond breaks. If constant-gap scaling is followed, the quantities m_q would scale all with the same exponent with respect to L for each value of q. Fig.9b shows the finite size scaling plot for the m_q's. They do not exhibit constant-gap scaling and the values of the critical exponents are universal for the different distribution considered.

If the same analysis is made at the maximum of the characteristics just before the catastrophic regime sets in, the results are instead in striking contrast (Fig. 9a). At this point, in fact, the m_q's are all parallel straight lines, following the constant-gap scaling with L. This result can be interpreted as follows: in the first regime, where small isolated craks are made, the system still behaves as a macroscopically homogeneous object and only when correlations dominate and a large spanning crack is formed, the system exhibits multifractal properties. This physically means that the regions with a very high local stress lie on fractal subsets of the whole system and the fractal dimensions of these sets themselves depend on how the stress increases with the system size.

For the multifractal case, just before failure, we determine for each value of q the asymptotic value of the critical exponent of the moment $M(q)$, by extrapolating its value $p(q)$ to infinite system size. We find that the exponents $p(q)$ are a monotonic function of q (Fig. 10a), that for $q \to \infty$ goes to a straight line of slope ~ 0.28 and

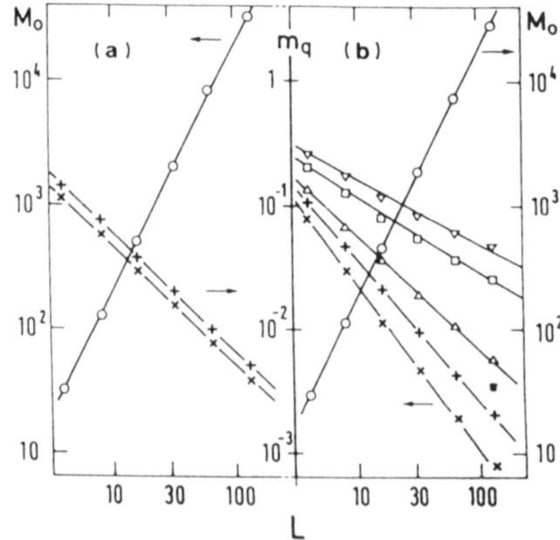

Figure 9 Moments of the current distribution for $m = 2$ (a) at the maximum of the characteristics and (b) just before the system falls apart in a log-log plot against L. The symbols are o for M_o, × for m_1, + for m_2, $\triangle$ for m_3, $\square$ for m_6 and $\triangledown$ for m_9 where $m_q = (M_q/M_0)^{1/q}$ (taken from Ref.[24]).

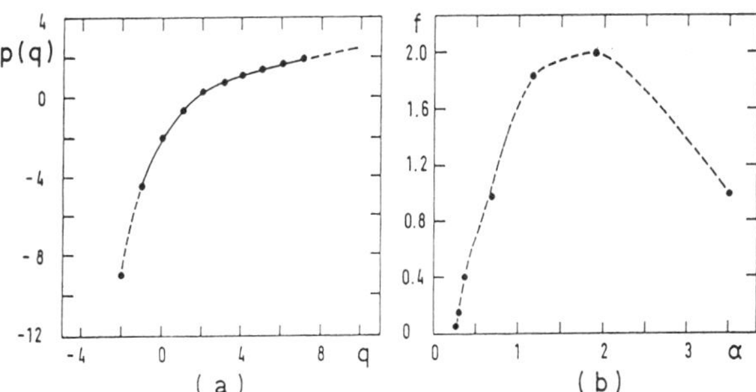

Figure 10 Line of critical exponents $p(q)$ versus q for the random fuse network model in the case of the uniform distribution (a). For the same distribution the spectrum of fractal dimensions $f(\alpha)$ (b).

vertical intercept at zero. For negative moments it is rather difficult to determine the value of $p(q)$ within our range of accuracy and we cannot exclude that they behave exponentially with L.

Finally, we analyse the spectrum of fractal dimensions for the different fractal subsets of the system. We follow the standard procedure by considering the Legendre transform of the critical exponents $p(q)$ $f(\alpha) = q\alpha - p(q)$, where $\alpha = \frac{\partial p(q)}{\partial q}$ represents how a given value of the current i goes to zero as $L \to \infty$ $(i \sim L^{-\alpha})$ and $f(\alpha)$ is the fractal dimension of the set of bonds characterized by that value of i.

The so calculated function $f(\alpha)$ (Fig.10b) has its maximum at the value of the fractal dimension of the system $(d_f = 2)$ and it goes to zero as $q \to +\infty$ at a value of $\alpha_{\min} \sim 0.28$, the asymptotic slope of the curve $p(q)$. These results are consistent with the picture that at the final point of breaking, one fuse only holds the system together carrying the whole current flowing in the network, its value dominating the moments $M(q)$ as $q \to \infty$. Moreover, in such a configuration and for a unit external voltage, the current in this fuse is equal to the conductance of the network, therefore we expect $\alpha_{\min}$ to be equal to the value of the critical exponent for the second moment $(M(q) \equiv G)$. We find, in fact, that the critical exponent $p(2) \sim 0.31$ is in agreement with $\alpha_{\min}$ within our accuracy. The behaviour of $f(\alpha)$ for higher α's represent the subsets of bonds carrying a very small current; it is therefore more difficult to accurately determine them due to the uncertainty on the negative moments.

Physically, the finding that the system has multifractal scaling properties means that the regions with highest variation in local strains, *i.e.* the regions that are finally responsible for rupture, lie on a fractal subset of the system. The fractal dimension of this subset depends on the strength of the local variations. In practical terms this means that the larger the system the more pronounced becomes the contrast between highly strained and practically unstrained regions. This effect only occurs just before the system breaks and not during the whole process as seen in Fig.9a. Our data permit to quantify this statement. The appeearence of multifractality is more astonishing if one considers that only a negligible number of bonds $(n \sim L^{1.7})$ has been cut, in contrast to the case of percolation where multifractality[24] only appears at the percolation threshold p_c (*i.e.* $n = p_c L^2$). Local strains can be studied by photoelasticity and this might also be the best mean to verify the multifractal properties.

5. Conclusions

In conclusion, we have discussed some selected problems of crack propagation on finite square lattice samples. Even without introducing any noise in the breaking criterion, the patterns of cracks can become very complex and in particular they can be fractal when a memory parameter is turned on, with a fractal dimension that depends on the exponent η in the breaking criterion. This phenomenon is due to the competition between the direction of global stress and the direction of local growth imposed by the lattice anisotropy. Therefore the *vectorial* nature of the elastic medium leads to crucially different results from what is known to occur in the *scalar* case of DLA. These differences are likely to subsist also when randomness is introduced either quenched in the form of random elastic constants[25] or annealed in the form of breaking probabilities.[6,7]

Moreover, we have presented some new scaling results for fracture models in the presence of quenched disorder. The breaking characteristics has provided novel universal exponents, whose experimental verification is under way. It would be very interesting and helpful to have a similar analysis made in the three dimensional case, more relevant from the experimental point of view.

Finally, the exciting result is that all the scaling properties found either for the breaking characteristics or for the total number of broken bonds and the multifractality of the local strain distribution, are independent, within error bars, on the particular distribution of quenched disorder and on the nature of the model.

Acknowledgements

We acknowledge the contributions of Janos Kertész.

References

[1] See e.g. H. Liebowitz (ed.), *Fracture*, Vols. I-VII, (Academic Press, New York, 1984)

[2] H.J. Herrmann in *Random Fluctuations and Pattern Growth*, eds. H.E. Stanley and N. Ostrowsky (Kluwer, Dordrecht, 1988), p.149

[3] B.K. Chakrabarti, Rev. Sol. State Sci. **2**, 559 (1988)

[4] L. de Arcangelis, S. Redner and H.J. Herrmann, J. Physique Lett. **46**, L585 (1985);
P.M. Duxbury, P.D. Beale and P.L. Leath, Phys. Rev. Lett. **57**, 1052 (1986);

P.M. Duxbury and P.L. Leath, J. Phys. A **20**, L411 (1987);
P.D. Beale and D.J. Srolovitz, Phys. Rev. B **37**, 5500 (1988) and references therein

[5] T.A. Witten and L.M. Sander, Phys. Rev. Lett. **47**, 1400 (1981) ;
L. Niemeyer, L. Pietronero and H.J. Wiesmann, Phys. Rev. Lett. **52**, 1033 (1984)

[6] E. Louis, F. Guinea and F. Flores in *Fractals in Physics*, eds. L. Pietronero and E. Tossatti (Elsevier, Amsterdam, 1986), p.117;
E. Louis and F. Guinea, Europhys. Lett. **3**, 871 (1987);
P. Meakin, G. Li, L.M. Sander, E. Louis and F. Guinea, J.Phys. A **22**, 1393 (1989)

[7] E.L. Hinrichsen, A. Hansen and S. Roux, Europhys. Lett. **8**, 1 (1989)

[8] L.D. Landau and E.M. Lifshitz, *Theory of Elasticity* (Pergamon, Oxford, 1986)

[9] S. Roux and E. Guyon, J. Physique Lett. **46**, L999 (1985)

[10] W. Nowacki, *Theory of Asymmetric Elasticity*, (Pergamon, Warsaw, 1986)

[11] G.G. Batrouni and A. Hansen, J. Stat. Phys. **52**, 747 (1988)

[12] C. Tang, Phys. Rev. A, **31**, 1977 (1985);
J. Szép, J. Cserti and J. Kertész, J. Phys. A **18**, L413 (1985);
J. Nittmann and H.E. Stanley, Nature **321**, 661 (1986);
J. Kertész and T. Vicsek, J. Phys. A **19**, L257 (1986)

[13] J. Fernandez, F. Guinea and E. Louis, J. Phys. A **21**, L301 (1988)

[14] H.J. Herrmann, J. Kertész and L. de Arcangelis, Europhys. Lett. **10**, 147 (1989)

[15] B.B. Mandelbrot and T. Vicsek, J. Phys. A **22**, L377 (1989)

[16] F. Family, D.E. Platt and T. Vicsek, J. Phys. A **20**, L1177 (1987)

[17] J.P. Eckmann, P. Meakin, I. Procaccia and R. Zeitak, Phys. Rev. A, **39**, 3185 (1989)

[18] M.J. Blackburn, W.H. Smyrl and J.A. Feeney, in *Stress Corrosion in High Strength Steels and in Titanium and Aluminium Alloys*, ed. B.F. Brown (Naval Res. Lab., Washington, 1972), p.344

[19] B. Kahng, G.G. Batrouni, S. Redner, L. de Arcangelis and H.J. Herrmann, Phys. Rev. B **37**, 7625 (1988)

[20] H.J. Herrmann, A. Hansen and S. Roux, Phys. Rev. B **39**, 637 (1989)

[21] L. de Arcangelis and H.J. Herrmann, Phys. Rev. B **39**, 2678 (1989)

[22] A. Hansen, S. Roux and H.J. Herrmann, J. Physique **50**, 733 (1989)

[23] L. de Arcangelis, A. Hansen, H.J. Herrmann and S. Roux, Phys. Rev. B (1989)

[24] G. Paladin and A. Vulpiani, Phys. Reps. **156**, 147 (1987);
L. de Arcangelis, S. Redner and A. Coniglio, Phys. Rev. B **31**, 4725 (1985)

[25] H. Takayasu in *Fractals in Physics*, eds. L. Pietronero and E. Tossatti (Elsevier, Amsterdam, 1986), p. 181;
H. Takayasu, Phys. Rev. Lett. **54**, 1099 (1985)

IV Rheology and Fracture

The first lecture, by Krajcinovic, recapitulates author's work on the micromechanical modelling of the brittle deformation processes in materials such as ceramics and concrete. The essential aspects of the deformation process, *i.e.* , the nucleation and growth of microcracks, are related to the microstructural fabric. The effectiveness of the proposed modelling is illustrated on several examples. A close fit of experimentally measured stress-strain curves is obtained without introduction of physically unidentifiable "fitting" parameters.

The next lecture, by Maugis, treats the case of solid adhesion in the framework of brittle fracture theory in presence of viscous losses. The separation of two adherent solids never occurs as a whole but the area of contact is progressively reduced until complete separation is achieved. This reduction in contact area can be considered as resulting from the propagation of a crack, and the concepts of fracture mechanics can be used. The mechanism of the "stick-slip" instability during the peeling process is discussed for viscoelastic solids.

A common thread through the next two lectures is provided by the tendency for deformation in ductile solids to localize into narrow bands. Localization of deformation occurs in a wide variety of solids, in rocks and concrete as well as in structural metals. Strain localization plays an important role in limiting ductility and, accordingly, the phenomenon of strain localization has received a great deal of attention over the last decade. Localization of deformation is associated with a change in type of the governing differential equations, and, in general, the theoretical framework used to predict the onset of localization is inadequate for determining the post-localization response.The two following lectures are devoted to a good description of this latter stage.

Needleman illustrates how accounting for the inevitable time and rate dependence of inelastic processes, for non-linear elastic solids or rate independent plastic solids, acts to regularize the governing equations and provides a framework for a meaningful investigation of post-localization response. The application to deformation and failure in porous plastic solids and the direct calculation of fracture toughness from an analysis of the microscale ductile fracture mechanism are discussed.

The following lecture, by Aifantis, shows that introduction of higher order spatial gradients of the macroscopic variables for capturing the evolution of the system in the post-instability regime is another feasible approach. Their presence may be justified on the basis of the "adiabatic elimination" of internal variables with diffusive transport, such as point defects and dislocation ensembles, characterized by short life-times and evolving at scales smaller than the macroscopic scale. Emphasis is put on illustrating the origin of plastic instablities at various levels of observation.

Chapter 10

Micromechanics of Brittle Deformation Processes

Dusan Krajcinovic

Mechanical and Aerospace Engineering
Arizona State University, Tempe, AZ 85287, USA

The present study recapitulates author's work on the micromechanical modelling of the brittle deformation processes in materials such as ceramics and concrete. The essential aspects of the deformation process, i.e. the nucleation and growth of microcracks, are related to the microstructural fabric. The effectiveness of the proposed modelling is illustrated on several examples. A close fit of experimentally measured stress-strain curves is obtained without introduction of physically unidentifiable "fitting" parameters.

1. Introduction

The objective of the present study is to recapitulate and summarize some of the work done by the author and his associates on the micromechanical modelling of brittle deformation processes. After a short discourse on thermodynamic aspects of the constitutive modelling of processes dominated by nucleation and growth of microcracks, the study focuses on the salient aspects of the mechanical response of particulate composites (concrete) and semi-brittle ceramics.

The adopted strategy emphasizes analytical description of the dominant mechanism of the irreversible changes of the microstructure and its homogenization into a relation mapping macrostresses on macrostrains. The latter task is accomplished through a micro-to-macro transition within the framework of a suitable effective continua approximation.

One of the objectives of this study is to underline the essential dependence of the deformation process (damage evolution) on the microstructural fabric. In particular, the gradual growth of the inelastic deformation depends on the stochastics distribution of the energy barriers on the microscale. Knowing, for example, the statistics of the grain size distribution, it is possible to determine the salient aspects of the kinetics of the damage process in contrast to making *a priori* postulates characteristic of most phenomenological theories.

Disorder and Fracture, Edited by J. C. Charmet *et al.*
Plenum Press, New York, 1990

The illustrative examples clearly demonstrate the effectiveness of the proposed model and its accuracy in analytical replication of the experimental data. Moreover, this is accomplished without use of additional, physically unidentifiable, "material" parameters commonly utilized in practice. Finally, the proposed model allows for a rational study of the optimal microstructure for a given deformation process.

The present study underlines inherent shortcomings of the micromechanical studies typically having their root in computational complexities which can become substantial in the case of complex geometries and arbitrary stress fields. In this sense, the micromechanical and phenomenological theories are viewed as complementary rather than competing with each other.

2. Scale

The computational efficiency in analyses of engineering problems to a large degree depends on the smoothness and continuity of the computed fields. Persisting infatuation with homogeneous continua is but a natural consequence of that fact. The underlying logic is that the knowledge of the expected values of computed variables suffice to predict the degree of the functional and structural reliability of the analyzed structural or machine component. This is, however, true only at the low energy levels, associated with the design conditions, at which the structure of the material changes only slightly and homogeneously. On the other hand, the onset of failure is commonly attributed to the nucleation, growth and coalescence of microdefects (*i.e.* decrease and ultimately the loss of homogeneity in the specimen). Prompted by a variety of microstructural mechanisms, the ultimate failure is typically characterized by a localization process during which the energy dissipating sources in a synergistic manner organize into clusters (macrocracks, shear bands, *etc...*). Hence, a rational prediction of the events on scale recognizing the heterogeneity of the material and its influence on the kinetics of the defects leading to the eventual failure (loss of integrity).

In the case of brittle processes in engineering materials, it is customary to adopt a two-scale model incorporating:

• microscale (or sometimes called mesoscale) recognizing local heterogeneities being in size commensurable to crystal grains or second phase particles, and
• macroscale on which the material can be considered as continuous and homogeneous

Modelling of this type typically involves (see *e.g.* Refs.[1-4]):

• analytical description of the local fields within the microstructural element, and
• a transition process during which these micromechanical fields are embedded (homogenized or averaged) into corresponding macrofields.

The continuum modelling requires the description of the material structure and recorded history (qualifying and quantifying the accumulated defects in a physically appropriate sense) in each point of the continuum using, whenever possible, continuous functions with moderate spatial and temporal gradients. Since the probability that a microdefect that is actually located in the observed material point is zero, the micro-to-macro transition inherently requires determination of an influence volume enveloping the material point. The defects within this volume, commonly labeled as representative volume element, are considered to have substantial influence on the thermodynamic state of the point of the continuum and the rate of its change. The rate of change of the microstructure within this volume is described by the requisite thermodynamic fluxes (rates of change of the internal variables) comprising the configurational space attached to its material point of the continuum. The skill in constitutional modelling is directly related to the minimization of the configurational space, *i.e.* ability to single out the salient aspects of the deformation process and blur the less important details changing from one experiment to the other. Since the response and, ultimately, the failure of brittle solids depends on the size and distribution of defects in the material, the scatter in experimental results is typically rather large. Consequently, sophisticated models based on complex analyses are sometimes hard to justify.

The representative volume element is, roughly speaking, the smallest volume of the material which in a given sense responds as a macro-continuum. More specifically, the size of the representative volume element is determined by the choice of the tolerance levels for the difference between the statistical momenta (of desired order) of a given field computed for the macro-continuum and the representative volume element.

Once determined the representative volume element maps on the material point of the continuum and the statistical momenta of the defect population within the volume are then assigned to the relevant material point of the macro-continuum. By definition, a crack will be referred to as a microcrack if its size is comparable to that of the source of the inhomogeneity on the microscale (grain, second-phase particle, *etc...*).

3. Micro-Response

Consider a single microcrack embedded in a perfectly elastic matrix occupying volume V of the representative volume element. Assuming, for convenience, the process to be isothermal, the inelastic change of the Gibbs (complementary) energy of the volume associated with the increase of the crack size[5] is

$$d^i\psi = \int_L ((G - 2\gamma)\delta\ell)dL \geq 0 \qquad (1)$$

where L is the curvilinear coordinate measured along the crack front, $\delta\ell$ the advance of the crack front in the direction to its normal, G the energy release rate and γ the surface energy. The sign of the inequality reflects the entropy production rate (according to the second law of thermodynamics). The superscript (i) in the expression

$$d^i\psi = \psi(\boldsymbol{\sigma}, H + dH) - \psi(\boldsymbol{\sigma}, H) \qquad (2)$$

is used to label the irreversible changes reflected in the change of recorded history denoted symbolically by letter H.

In general, the surface energy is on the microscale a random variable defining preferred planes of crack propagation (cleavage planes in crystals, transition zones in concrete, glassy grain boundary phases in vitreous ceramics, *etc...*). The existence of energy barriers (separating zones of low surface energy from the rest of the solid) explains the fact that a microcrack can be arrested in a brittle material subjected to uniform stress fields.

The energy release rate is a homogeneous function of second order in stress intensity factors $\boldsymbol{k}$ (arranged into a vector)[5-7]

$$G = \frac{\pi}{4} C_{ij} k_i k_j \qquad (3)$$

where $\boldsymbol{C}(H)$ is a second rank tensor depending on the material parameters of the surrounding matrix (including possibly the adjacent defects).

Assuming the macrostrains to be infinitesimal during a brittle deformation process, the Gibbs energy is a homogeneous function of second order in macrostresses

$$\psi(\boldsymbol{\sigma}, H) = \frac{1}{2} V \boldsymbol{\sigma} : S(H) : \boldsymbol{\sigma} \qquad (4)$$

with $S(H)$ being the fourth rank compliance tensor. Thus, in view of Eq.(2)

$$d^i S = \frac{1}{V} \frac{\partial^2 (d^i \psi)}{\partial \sigma \partial \sigma} \qquad (5)$$

Differentiating Eq.(1) twice respect to stresses and using Eqs.(3) and (5), the following expression is obtained for the compliance tensor

$$d^i S = \frac{\pi}{2V} \int_L \left(\left(\frac{\partial \boldsymbol{k}}{\partial \boldsymbol{\sigma}} \cdot \boldsymbol{C} \cdot \frac{\partial \boldsymbol{k}}{\partial \boldsymbol{\sigma}} \right) \delta \ell \right) dL \qquad (6)$$

The expression (6) can be formally extended to the case of an ensemble of a large number of cracks within the volume V. Using the principle of superposition, the expressions (6) can be modified to the form

$$d^i S = \frac{\pi}{2V} \sum_a \int_{L_a} \left(\left(\frac{\partial \boldsymbol{k}(\hat{\boldsymbol{\sigma}})}{\partial \boldsymbol{\sigma}} \cdot \boldsymbol{C}(H) \cdot \frac{\partial \boldsymbol{k}(\hat{\boldsymbol{\sigma}})}{\partial \boldsymbol{\sigma}} \right) \delta \ell \right) dL \qquad (7)$$

In Eq.(7), the summation is extended over all active cracks (increasing their size) in V. The cap above the stress $\hat{\boldsymbol{\sigma}}_{(a)}$ refers to the external field of the ath crack. This stress is nor necessarily equal to the expected — or macrostress — $\boldsymbol{\sigma}$. The two stresses are related through the influence tensor $\boldsymbol{M}$[8] of fourth rank

$$\hat{\boldsymbol{\sigma}} = \boldsymbol{M}(H) : \boldsymbol{\sigma} \qquad (8)$$

Naturally, the stress intensity factors k are also functions of the local stresses reflecting the direct interaction of cracks, while the tensor $M(H)$ depends on the elastic properties of the weakened matrix (distribution, size and orientation of defects).

In principle, the external stress fields $\hat{\sigma}$ can be determined for all N cracks solving a system of N coupled integral equations.[9] This problem, in general, admits an analytical solution only in very special cases of isotropic, elastic matrices weakened by regular en-echelon arrays of slits of equal size and orientation. Approximate solutions, such as one suggested by M Kachanov[10], are available as well but had still not been tested in applications. Naturally, the actual problem of a large number of irregularly spaced and oriented microcracks of irregular shape embedded in a solid comprised of anisotropic grains joined at irregularly oriented grain boundaries would tax the patience of most ardent numerical analysts.

An approximate solution to this problem is often sought within the context of the effective continua models.[9] The most popular among them, known as the self-consistent model, is based on the assumptions of:

• weak dependence of the external fields of each microdefect on the exact disposition of adjacent defects within the representative volume element, and
• equality of the local and macrostress tensors $\hat{\sigma} = \sigma$

The first of two assumptions reduces the problem of N interacting cracks in the actual medium to a sum of N problems of single cracks embedded in an effective medium.

The parameters of the effective continuum are computed such as to approximate the actual continuum in an appropriate sense. The second of two assumptions eliminates the need to solve the system of N integral equations significantly enhancing the tractability of the problem.

Consequently, within the context of the self-consistent model, the irreversible change of the compliance Eq.(7) can be determined at each stress and damage level of the loading cycle:

• if the analytical expressions for the stress intensity factors k and tensor C are available for a crack embedded in the effective continuum, and
• if the crack front advance is a known function of the stress.

Since a solid becomes anisotropic when weakened by an irregular distribution of microcracks, neither of the two tasks is trivial. In fact, closed-form expressions for k and C are available only for a handful of crack geometries embedded in isotropic solids, and for two-dimensional problems involving rectilinear slits in anisotropic solids. In all other cases, the determination of k and C requires a complicated numerical effort.[6,11] The approximate expressions for the stress intensity factors are available (see *e.g.* Refs.[12,13]) in the case of irregular shapes rather than the lack of material symmetries. In either case, the relative paucity of analytical solutions for k and C places a serious limitation on the range of applicability of the micromechanical models. It should be pointed out that expression (7) can be, in conjunction with the formulae derived by Hoenig[6], reformulated in terms of eigenstrains.[9] Naturally, the same limitations on the tractability are now imposed by the lack of the analytical solutions for the crack opening displacements for cracks of arbitrary shapes embedded in anisotropic media.

Finally, the relatively primitive effective continua models such as the conventional self-consistent theory impose another limitation on the range of applicability. Eliminating from the considerations the distance between the adjacent cracks, the models based on the self-consistency are inherently incapable of predicting failure modes characterized by localization of crack-like defects into cracks-bands.

In conclusion, the self-consistent theory yields applicable and reliable results for small to moderate crack densities and relatively homogeneous fields. In all other cases, it might be necessary to introduce higher order statistical momenta of the crack distribution which would seriously reduce the simplicity of the analytical model making it much less attractive for practical applications.

4. Macro-Response

Consider next the macrofields in the effective continuum. The already formed micro-cracks, assumed to be distributed in a reasonably homogeneous manner over a large portion of the volume, are reflected through the compliance tensor. The rate of change of the compliance tensor, in function with the extent of the lacunity, can then be determined from expression (7).

From Eq.(4), the expression mapping the macrostresses on macrostrains is

$$\varepsilon^e = \frac{1}{V}\frac{\partial \psi}{\partial \boldsymbol{\sigma}} = \boldsymbol{S} : \boldsymbol{\sigma} \tag{9}$$

where $S = S(H)$ reflects not only the properties of the original (virgin) matrix but also the damage induced anisotropy. Thus the total strain is

$$\varepsilon = \varepsilon^e + \varepsilon^p = \boldsymbol{S} : \boldsymbol{\sigma} + \varepsilon^p \tag{10}$$

The plastic strain ε^p is defined as

$$\varepsilon = \varepsilon^p \quad \text{for} \quad \boldsymbol{\sigma} = 0 \tag{11}$$

The incremental form of Eq.(10) is for the case of loading

$$\dot{\varepsilon} = \boldsymbol{S} : \dot{\boldsymbol{\sigma}} + \dot{\boldsymbol{S}} : \boldsymbol{\sigma} + \dot{\varepsilon}^p \tag{12}$$

with dots above symbols denoting increments. The second term on the right-hand side of Eq.(12) represents the inelastic strain increment attributable to the damage evolution (increase of the surface area of microcracks within the representative volume element mapping on the material point). The loading process is here defined as one during which the recorded history changes, *i.e.* during which at least one microcrack increases it surface area, and/or for which the plastic strain rate is non-zero.

In unloading $(dH = 0)$

$$\dot{\varepsilon} = S : \dot{\sigma} \tag{13}$$

However, since $S = S(H)$ the slope of the unloading segment of the stress-strain curve is not identical to the slope of the initial segment of the loading curve ($H = 0$). The difference in slopes reflects the damage accumulated during the loading implying that the change in the compliance is a rational macroscopic measure of the microcrack density.[14] In the case of some brittle materials, such as concrete, the residual strain measured upon unloading is a result of rough crack surfaces and aggregate interlock[15] rather than plasticity in a conventional sense of this term.

Assuming that the inelastic change of the compliance can be determined from the micromechanical considerations Eq.(7), and that the plastic strain can be computed, equation (12) suffices for the solution of a given boundary value problem. The successful completion of this task requires availability of analytical expressions for the tensors C and k, for a given effective continuum and state of stress. Additionally, the kinetic equation relating the increase of the crack size and the state of stress must be known as well.

5. Kinetic Equations

The determination of the crack growth patterns is, perhaps, the least settled issue in fracture mechanics. A variety of empirical and phenomenological methods are employed in order to match the experimental data for a variety of materials, specimen shapes and states of stress.

The Griffith's criterion (obtained from Eq.(1) assuming that the inequality is valid for every point of the crack front)

$$G - 2\gamma \geq 0 \quad \text{for} \quad \delta\ell > 0 \tag{14}$$

and the condition

$$\frac{\partial G}{\partial l} \delta\ell > 0 \tag{15}$$

which is valid for a crack in a uniform tensile field would indicate that the crack cannot be arrested once it starts propagating. Implicit in this conclusion is the assumption that the surface energy is constant for a homogeneous continuum. This assumption, naturally, is oblivious to the fact that a crack typically grows along a tortuous path defined by interior surface energies. In an empirical sense, this is implicity recognized by the so-called R−curve approach which assumes that the resistance to the crack growth is not constant. Moreover, the notion of short cracks is another indication of the same phenomenon which is for obvious reason even more emphasized when the crack is of the size of source of the material heterogeneity.

The basic expression relating the increase of the crack length and the energy release rate and, possibly, the accumulated history is of the form[16]

$$\delta \ell = f(G, H) \tag{16}$$

The problem of the determination of a rational function $f(G, H)$ is significantly alleviated if the path of the crack growth is a priori known.[17] In other words, if the planes of inferior surface energy (such as cleavage planes in crystals, transition zones at the interface of different phases in particulate or laminated composites, weak grain boundaries, *etc...*) are known, the number of generalized coordinates defining the pattern of crack growth is reduced to a finite number. For example, in the case of a self-similar growth of a penny-shaped crack in its own plane, the kinematics of the crack growth is fully defined by a single generalized coordinate (increase of the crack radius).

As an example, consider the two-dimensional time-independent deformation of the perfectly brittle polycrystalline vitreous alumina specimen in uniaxial tension. Since the glassy boundary phase has a much smaller surface energy than the alumina crystals, all cracks will, in general, propagate along the grain boundaries until they acquire a size being critical in macrosense.

In view of Eq.(14), the initial slit (introduced during the processing) of length $2a_0$ will become unstable when

$$F_0(a, \beta, q) = G_0 - 2\gamma = (\frac{\pi a_0}{E} \cos^2(\beta))q_0^2 - 2\gamma_0 = 0 \tag{17}$$

where the Young's modulus E must be amended for the case of plane strain. Also, β is the angle subtended by the normal to the slit and the tensile axis with q being the tensile stress. Since the inequality (15) is clearly satisfied in the case of Eq.(17), as soon as the loading point intersects the hypersurface (17), the crack will start propagating in an unstable (runaway) fashion. Assuming the surface energy to be constant along the grain boundary, the crack can be arrested only at the triple grain junctions where the superior toughness of the grains forces it to kink to the sequent grain boundary in order to recommence its growth.

Using the Cotterell-Rice[18] criterion, it is possible to derive the kinking condition (see [19]) using Eq.(14) in the form

$$F_k(a_1, \beta, q) = G_1 - 2\gamma_0 = \frac{\pi a_1}{E} g(\beta, \phi)q^2 - 2\gamma_0 = 0 \tag{18}$$

where $2a_1$ is the size of the straight segment of the grain boundary (the distribution of which is considered to be known), while $g(\beta, \phi)$ is a trigonometric function of angles defining the orientation of the grain boundaries with respect to the tensile axis (or fixed specimen coordinate system). Naturally, the adjacent grain boundaries intersect each other at an angle of $\pi/3$.

The first crack to become unstable will, according to Eq.(17), be the largest crack which is perpendicular to the tensile axis. Thus the tensile stress at which the nonlinear response will commence is from Eq.(17)

$$q = q_0 = \left(\frac{2\gamma_0 E}{\pi a_{0M}} \right)^{\frac{1}{2}} \tag{19}$$

where a_{0M} is the half-length of the largest of initial cracks. The onset of kinking will follow when from Eq.(19)

$$q = q_k = \left(\frac{32\gamma_0 E}{9\pi a_1} \right)^{\frac{1}{2}} = q_0 \left(\frac{16 a_{0M}}{9 a_1} \right)^{\frac{1}{2}} \tag{20}$$

The cracking will be gradual $q_k > q_0$ only if

$$a_{0M} > \frac{9}{16} a_1 \tag{21}$$

In the case of very small initial defects, the response will, therefore, be brittle. This was noticed in the case of high strength concrete to which the modifiers are added to increase the toughness of the transition zone. Reduction of the difference in the height of the consecutive energy barriers makes the deformation less gradual and more susceptible to brittle failure.

The spasmodic growth of a crack of given orientation along a path intersecting zones of increasing toughness, similar to the jerky glide of dislocations, has been experimentally verified recording acoustic emissions during the loading process.

Naturally, in the case of a large specimen the crack and grain sizes and orientations form continuous distributions. The hypersurfaces (17) and (18) define the loci of:
- all cracks retaining their original size if $F_0 < 0, F_k < 0$,
- cracks occupying the entire straight segment of a grain boundary, $F_0 > 0, F_k > 0$, and
- kinked cracks $F_k > 0$.

In a complete analogy to the slip theory, the loading surface will develop vertices serving as attractors for the loading point. The change of the shape of the loading surface can be followed in a relatively straightforward fashion for the cases of monotonic and proportional loading programs. However, in the cases of non-proportional loads, the cracks will change their status from passive to active and *vice versa*. Consequently, the loading surface will change in a stepwise fashion making bookkeeping a significant problem. Even though the same problem exists in the case of the slip theory, the difficulties are further exacerbated as a consequence of the fact that the cracks impose unilateral constraint on the displacement field.

Conceptually simple, the described algorithm may become computationally unattractive in the case of complex expressions for the grain size distributions, large number of defects, *etc...*, involving significant computational effort.

The described damage evolution process, typically labeled as perfectly brittle or "cleavage 1" response, is but the simplest case dominated by a single class of microdefects. In this case, all cracks are present at the outset and the deformation process does not include the crack nucleation phase. The microcracking can, in addition, occur as a result of stress concentrations induced by dislocation pile-ups in isolated shear bands, relative motion of adjacent grains associated with grain boundary sliding, *etc....* In all of these cases, the Griffith's condition Eq.(14) is reached through the increase of the energy release rate G.

In addition, at constant (initially subcritical) G the condition (14) can be satisfied by a gradual decrease of the surface energy attributed to the reactions with chemically aggressive substances found in many ambients. Corrosion and hydrogen embrittlement of metals, alkali and sulphate reactions in cementitious composites, *etc...*, are just some of the examples often encountered in practice. The solution of this type of problems requires the determination of the rate of decrease of the surface energy as a function of the chemical potentials and the supply of the chemically aggressive substance.[20] Specifically, the solution of the problem requires solution of the mass transport equation through the porous solid, determination of the actual chemical process, determination of the stresses associated with the change in molar volume of the reactants and the reaction product, and the change of the surface energy as a result of the chemical reaction.

6. Macro-Compliance

The determination of the macro-compliance, based on expressions such as Eq.(7), is conceptually simple but involves too many details (see *e.g.* [2,21,3,7]) to allow for a comprehensive and self-sufficient discourse within this study. Nevertheless, it must be said that the quest for reducing the computational effort to manageable levels necessitates introduction of simplifying assumptions regarding the shape of cracks and grains, and defect interaction.

As demonstrated in the referenced literature, the change in the rate of the compliance can be, within the framework of the self-consistent model, on the basis of Eq.(7) written in the form

$$d^i S = \frac{N}{V} < a^2 \delta a \boldsymbol{B} > \tag{22}$$

where a is a characteristic linear dimension of the crack, N the number of active cracks in the volume element and $\boldsymbol{B}(\mathrm{H})$ a fourth rank tensor depending on the recorded history through the size, shape and orientation of cracks.

As an illustration, consider a two-dimensional case of a solid weakened by a large number of randomly oriented rectilinear slits of equal length. In the case of uniaxial tension, the onset of cracking will occur at q_0 determined from Eq.(19). As the load is further increased, more and more slits will change their size from $2a_0$ to $2a_1$. The fan of cracks which have already changed their length is confined within the fan, the limits of which are determined from Eqs.(17) and (20) as

$$\beta_{1,2} = \pm \cos^{-1}(\frac{q_0}{q})^{\frac{1}{2}} \tag{23}$$

Assuming, for the sake of this example, that the slits can be only of two sizes (which is the same as assuming that all grains are equal in size), the compliance attributable to the cracks is

$$\hat{S} = \frac{1}{A}\{N_0(\int_0^{\beta_1} a_0^2\boldsymbol{B}(\beta)d\beta + \int_{\beta_2}^{\frac{\pi}{2}} a_0^2\boldsymbol{B}(\beta)d\beta) + N_1\int_{\beta_1}^{\beta_2} a_0^2\boldsymbol{B}(\beta)d\beta\} \tag{24}$$

where A is the surface area of the surface element. Also,

$$N_0 = \frac{\beta_1 + \beta_2}{\pi}N_T, \quad N_1 = N_T - N_0 \tag{25}$$

where N_T, N_0 and N_1 are the total number of slits in A, number of slits retaining their original size and number of slits of increased size, respectively. The general case, characterized by a given distribution of crack sizes is discussed in [21] and [3].

The described procedure is exceedingly simple in the case of Taylor's model (totally neglecting the crack interaction, $i.e.$ assuming that each crack is embedded in the undamaged solid) allowing for a closed-form, analytical solution in the case of homogeneous stress fields. The application of the self-consistent model requires a modest iterative procedure which is still well within acceptable limits. Higher order statistical theories, allowing for direct crack interaction, involve substantial computations which are not entirely justifiable in practical applications.

7. Perfectly Brittle Response

The initial effort in the establishment of micromechanical models was concentrated on the study of the perfectly brittle, time-independent response of solids such as plain concrete. This class of problems is characterized by a single energy dissipating mechanism (microcracking) and the absence of the crack nucleation phase. All cracks are assumed to be already in the material (as a result of temperature gradients and moisture transport associated with the hardening process) and are typically located within the transition zones separating the aggregate from the surrounding cement paste. Occupying initially only a portion of the transition zone, these cracks increase their size in two discrete increments. The crack growth pattern is a function of the hierarchy of surface energies (transition zone, cement paste and aggregate in ascending order) and distribution of aggregate sizes (sieve grading curve).

The results for uniaxial tension[24] and compression[22] — Figs.(1) and (2) — clearly demonstrate the efficiency of the proposed algorithm in accurately replicating the experimental data without using fitting parameters. The macrofailure criterion is introduced in the case of tension considering the probability that a crack can be arrested once it starts propagating through the cement paste. The determination of the failure criterion in the case of the uniaxial compression represents a more complicated task since the growth of individual cracks is stable. The splitting failure[13] occurs as a result of the direct crack interaction which cannot, for already mentioned reasons, be easily incorporated into the analyses.

The accuracy of the closed-form, analytical solutions obtained neglecting the crack interaction (Taylor's model) was checked using the self-consistent model.[23] The results proved to be surprisingly accurate in the case when the stress at failure did not exceed the stress at the onset of cracking by a factor of 2.5 to 3.0. According to be available data, this would include most of the commonly used concretes.

The developed model proved to be efficient in considerations of the effects of the microstructural features (distribution of aggregate sizes, volume fraction of the aggregate, size and distribution of initial defects, changes in surface energy distributions, etc...) on the mechanical response and macrofailure. Additionally, these models allow for rational discussion of the size effect as well.

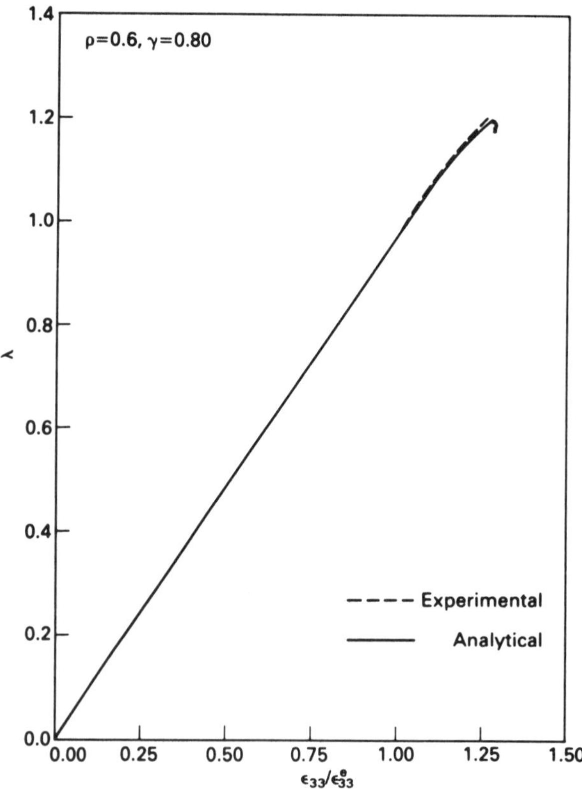

Figure 1 Stress-strain curve for concrete in uniaxial tension for a given sieve-grading.

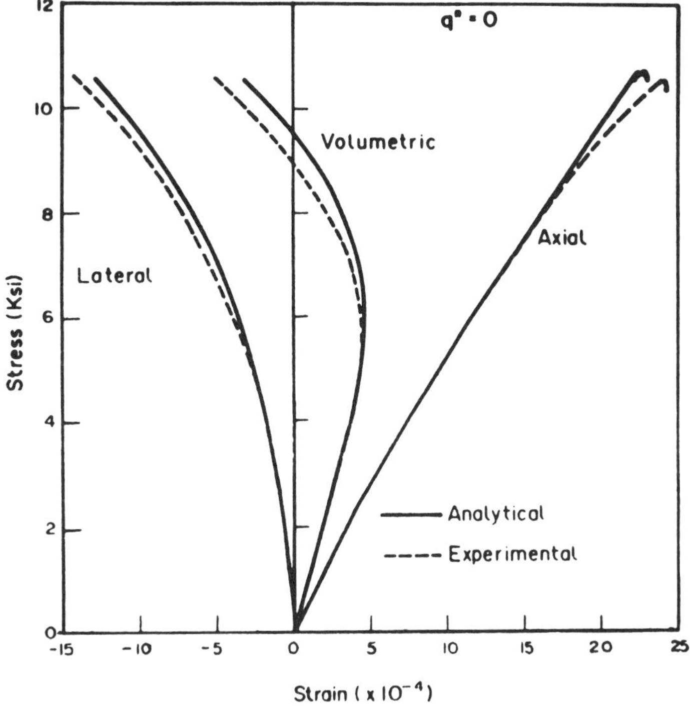

Figure 2 Stress-strain curve for concrete in unconfined uniaxial compression.

8. Semi-Brittle Deformation Processes

The deformation of semi-brittle materials, such as MgO ceramics, is characterized by the microplasticity induced microcracking. During the initial phase of the deformation in MgO crystals (at low stress levels), the inelastic strain is attributed to the easy glide of dislocations along two conjugate primary slip systems $\{110\}\ \langle 110\rangle$. The inability to accommodate a general inelastic state of strain with less than five independent slip systems and the existence of glide barriers (grain boundaries and to a lower extent intersection of slip bands) causes large stress concentrations which become potential sources of crack nucleation.

The stress concentration at the head of a dislocation pile-up, near which the so-called Zener-Stroh cracks are typically nucleated, was determined by Smith and Barnby.[24] In the case of uniaxial tension of an infinitely extended, elastic and homogeneous plate, the Smith-Barnby criterion may be written in the form (see also [19] and [25])

$$F_\eta = \frac{1}{2}q\sin(2\beta) - \left(\tau_i + \left(\frac{3\pi\gamma_s G_0}{2(1-\nu)D}\right)^{\frac{1}{2}} f(\phi)\right) = 0 \tag{26}$$

where D is the grain size, β the angle between the normal to the slip plane and

180   Dusan Krajcinovic

the tensile stress q, ϕ the angle subtended by the slip plane and slit, τ_i the lattice
resistance to the dislocation glide, G_0 the shear modulus and $f(\phi)$ a function of
angle ϕ available in the referenced literature.

On the basis of the experimental evidence, it was assumed that the considered
material favors intergranular cracks. Moreover, due to the presence of the residual stresses along the grain boundaries (associated with the thermal anisotropy of
crystals) once nucleated, a crack can be arrested only at the triple grain junction.
Further crack growth involves kinking on the sequent grain boundaries at one of
the crack tips.

Consequently, the entire population of intergranular slits can be divided into
two classes:

- slits occupying a single rectilinear stretch of the grain boundary $F_\eta > 0, F_k < 0$,
- slits kinked over two sequent grain boundaries $F_k > 0$.

The total inelastic deformation includes inelastic slip over the active slip bands
as well as the effect of intergranular slits. The macrofailure is assumed to occur
when the first crack kinks onto the third grain facet acquiring length too large to
allow trapping by triple grain joints.

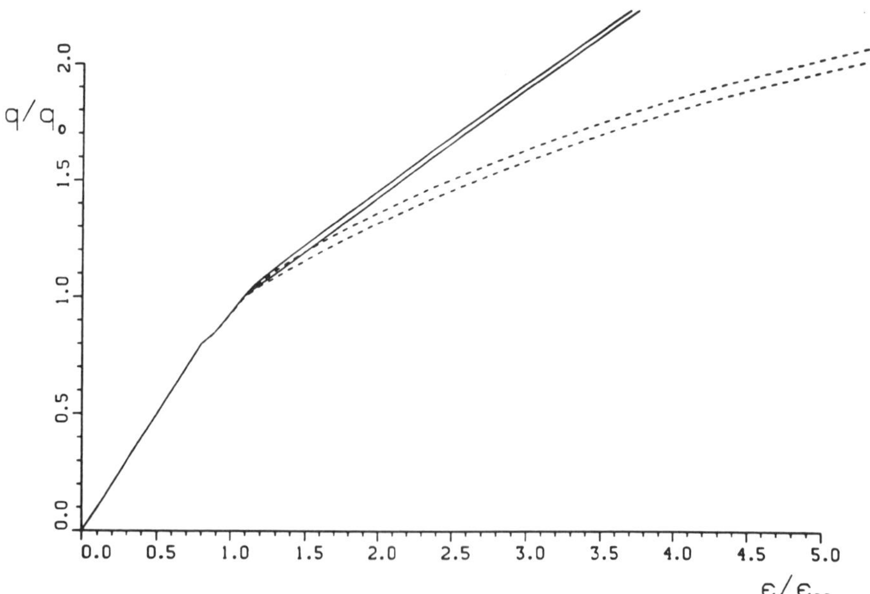

Figure 3 Stress-strain curve for an MgO specimen in uniaxial tension (solid lines
Taylor model, dashed lines self-consistent approximation). Upper lines uniform and
lower lines triangular distribution of grain sizes.

The computations were performed for both Taylor and self-consistent models[25] for uniform and triangular distributions of grain sizes Fig.(3). As in the case of perfectly brittle solids, the accuracy of the Taylor's model diminishes in the latter stages of the process. The influence of the grain size distribution (at indentical maximum grain size has a profound effect on the rupture stress.

9. Time-Dependent Deformation of Vitreous Ceramics

As a popular refractory material, ceramics are often used in high temperature applications. Consequently, the investigations of the creep deformation of ceramic specimens exposed to high temperatures for prolonged periods of time are of more than theoretical interest. The so-called creep deformation, at constant stress levels, reflects an array of different modes of irreversible changes of the microstructure such as dislocation glide, grain boudary sliding, grain boundary and bulk diffusion, nucleation and growth of spheroidal voids and microcracks, *etc.*

Naturally, it is unrealistic to expect that a single, comprehensive, analytical model can account for a substantial number of different deformation processes.[27] Consequently, it appears more promising to focus the attention initially to a relatively simple case of vitreous alumina at the stress and strain levels at which the grains themselves remain elastic. The inelastic deformation is in the case caused by a relatively low resistance of the glassy intergranular phase to viscous flow.

The microcrack nucleation and growth is attributed to the stress concentrations at the triple grain junctions caused by the slip over the grain boundaries exposed to high shear stresses. The shear stresses (associated with the externally applied tractions) are resisted by the viscosity of the intergranular phase (including the inhibiting effect of the hard particles along the grain boundary) and the resistance of the triple joint crack to the wedging action.[28]

The determination of the stress concentrations at the confluence of two sliding grain boundary and a grain boundary containing a crack is by no means a trivial problem. In absence of an anlytical solution, it was possible to determine the stress concentrations approximately using simplifying assumptions supported by numerical analyses. The problem was reduced to a system of two coupled differential equations in the crack length and the crack opening displacement (or average slip over the sliding grain boundary).[28] The numerical integration of these two equations proved to be routine.

The crack growth along the grain boundary was found to be stable for the considered circumstances. The growth is again arrested upon reaching the far end triple grain junction. At this point, the crack will kink when the kinking condition similar to (18) (see [3] and [28]) is satisfied. The driving force is supplied by the wedging action caused by the sliding over the grain boundaries exposed to shear stresses.

The computations were performed for two different sizes of the initial triple

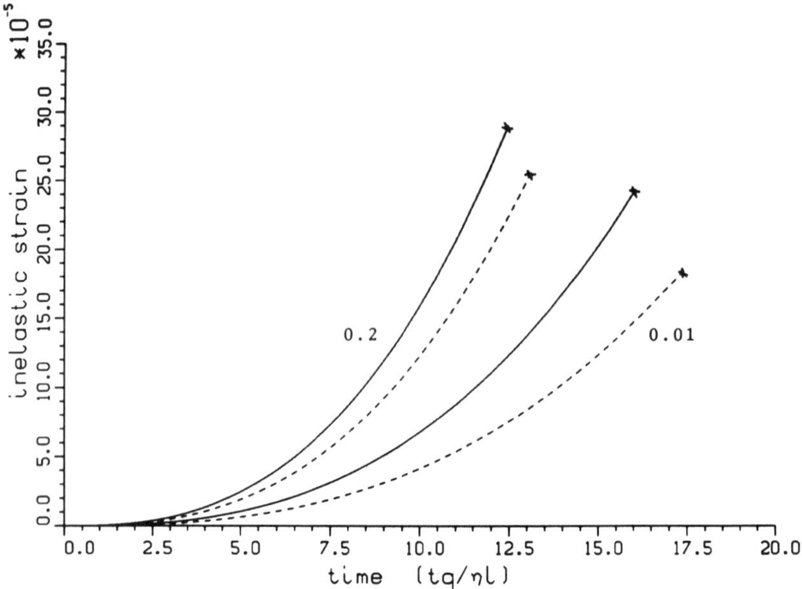

Figure 4 Inelastic strain for an Al_2O_3 specimen in creep. Solid lines for tension, dashed lines for compression. Numbers 0.2 and 0.01 refer to the initial crack grain boundary length ratio.

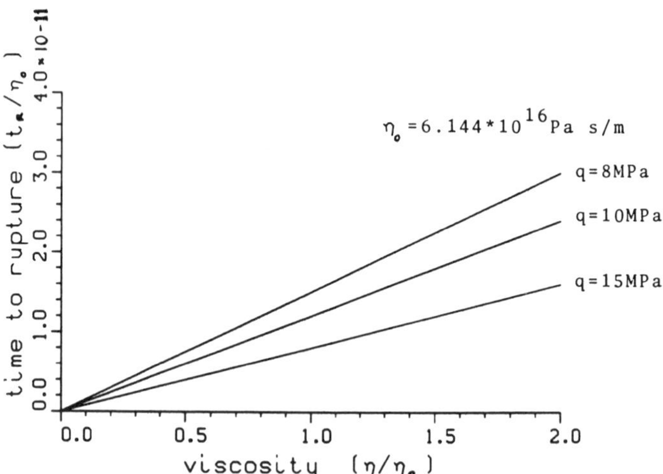

Figure 5 Linear relationship between the viscosity and the time to rupture (Monkman-Grant rule).

joints slits (assuming them to be either 0.01 or 0.2 of the grain boundary facet length). The general trends of the strain vs. time curves Fig.(4) is in agreement with the experimental observations. Just as importantly, the time to rupture indeed follows the Monkman-Grant rule since the rate of the inelastic strain in the stationary creep phase is inversely proportional to the viscosity Fig(5). It is interesting to note that the strain rate is linearly proportional to the stress and the nonlinearity of the strain vs. time curve is directly traceable to the increase of the compliance resulting from the damage evolution (*i.e.* increase in the lacunity).

Summary and Conclusions

The proposed modelling strategy based on the analytical description of the dominant mode of irreversible rearrangements of the microstructure has several important advantages when compared to traditional phenomenological theories. Specifically it:

• does not require introduction of physically unidentifiable fitting parameters.
• relates the salient aspects of microstructure (grain size distribution, volume fraction of second phase, *etc...*) to the mechanical response and circumstances leading to macrofailure.
• does not require a priori postulates of kinetic equations, and
• establishes physically justifiable relations between kinematic and force variables even when a "rigorous" analytical description of microscale events is not possible.

However, some limitations of the micromechanical modelling are quite apparent as well. In particular:

• analytical descriptions of complex microstructural mechanisms are unlikely to be feasible,
• considerations of non-uniform states of stress and strain involve horrendous book-keeping problems and attendant computational effort will be difficult to justify,
• macrofailure analyses will in most cases involve introduction of higher order statistical momenta of defect distribution,
• incorporation of direct defect interactions, grain anisotropy, residual stresses, three-dimensional effects (such as keying of adjacent grains in relative motion), *etc...*, will significantly increase the complexity of these methods.

All of these problems will be exacerbated in the case of non-proportional and cyclic loads during which the status of individual cracks will change from passive to active and vice versa.

Consequently, the micromechanical modelling should be regarded as complementary to phenomenological theories. Once the essential structure of the governing equations is established from microstructural considerations, it should be possible to formulate a rational phenomenological model preserving these relations in an appropriately smoothed sense. This is especially important in the case of the damage theory which has been plagued by an inordinately large number of different and contrasting models introducing confusion in applications.

Initial effort to establish a continuum damage model based on micromechanical considerations has been undertaken by the author in order to determine a rational and physically justified choice of internal variable and the conditions under which the kinetic equations can be derived from the potential.

Acknowledgement

It is pleasant duty to acknowledge the financial support in the form of research grants from the U.S. Department of Energy, Office of Basic Energy Sciences, Engineering Research Program (managed by Dr. Oscar Manley) and Air Force Office of Scientific Research, Directorate of Aerospace Sciences, Civil Engineering Program (managed by Dr. Spencer Wu) which made this work possible.

References

[1] B. Budiansky and T.T. Wu, *Theoretical prediction of plastic strains of polycrystals*, Proc. Fourth U.S. National Congress of Applied Mechanics, ASME Publ., New York, 597, (1962)

[2] H. Horii and S. Nemat-Nasser, *Overall moduli of solids with microcracks: load induced anisotropy*, J. Mech. Phys. Solids, **31**, 155, (1983).

[3] D. Krajcinovic, and D. Sumarac, *Michromechanics of damage processes* in *Continuum Damage Mechanics: Theory and Applications*, D. Krajcinovic and J. Lemaitre eds., Springer-Verlag, Wien, (1987).

[4] H. Kitagawa, *Macro- and microaspects in computational mechanics*, ASME Inter. Journal, **30**, 1361, (1987).

[5] J.R. Rice, *Continuum mechanics and thermodynamics of plasticity in relation to microscale deformation mechanisms* in *Constitutive Equations in Plasticity*, A.S. Argon ed., The MIT Press, Cambridge MA, (1975)

[6] A. Hoenig, *The behavior of a flat elliptical crack in an anisotropic body*, Int. J. Solids Structures., **14**, 925, (1978).

[7] D. Krajcinovic, and D. Sumarac, *A mesomechanical model for brittle deformation processes - Parts I and II*, J. Appl. Mech., **56**, 51, 57, (1989).

[8] R. Hill, *The essential structure of constitutive laws for metal composites and polycrystals*, J. Mech. Phys. Solids, **15**, 79, (1967).

[9] I.A. Kunin, *Elastic Media with Microstructure II, Three-dimensional Models*, Springer-Verlag, Berlin, (1983).

[10] M. Kachanov, *Elastic solids with many cracks: a simple method of analysis*, Int. J. Solids Structures, **23**, 23, (1987).

[11] T. Mura, *Michromechanics of Defects in Solids*, Martinus Nijhoff,The Hague, The Netherlands, (1982).

[12] H. Kitagawa, R. Yuki and T. Ohira, *Crack morphological aspects in fracture mechanics*, Eng. Fract. Mech, bf 7, 515, (1975).

[13] H. Horii and S. Nemat-Nasser, *Brittle failure in compression: splitting, faulting and brittle-ductile transition*, Phil. Trans. Royal Soc. London., **A319**, 337, (1986).

[14] J. Lemaitre and J.L. Chaboche, *Mécanique des Matériaux Solides*, Dunod, Paris, (1985).

[15] S. Nemat-Nasser and M. Obata, *A microcrack model of dilatancy in brittle materials*, J. Appl. Mech., **55**, 24, (1988).

[16] G.P. Cherepanov, *Fracture Mechanics of Composite materials*, Nauka, Moscow, (1983).

[17] V.V. Bolotin, *Multiparametric fracture mechanics* in *Strength Analysis*, Vol.25, Mashinostroenie, Moscow, (1984).

[18] B. Cotterell and J.R. Rice, *Slightly curved or kinked cracks*, Int. J. of Fracture, **16**, 155, (1980).

[19] A. Stojimirovic, D. Krajcinovic and T. Sadowski, *Constitutive model for polycristalline MgO ceramics* in *Constitutive Modelling for Non-traditional materials*, V.K. Stokes and D. Krajcinovic eds, ASME Publ. AMD, Vol.85, (1987).

[20] J.R. Rice, *Thermodynamics of the quasi-static growth of Griffith cracks*, J. Mech. Phys. Solids, **26**, 61, (1987).

[21] D. Krajcinovic and D. Fanella, *A micromechanical damage model for concrete*, Eng. Fract. Mech., **25**, 585, (1986).

[22] D. Fanella, and D. Krajcinovic, *A micromechanical model for concrete in compression*, Eng. Fract. Mech. **29**, 49, (1988).

[23] D. Sumarac and D. Krajcinovic, *A self-consistent model for microcrack weakened solids*, Mech. Mater., **6**, 39, (1987).

[24] E. Smith and J.T. Barnby, *Crack nucleation in crystalline solids*, Metal Sci. J., **1**, 56, (1978)

[25] D. Krajcinovic and A. Stojimirovic, *Deformation processes in semi-brittle polycrystalline ceramics*, Int. J. of Fracture, to appear.

[26] M.F. Ashby and B.F. Dyson, *Creep damage mechanics and micromechanims*, National Physics Lab Teddington, Report DMA(A) 77, (1984).

[27] A.C.F. Cocks and T.A. Leckie, *Creep Constitutive Equations* in *Advances in Applied Mechanics*, J.W. Hutchinson and T.Y. Wu eds., Academic Press, New York, Vol.25, 597, (1987).

[28] A. Stojimirovic and D. Krajcinovic, *Time dependent response of semi-brittle ceramics*, Mech. Mater., to appear.

Chapter 11

Fracture Mechanics and Solid Adhesion

Daniel Maugis

C.N.R.S./L.C.P.C., 58 Bld Lefebvre
F-75732 Paris Cédex 15, France

1. Introduction

The forces insuring the adhesion between two solids are the same as those insuring cohesion of solids *i.e.* Van der Waals, metallic, covalent, ionic. Fig.(1) displays the potential and the force of interaction between two half-spaces for Van der Waals forces. When the separation increases from the interatomic equilibrium distance Z_0, the attraction force increases up to a maximum, the theoretical strength σ_{th}, and then decreases. The work needed to completely separate the two half-spaces is two times the surface energy γ if they are identical, or the Dupré energy of adhesion $w = \gamma_1 + \gamma_2 - \gamma_{12}$ if they are different (γ_{12} is the interfacial energy of the two solids in contact).

In fact the only knowledge of w, or even the precise law of interaction, is insufficient to predict the force to separate two solids (pull out force, or adherence force). The reason is, first that it is generally not possible to stretch the bonds between the two half-spaces without stretching the bonds near the interface and storing elastic energy, second, that separation never occurs as a whole but the area of contact is progressively reduced until complete separation is achieved. This reduction in contact area can be considered as resulting from the propagation of a crack, and the concepts of fracture mechanics can be used.

The work needed to break the bonds is taken from the elastic energy stored in the solid, or from the potential energy or both. Consider for example a double cantilever beam with a wedge inserted (fixed grips experiment). An elastic energy U_E is stored in the bended arms. If the length of the arms is allowed to increase by dc , the stored elastic energy decreases by $(\frac{\partial U_E}{\partial c})dc$. The quantity $G = -(\partial U_E/\partial c)$ is called "strain energy release rate". The crack can advance if the available energy Gdc is larger than the work $2\gamma dc$ or wdc to break the bonds. If $G = w$ the crack is in equilibrium, but this equilibrium can be stable, unstable, or neutral. This is the basis of fracture mechanics.

Disorder and Fracture, Edited by J. C. Charmet *et al.*
Plenum Press, New York, 1990

2. Adherence of Elastic Solids

Let us consider the system made of two elastic solids in contact over an area
A. This system can exchange work and heat, but not matter, with the exterior. A
force P (compressive or tensile) can be applied to the two elastic bodies, either by
a dead load as in Fig.(2a), or by a spring of stiffness K_m as in Fig.(2b). The area of
contact is allowed to vary at fixed load P, or at fixed displacement δ for K_m infinite
(or more generally at fixed displacement Δ), so that the state of the system depends
in general on two independent variables P, A or δ, A. The edge of the contact area
can be considered as an interface crack tip in Mode I that recedes or advances as the

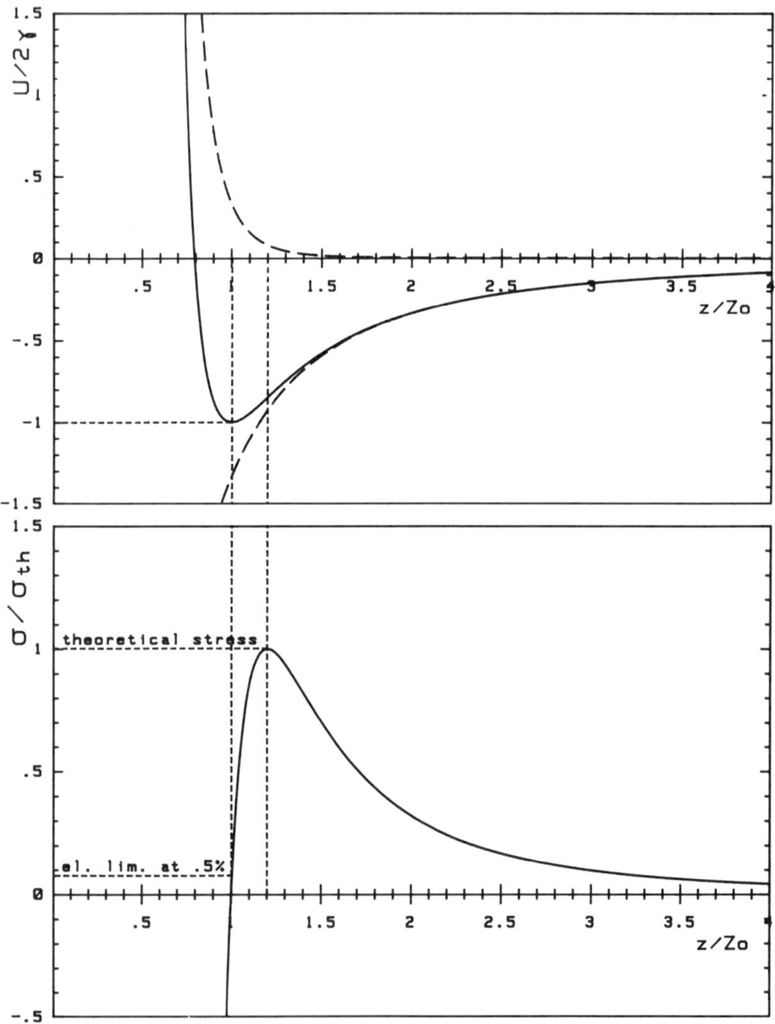

Figure 1 Lennard-Jones potential between two half-spaces.

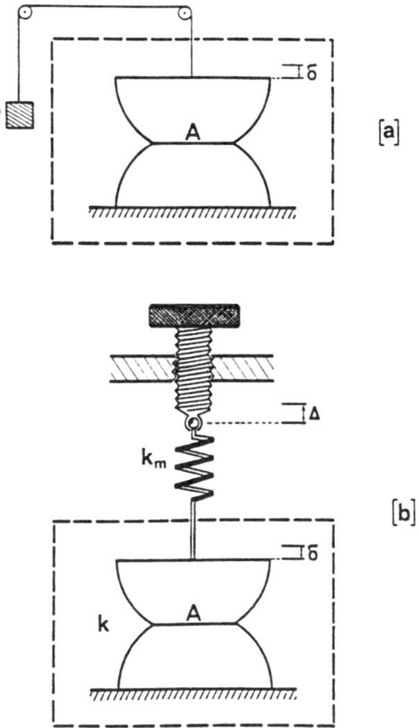

Figure 2 Equilibrium state of the system with force applied by dead load, Fig.(2a), or by a setup of finite stiffness k_m, Fig.(2b). The case of imposed displacement corresponds to $k_m = \infty$.

area of contact increases or decreases. We will neglect any interfacial shear stress so that the system is the same as for a notched solid, but with an imposed crack path.

The energy of the system $U = U(S, \delta, A)$ is a function of the extensive variables S (entropy), δ, A; it can be decomposed into elastic energy U_E and interface energy $U_S = -wA$. The first differential of the energy can be written in the form

$$dU = TdS + Pd\delta + (G - w)dA \qquad (1)$$

with

$$\left(\frac{\partial U}{\partial S}\right)_{\delta,A} = T$$

$$\left(\frac{\partial U}{\partial \delta}\right)_{S,A} = \left(\frac{\partial U_E}{\partial \delta}\right)_{S,A} = P$$

$$\left(\frac{\partial U}{\partial A}\right)_{S,\delta} = \left(\frac{\partial U_E}{\partial A}\right)_{S,\delta} + \left(\frac{\partial U_S}{\partial A}\right)_{S,\delta} = G - w$$

G, which describes the variation of elastic energy with A, at constant δ, is the strain energy release rate. Note that wA is a kind of potential energy which can be recovered by crack healing. The three relationships

$$T = T(\delta, A, S) \tag{2a}$$

$$P = P(\delta, A, S) \tag{2b}$$

$$G = G(\delta, A, S) \tag{2c}$$

expressing intensive parameters in terms of the independent extensive parameters, are the equations of state of the system. The knowledge of these three equations of state is equivalent to that of the fundamental equation $U = U(S, \delta, A)$ and gives a thermodynamically complete description of the system.

A system is in equilibrium if virtual perturbations of the extensive variables leave its energy constant. However, equilibrium is often studied in the presence of constraints such as constant pressure, constant volume or constant temperature. In this case, the Legendre transformation of the energy U is used to exchange any variable X_j with its derivative $P_j = (\partial U / \partial X_i)$. The equilibrium of the system at constant P_j corresponds to the extremum of the function $\Psi = U - P_j X_j$. Ψ, which is the Legendre transform of the energy U, is called the thermodynamic potential. In the present case we wish to study equilibrium at constant temperature, by allowing perturbations of the area of contact at constant load P, or at constant displacement δ (fixed grips conditions). Of interest are thus the Helmholtz free energy $F = U - TS$, and Gibbs free energy $G = U - TS - P\delta$, whose differentials are

$$dF = Pd\delta + (G - w)dA - SdT \tag{3}$$

$$d\mathcal{G} = -\delta dP + (G - w)dA - Sdt \tag{4}$$

Noting that $-P\delta$ is the potential energy U_P of the load, these expressions show that

$$G = \left(\frac{\partial U_E}{\partial A}\right)_{\delta, S} = \left(\frac{\partial U_E}{\partial A} + \frac{\partial U_P}{\partial A}\right)_{P, S} \tag{5}$$

Equilibrium at fixed temperature and fixed load conditions ($dT = 0, dP = 0$) corresponds to an extremum of $\mathcal{G}$, and equilibrium at fixed temperature and fixed grips ($dT = 0, d\delta = 0$) to an extremum of F. In either case equilibrium is given by

$$G = w \tag{6}$$

(The same equilibrium condition would be obtained at fixed Δ with a spring of stiffness k_m). Eq.(6) is the Griffith criterion which links two of the three variables δ, P, A, of the equations of state (Eq.(2)), so that the equilibrium curves $\delta(A), A(P), P(\delta)$, are functions of w.

If $G \neq w$ the area of contact will spontaneously change so as to decrease the thermodynamic potential. If $G < w$, Eq.(3-4) show that A must increase, and the crack recedes. Conversely, if $G > w$ the area of contact must decrease to give $d\mathcal{G} < 0$

or $dF < 0$, and the crack extends. GdA is the mechanical energy released when the crack extends by dA. The breaking of interfacial bonds requires an amount of energy wdA, and the excess $(G - w)dA$ is changed in kinetic energy if there is no dissipative factor. $G - w$ is the crack extension force, which is zero at equilibrium.

The equilibrium given by $G = w$ can be stable, unstable or neutral. A thermodynamic system under a given constraint is stable if the corresponding thermodynamic potential is minimum, *i.e.* if its second derivative is positive. Thus, from Eq.(3-4) stability is defined by

$$(\frac{\partial G}{\partial A})_\delta > 0 \text{ at fixed grips;}$$

$$(\frac{\partial G}{\partial A})_P > 0 \text{ at fixed load,}$$

or more generally by $(\partial G/\partial A)_\Delta > 0$ if the machine has a finite stiffness k_m (or compliance C_m) defined by

$$\Delta = \delta + \frac{P}{k_m} = \delta + C_m P \tag{7}$$

In this last case the stability depends on the stiffness of the testing machine, since elastic energy stored in the spring can be used for crack propagation.

If, under a stable equilibrium, a fluctuation decreases A, G incrementally decreases and one has $G < w$: the crack recedes to its equilibrium position. It can only advance if the load P or the displacement δ is varied, bringing back G to the value w: one is dealing with controlled rupture of an adhesive joint. In this case one has

$$dG = (\frac{\partial G}{\partial A})_\Delta dA + (\frac{\partial G}{\partial \Delta})_A d\Delta = 0 \tag{8}$$

Starting from a stable equilibrium with the two bodies compressed ($P > 0, \delta > 0$) let us decrease the cross-head displacement Δ: one generally encounters a progressive reduction of area of contact, *i.e.* a controlled rupture with $(\frac{\partial G}{\partial A})_\Delta > 0$ up to a point where $(\frac{\partial G}{\partial A})_\Delta = 0$; the equilibrium becomes unstable and the crack spontaneously extends toward the rupture under this given cross-head displacement. It extends with acceleration, for the crack extension force $G - w$ increases as A decreases.

Eq.(8) shows that the limit of stability corresponds to $d\Delta = 0$, i.e.

$$k_m + \frac{dP}{d\delta} = 0. \tag{9}$$

On the equilibrium curve $P(\delta)$, the stability ends at the horizontal tangent at fixed load ($k_m = 0$), or at the vertical tangent at fixed grips ($k_m \to \infty$).

The load P_c corresponding to the limit of stability will be called the *adherence force* of the two elastic bodies, and thus will generally depend on the stiffness of the measuring apparatus. In some geometries (unstable geometries) equilibrium is always unstable, and thus the criterion for adherence force is simply $G = w$. It is the case for the double cantilever beam at fixed load, or for a flat punch on an elastic half-space. It is also the classical Griffith case of a crack in an infinite solid.

2.1 Two Examples

The calculation of G for a number of geometries such as peeling, double cantilever, double torsion, blister test can be found in text books. We will concentrate on the case of adherence of punches which is conceptually an important topic to understand the connection between adherence, mechanics of contact and fracture mechanics.

The case of an axisymmetric flat punch of radius a on an elastic half-space was solved by Kendall[1] by evaluating the elastic energy U_E and the potential energy $U_P = -P\delta$ from the elastic displacement under the load P:

$$\delta = \frac{2P}{3aK},\tag{10}$$

where

$$\frac{1}{K} = \frac{3}{4}\frac{1-\nu^2}{E},$$

The result is:

$$G = \frac{P^2}{6\pi a^3 K} = \frac{3K\delta^2}{8\pi a}.\tag{11}$$

The equilibrium corresponding to $G = w$ is unstable (as for a crack of radius a in an infinite body) and the load P given by $G = w$ is the adherence force at both fixed load and fixed grips.

The case of a sphere of radius R is more subtle, since the contact is no longer conformal and elastic energy is stored under the action of molecular forces. To evaluate it, Johnson, Kendall and Roberts[2] (JKR theory) first apply the Hertzian load

$$P_1 = \frac{a^3 K}{R},\tag{12}$$

giving the elastic displacement

$$\delta_H = \frac{a^2}{R},$$

and then reduce the load from P_1 to P at fixed radius of contact. In this flat punch displacement under the load $P_1 - P$:

$$\delta_F = -2\frac{(P_1 - P)}{3aK}$$

tensile stresses appear in an annulus inside the area of contact, with singularities and discontinuity of displacement at the edge. The elastic displacement for a radius of contact a and a load P is thus $\delta_H + \delta_F$, $i.e.$.:

$$\delta = \frac{a^2}{3R} + \frac{2P}{3aK} \tag{13}$$

with the relation between a and P still unknown. Evaluation of elastic and potential energies leads to:

$$G = \frac{(P_1 - P)^2}{6\pi a^3 K} \tag{14a}$$

$$= \frac{3a^3 K}{8\pi R^2} (1 - \frac{R\delta}{a^2})^2 \tag{14b}$$

Eq.(13-14) are the equations of state (2b-2c) of the system. The Griffith criterion $G = w$, gives now the equilibrium relations $a(P), P(\delta)$ and $\delta(a)$:

$$a^3 = \frac{PR}{K} [1 + \frac{3\pi w R}{P} + \sqrt{\frac{6\pi w R}{P} + (\frac{3\pi w R}{P})^2}] \tag{15}$$

$$\delta = \frac{a^2}{R} - (\frac{8\pi w a}{3K})^{\frac{1}{2}} \tag{16}$$

In Eq.(15) the brackets represent the correction to the Hertz theory for solids with surface energy. Under zero load the radius of contact is given by:

$$a_0^3 = \frac{6\pi w R^2}{K}$$

By slowly decreasing the load P or the displacement δ these equilibrium curves can be followed up to the unstability point. This point depends on the stiffness of the testing machine. At fixed load ($k_m = 0$) it is given by:

$$(\frac{\partial G}{\partial A})_P = \frac{P_1^2 - P^2}{4\pi^2 a^5 K} = 0 \tag{17}$$

$$P = -P_1 = -\frac{3}{2}\pi w R \tag{18}$$

At fixed grips ($k_m \to \infty$) it is given by :

$$(\frac{\partial G}{\partial A})_\delta = \frac{3K}{16\pi^2 R^2 a^3} (a^2 - \delta R)(3a^2 + \delta R) \tag{19}$$

$i.e.$

$$P = -5P_1 = -\frac{5}{6}\pi w R \tag{20}$$

Figs (3-4) display the curves $\delta(P, a)$, Eq.(13), independent of w, with super-imposed equilibrium curves for $w = 0$ (Hertz) and $w \neq 0$. In Fig.(3) the curves $\delta(P, a)$ at constant a are straight lines tangent to the Hertz curve. Elastic energy and total surface energy $U_S = -\pi a^2 w$ at an equilibrium point B are marked as hatched zones. From the slope of the equilibrium curve

$$\frac{dP}{d\delta} = \frac{9Ka}{2} \frac{P_1 + P}{5P_1 + P} \tag{21}$$

one can see that, in agreement with Eq.(9) the adherence force at fixed load (point C) corresponds to an horizontal tangent, that at fixed grips (point D) corresponds to a vertical tangent. For a finite stiffness k_m, the adherence force corresponds to the point on the equilibrium curve (between C and D) where the slope is $-k_m$.

In Fig.(4) the curves $\delta(P, a)$, at constant P have their minimum (when it exists) on the Hertz curve. The dotted line is for $\left(\frac{\partial G}{\partial A}\right)_P = 0$ and goes through the point C for adherence at fixed load. Starting from an equilibrium point L under a compressive load P, and using a quasistatic unloading, the equilibrium curve $G = w$ is followed up to the point C, from where the curve at constant load is followed with G increasing as a decreases.

At fixed grips the equilibrium curve would be followed up to the point D, where $\frac{d\delta}{da} = 0$. If instead of a quasistatic unloading, a new load P' is instantaneously

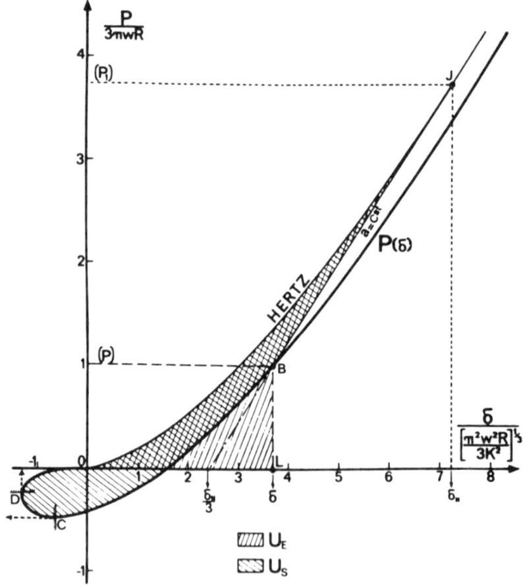

Figure 3 Curves $\delta(P, a)$ as defined by Eq.(13), with superimposed equilibrium curves for $w = 0$ (Hertz) and $w \neq 0$. Elastic energy and total surface energy $U_S = -\pi a^2 w$ at an equilibrium point B are marked as hatched zones.

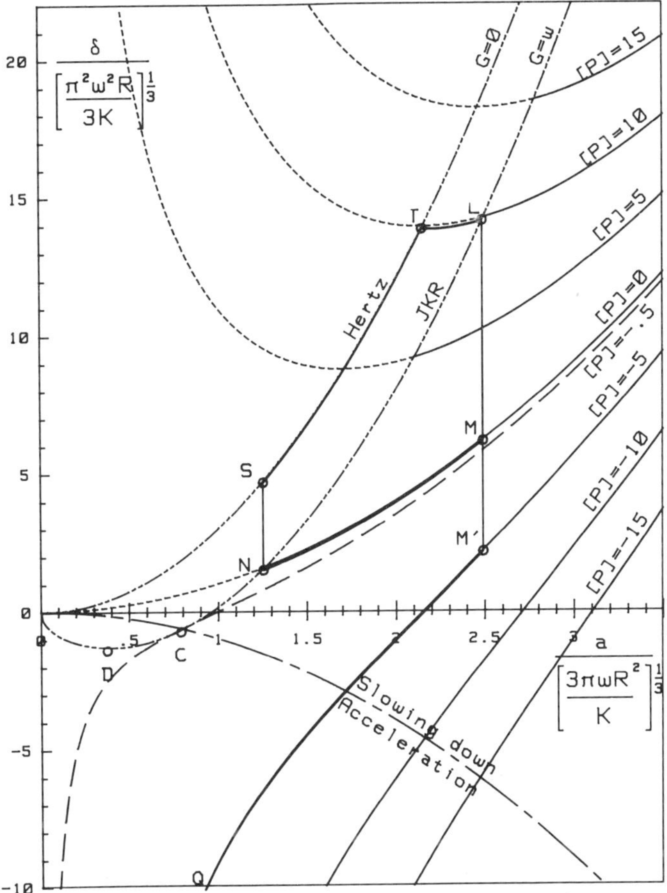

Figure 4 Curves $\delta(P, a)$, at constant P have their minimum (when it exists) on the Hertz curve. The dotted line is for $\frac{\partial G}{\partial A})_P = 0$ and goes through the point C for adherence at fixed load.

applied at point L, one must observe an instantaneous displacement at constant a (branch LM or LM') followed by a crack propagation at the constant load P' toward a new equilibrium if $P' > \frac{-3}{2}\pi wR$ (branch MN) or towards rupture if $P' < \frac{-3}{2}\pi wR$ (branch $M'Q$) with the crack first slowing down and then accelerating. Conversely, if one applies an instantaneous loading from P' to P, G immediately decreases at constant a, but as the stress intensity factor cannot be negative, branch ST of the Hertz curve ($G = 0$) is followed, and then the crack recedes at constant load P towards its new equilibrium (branch TL). All the theoretical predictions have been verified by Maugis and Barquins[3] for glass ball on polyurethane with a precision of about 2%.

2.2 Stresses at the Crack Tip

The stresses around a crack can be computed from the theory of elasticity by solving the biharmonic equation $\Delta\Delta\Phi = 0$ with the imposed boundary conditions. One of these boundary conditions is that the crack faces are stress-free, which leads to infinite stresses at the crack tip instead of the theoretical stress. The stress tensor for a crack of length a traversing a plate was computed by Inglis[4] in case of a uniform stress σ applied at large distance, and by Chang and Wu[5] in case of a biaxial loading (σ along Oy and $k\sigma$ along Ox). Figs.(5-7) display these stress tensors for a crack along Ox subjected to uniaxial ($k = 0$), uniform ($k = 1$) or biaxial loading. The principal stresses σ_1, and σ_2, the isochromatic lines are in agreement with the figures given by Sneddon[6] for $k = 0$ and by Kuppers[7] for $k = 1$. These pictures show how the stress distribution is sensitive to boundary conditions.

Another example of completely determined stress tensor around a crack is the stress tensor associated with the JKR theory of contacting spheres.[8]

However, the hypothesis of stress-free crack surface has a number of drawbacks. The interaction energy (w or 2γ) abruptly appears as a step function at the crack tip, and the theory of elasticity shows that under loading, whatever its magnitude, an initially sharp crack becomes rounded and stresses become infinite at the crack tip, Fig.(8). More precisely, for a mode I loading (opening mode) the leading term of the power expansion of the stress σ_z normal to the crack plane ahead of the crack tips is $\sigma_z = K_1/\sqrt{2\pi\rho}$ where ρ is the distance to the crack tip, and K_1 is the stress intensity factor which describes the force of the singularity. Physically, infinite stresses cannot exist in a material, they cannot exceed the theoretical strength of the bond, or for an elastoplastic material the yield stress σ_0. Secondly, it is not easy to see by what mechanism the crack recedes when $G < w$ and by what mechanism a reduction of w by adsorption or segregation increases the crack velocity when $G > w$. The difficulties arise from the fact that cohesion or adhesion forces $g(z)$ are only taken into account by their integral

$$w = \int_{Z_0}^{\infty} g(z)dz$$

where Z_0 is the interatomic equilibrium distance, and that only stresses due to external loading are studied. In reality, as shown by Barenblatt[9], cohesive forces act inside the crack on a length d. They elastically deform the crack, as would do any external force, and give rise to a stress intensity factor

$$K_m = -\sqrt{\frac{2}{\pi}} \int_0^d \frac{g(t)dt}{\sqrt{t}}$$

which tends to close the crack, and cancels with the stress intensity factor K_1 due to external loading. Any singularity is thus removed, and the crack has a cusp-shaped tip, varying as $\rho^{3/2}$ irrespective of the laws of molecular forces, Fig.(8). At the crack tip the stress reaches the theoretical strength, much higher than the

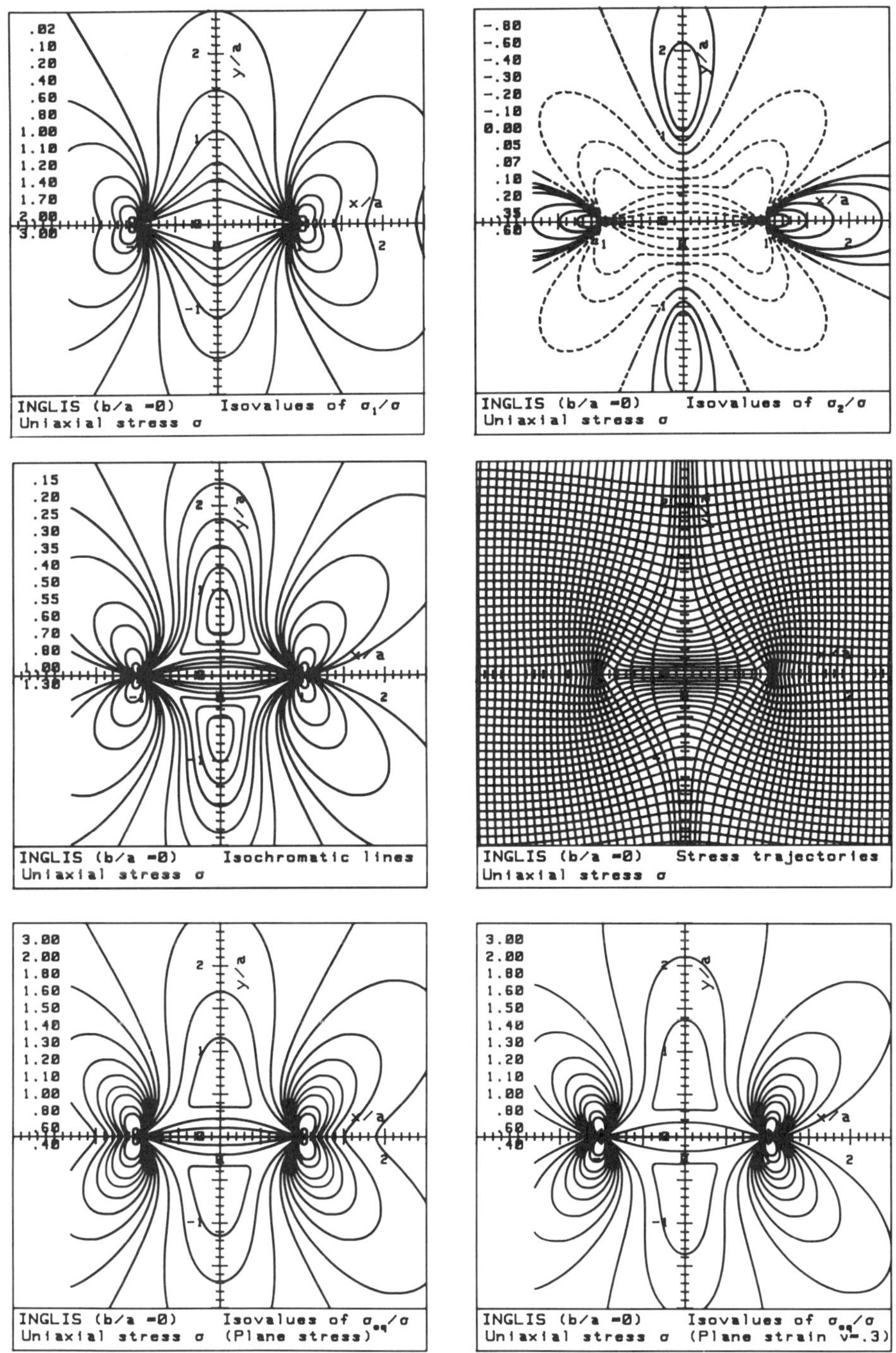

Figure 5 Stress field around a crack oriented along Ox and subjected to uniaxial ($k = 0$) loading.

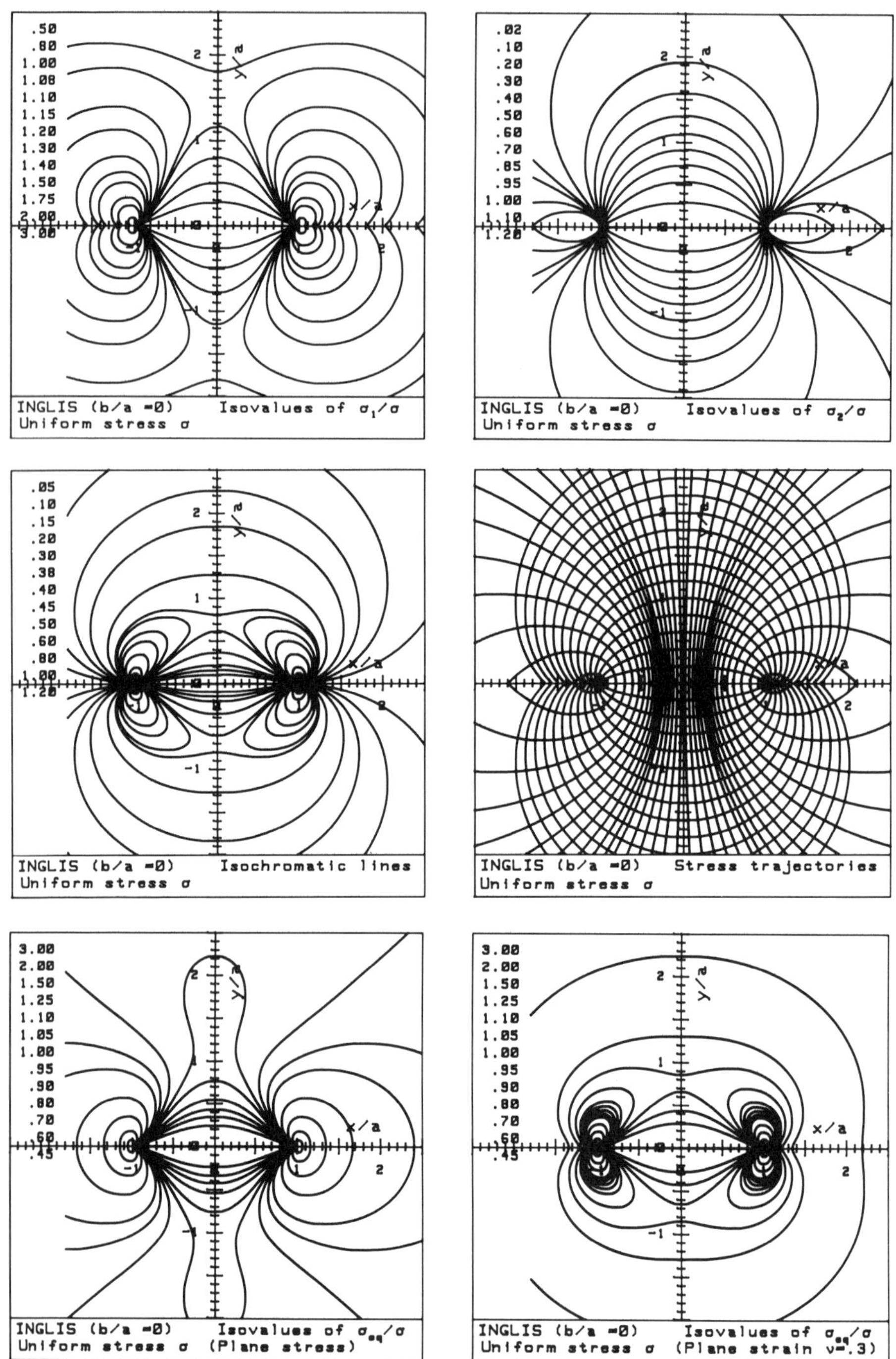

Figure 6 Stress field around a crack oriented along Ox and subjected to uniform ($k = 1$) loading.

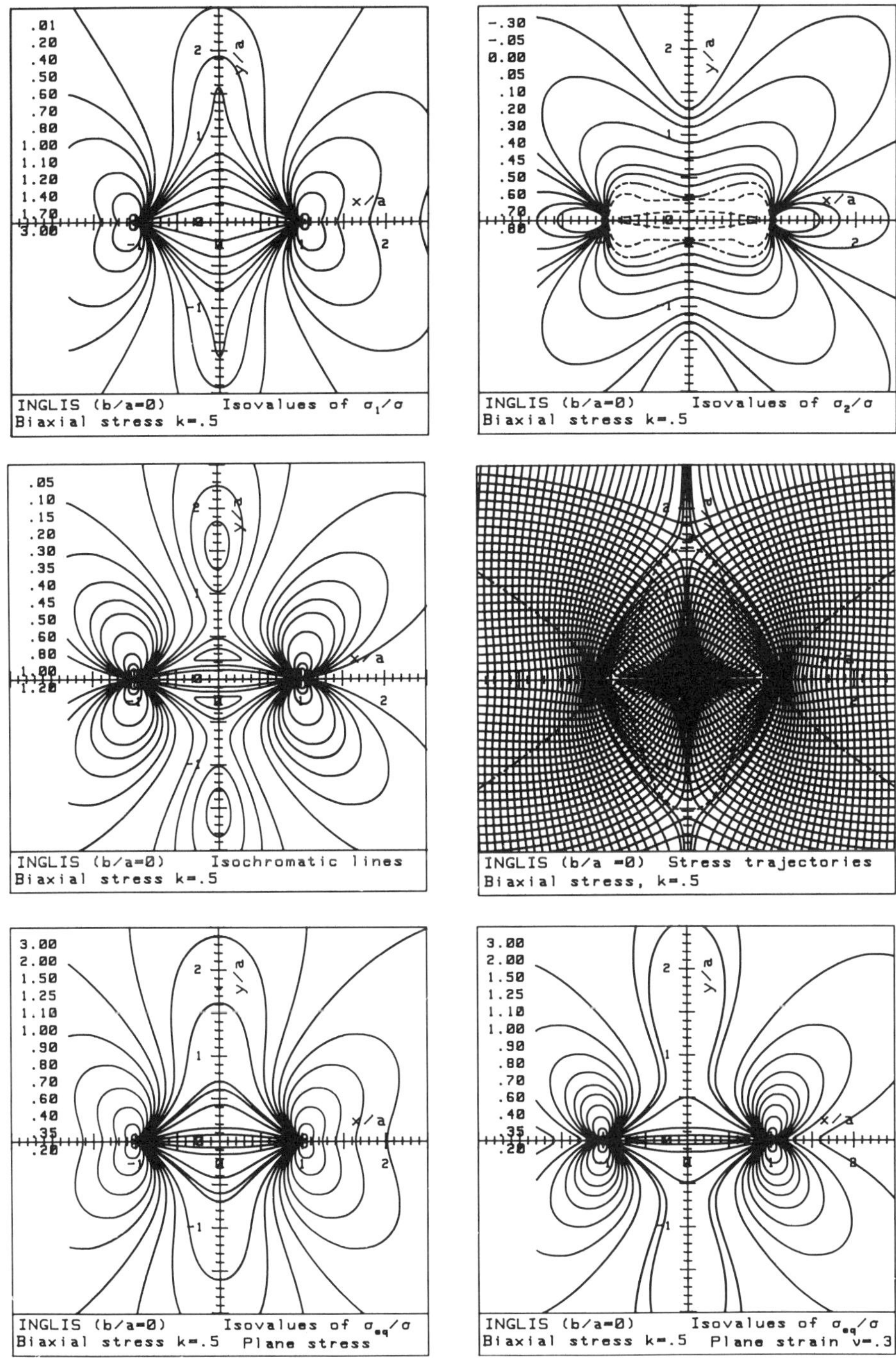

Figure 7 Stress field around a crack oriented along Ox and subjected to biaxial loading.

remote stresses, so that two solids in adhesive contact are essentially sustained by the stresses, near the crack tip, regardless of stresses in the contact area. The case of peeling is particularly instructive, since the load which can be hung is independant of the contact area.

3. Various Losses

$G - w$ is the generalized force applied per unit length of the crack and moving it. Indeed this force is zero at equilibrium. In absence of any dissipation a crack subjected to a constant force $G - w$, would accelerate up to the Rayleigh waves velocity. However, such non-dissipative materials do not exist. The dissipated energy is taken to the available energy $G \rightarrow w$, and as a ball falling in oil, the crack will take a limit velocity v depending on the dissipative phenomena. It will be possible to draw $G(v)$ curves characteristic of the dissipation process. Of course, if $\partial G / \partial A \neq 0$, G varies during the propagation, the crack speed will vary either.

A first kind of dissipation, very general, follows from the Barenblatt model. When a crack moves at velocity v the stress peak at its tip moves with it, and an element of volume near the trajectory undergoes a cycle of stress when the crack tip comes on and then comes off, whose characteristic time is of the order of d/v (where a is the width of the stress peak) and the magnitude proportional to the theoretical strength, hence to w. Energy is lost in such a cycle, and the crack

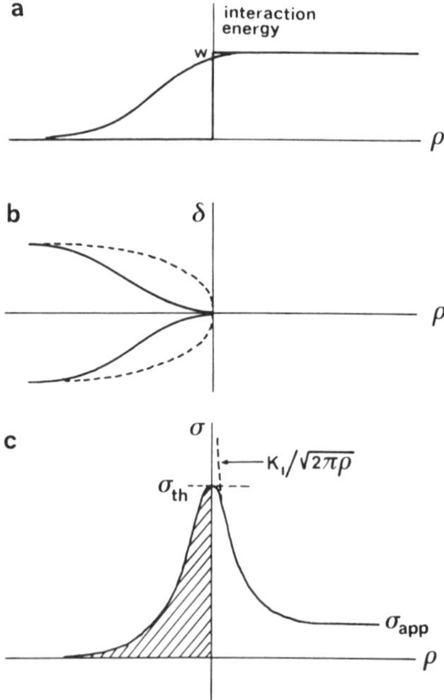

Figure 8 Griffith and Barenblatt models of crack.

undergoes a drag in proportion to w, as pointed out by Mc Lean[10] for the plastic dissipation in intergranular fracture, and function of the crack velocity. We can now understand, at least qualitatively, why adsorption increases the crack velocity. Adsorbed species in the Barenblatt tip screen the cohesion or adhesion forces, lower the peak stress and reduce the drag. But if the size of molecules is too large, or if the crack is running too fast for the medium to follow it, there is no effect, although the crack walls at some distance have their surface energy reduced by adsorption.

Note also that the dissipation does not generally cancel with the crack velocity. There is then a static friction term ψ, and the crack can only advance if $G - w > \psi$, giving the appearance of a very large energy of adhesion. In metals, for example, when the dissipation is mainly due to the motion of dislocations, there is a threshold stress to move these dislocations. For polymers, work is needed to extract molecular chains even at very low crack velocity. In presence of such a static friction, the crack is no longer reversible, and it appears a hysteresis between advancing crack and receding crack, as the crack advances when $G > w + \psi$ and recedes when $G < w$. Such hysteresis have been observed on glass and mica.[11]

There are other types of dissipation that can be added on. When a crack propagates in a liquid medium, the viscosity can impede the crack opening, and causes a newtonian drag which can be higher than the drag by internal losses above a given velocity. There can also exist a dissipation, not only at the crack tips on inside a liquid filled crack, but in the deformed material. In case of peeling a viscoelastic strip, for example energy can be dissipated in the transient flexure of the strip when the crack advances, the amount of which depends on the thickness of the strip.

4. Adherence of Viscoelastic Solids

As discussed above a crack propagating under a constant crack extension force $G - w$, undergoes a drag proportional to w and function of its velocity v and one can write.[3,12,13]

$$G - w = w\phi(a_T v) \tag{22}$$

which generalizes for any geometry the equation given by Gent and Schultz[14] to account for peeling in various liquids. a_T is the Williams-Landel-Ferry shift factor for frequency-temperature equivalence, given by:

$$\log a_T = -\frac{8.86(T - T_S)}{101.6 + T - T_S} \tag{23}$$

where $T_S = T_g + 50$ (T_g is the glass transition temperature), and first used in fracture mechanics by Mullins[15] for tearing.

Thus in Eq.(22) surface properties (w) and viscoelastic properties $\varphi(v)$ are completely decoupled from elastic properties, geometry and loading conditions included in G. The dimensionless function $\varphi(a_T v)$ is a characteristic of crack propagation in Mode I in the material. Once $\varphi(a_T v)$ is known, Eq.(22) allows one to predict any feature such as kinetics of detachment at fixed load, fixed grips or fixed cross-head velocity.

Eq.(22) was verified for glass on polyurethane for various geometries where G can be computed (sphere, flat punch, flat ended spheres, peeling) by studying crack propagation at a fixed load[3,16] or at a fixed displacement[17,18] at various temperatures. Fig.(9) displays the kinetics of crack propagation for the unloading of a glass ball on polyurethane from $P = 50mN$ to $P' = -30mN$, for two temperatures. The crack first slows down and then accelerates. The inflexion point corresponds to $(\partial G/\partial A)_P = 0$ $i.e.$ $a^3 = -P'R/K$. At any point of the curves , G can be computed from the values of a and P', and the crack velocity by the slope, so that the G vs $a_T v$ curve can be constructed. The same master curve is obtained whatever the geometry studied, which is a proof of the correctness of the formulae for G. Fig.(10) shows the function $\varphi(a_T v) = (G - w)/w$ for polyurethane, with w deduced from the equilibrium contact area of a glass ball. This function varies as $(a_T v)^{0.6}$ over about five decades, a result often found for the peeling of rubber-like materials.[19,20]

Water adsorption decreases the Dupré energy of adhesion, and hence the viscoelastic losses according to Eq.(22)). This point was verified by measuring the rolling resistance $\mathcal{R}$ of a glass cylinder rolling on an inclined sample $G \simeq \mathcal{R}/l$) as a function of velocity for various humidity conditions.[20] The result is a translation of the $G(v)$ curves with water adsorption. As $G \gg w$ in these experiments, the shift

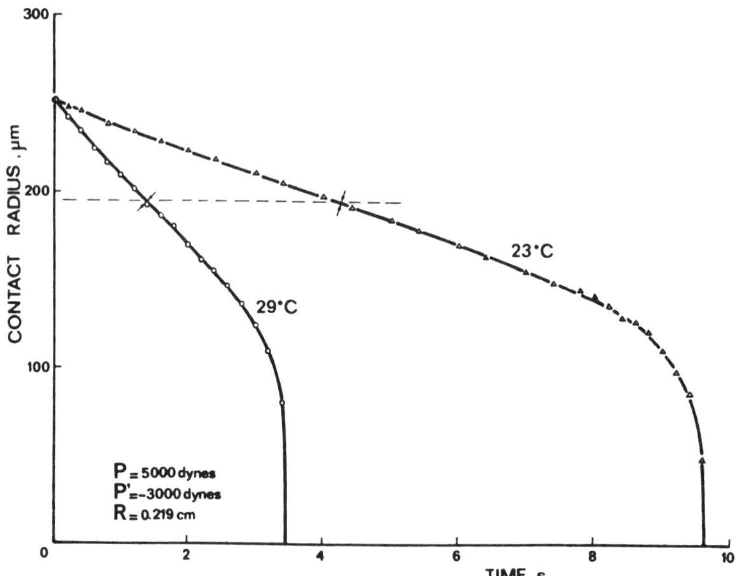

Figure 9 Kinetics of crack propagation for the unloading of a glass ball on polyurethane from $P = 50mN$ to $P' = -30mN$, for two temperatures.

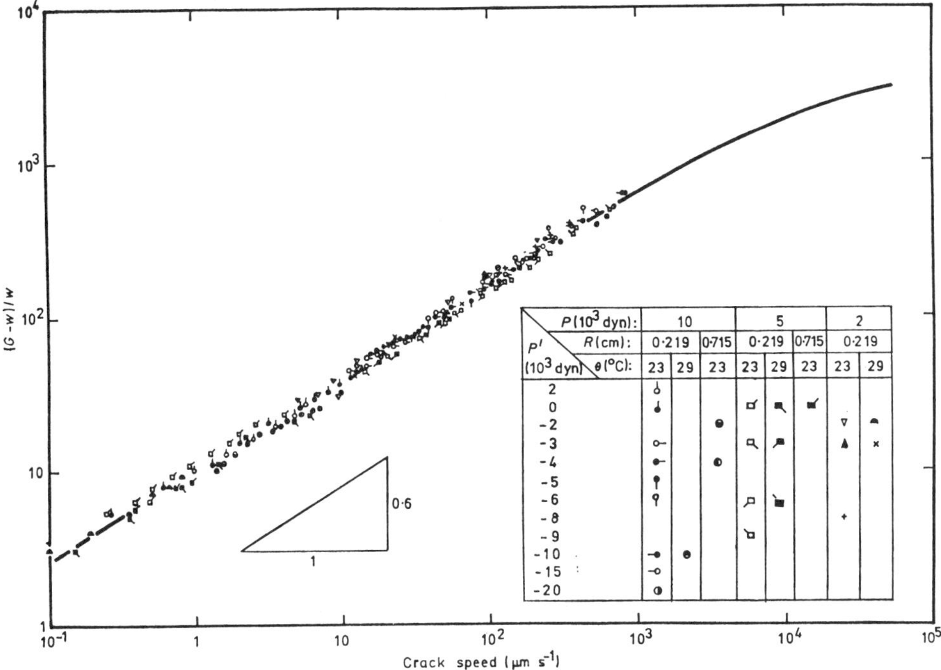

Figure 10 The function $\varphi(a_T v) = (G - w)/w$ for polyurethane, with w deduced from the equilibrium contact area of a glass ball.

clearly arises from the multiplicative term w on the right hand side of Eq.(22) as previously deduced from peeling in various liquids[14] or on various substrates.[13]

4.1 Tackiness

The adherence of solids is more often studied with a tensile test machine at constant cross-head velocity than at constant load, but the kinetics of separation is less easy to interpret, due to the competition for increasing G with time between increasing δ with time at constant a, and decreasing a at constant δ. If the machine has an infinite rigidity k_m, one has $\Delta = \delta$ and the variation of G with time is given by:

$$\frac{dG}{dt} = (\frac{\partial G}{\partial \delta})_a \dot{\delta} + (\frac{\partial G}{\partial a})_\delta \dot{a} \tag{24}$$

The recorded force first increases, then dicreases. The maximum value, termed the tack force, is a measure of the adherence under this particular experimental condition, and has no clear physical significance. The area under the curve, termed tack energy, is equal to the work $\int G da$ of the cohesive stress at the crack tip. "Tackiness" refers to the ability of an elastomer to adhere instantaneously to a

solid surface, or to itself, after a brief time of contact under low pressure. Probe
tack testing can be analysed by Eq.(22), and tack curves obtained by computer
integration closely coincide with experimental ones.[21] Fig.(11) is for a spherical
probe and shows that even at very low cross-head velocity the viscoelastic effects
considerably increase the adherence force compared to the elastic (or quasistatic)
adherence force at fixed displacement (Point D).

4.2. Viscoelastic Losses and Negative Resistance Branch

As pointed out by Mullins[15] viscoelastic losses at the crack tip are related
to the frequency dependence of the loss modulus E'' (the imaginary part of the
Young modulus). As a matter of fact, E'' varies as $\omega^{0.6}$ for polyurethane at low
frequency.[22] Such a correlation between $\varphi(a_T v) \sim (a_T v)^n$ and $E'' \sim (\omega \tau)^n$ where τ
is a relaxation time, is not limited to polyurethane. It can be found for PMMA, for
which $\varphi(v) \sim v^{0.17}$ is observed for crack propagation[23] and $E'' \sim (\omega \tau)^{0.17}$ at low

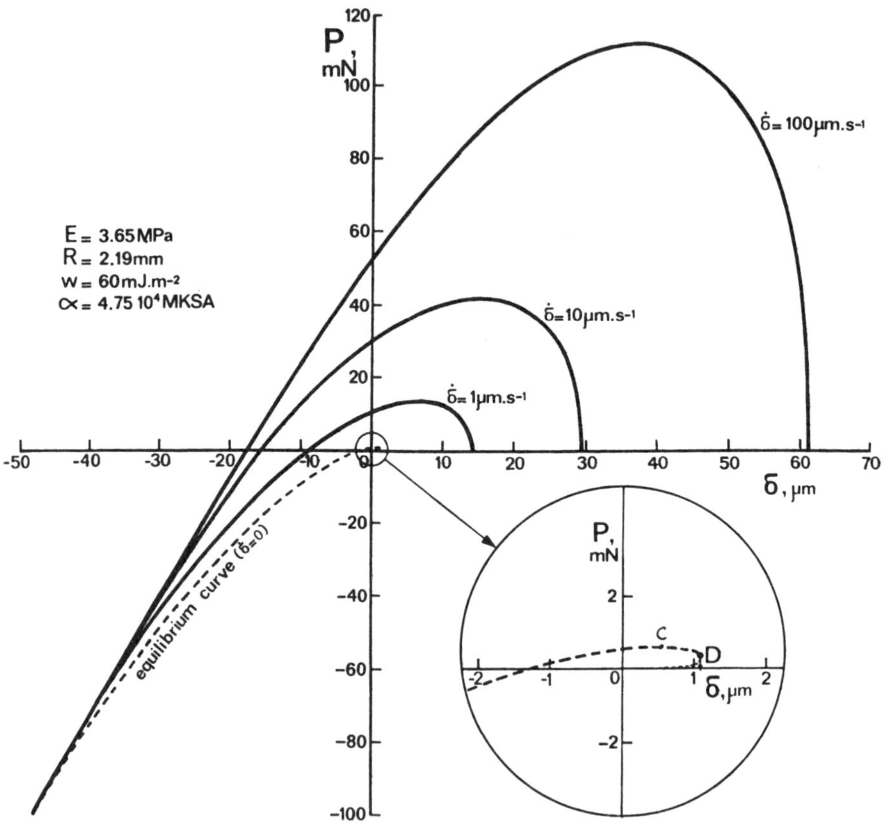

Figure 11 Glass ball on polyurethane: influence of withdrawal speed on the
recorded force (tack force): $E = 3,65\ Mpa, R = 2,19\ mm, w = 60\ mJm^2, \alpha = 4,75\ 10^4\ SIunits$ (at 295K) in Eq.(22).

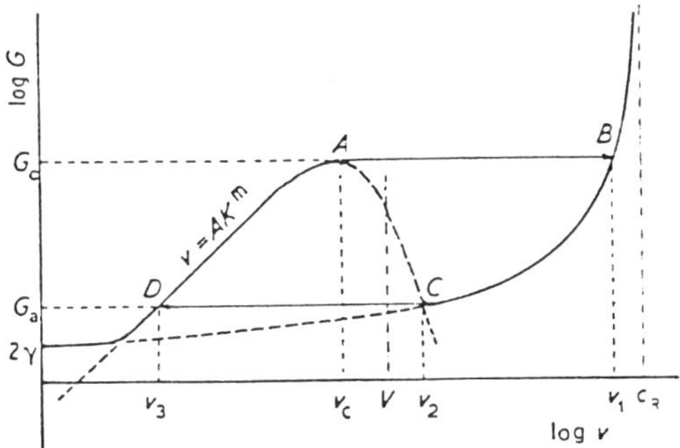

Figure 12 Curve $G(v)$ vs v. It can be seen as the superposition of the elastic solution with G increasing as v approaches the Rayleigh velocity c_R and a broad peak due to viscoelastic losses.

frequency.[24] However, one must take care that the presence of a "static friction" term ψ can mask the correlation, by adding to the $G(v)$ curve a   horizontal line in the $\log G, \log v$ diagram. A viscoelastic model such as the parabolic Zener model:

$$E^*(iw) = \varepsilon_0 + \frac{E_\infty - E_0}{1 + (i\omega\tau)^{-n}} \tag{25}$$

where E_0 is the relaxed modulus, and E_∞ the instantaneous modulus, can be used to have $E''(\omega) \to (\omega\tau)^n$ when $\omega\tau \to 0$. Such a model corresponds to a continuous spectrum of relaxation times, and leads to symmetrical Cole-Cole diagram with $E''(\omega) \to (\omega\tau)^{-n}$ when $\omega\tau \to \infty$. For such a material the loss modulus decreases and the behaviour tends to become elastic at sufficiently high frequencies or low temperature, so we can think that the function $\varphi(v)$ would decrease with v at higher speeds, giving a negative resistance to crack propagation. The implications of such a negative resistance branch are discussed below.

4.3. Velocity Jumps and Stick-Slip

The curve $G(v)$ can be seen as the superposition of the elastic solution with G increasing as v approaches the Rayleigh velocity c_R and a broad peak due to viscoelastic losses, Fig.(12). The negative resistance branch AC corresponding to unstable stationary equilibrium cannot be observed, and the behaviour will depend on the stability of the geometry studied.

If the geometry and the loading are such that $(\partial G/\partial A) < 0$, G increases with the crack length up to a critical value G_c (point A) where the velocity jumps from v_c to v_1 on the second branch (point B) with acoustic emission. G_c is the critical energy release rate for catastrophic failure, which is thus preceded by slow crack growth (or subcritical crack growth) as clearly described by Rivlin and Thomas[25] for rupture of rubber. The criterion $G = G_c$ for catastrophic failure must not be confused with a Griffith criterion, although for unstable geometries, once the Griffith criterion is reached, crack propagation starts until catastrophic failure at fixed load. The value of G_c must not be deduced from the maximum tensile force recorded with a tensile machine at constant cross-head velocity, since the maximum can correspond to subcritical crack growth.[26,27] In Fig.(11) for example, the whole curves are for subcritical crack growth. Velocity jumps over several decades have been observed by many investigators as discussed in Ref.[27] and hysteresis between the velocity jump AB for accelerating cracks and the velocity jump CD for decelerating cracks was observed by Kobayashi and Dally[28] on epoxy resins.

If the geometry and the loading are such that $(\partial G/\partial A) \geq 0$, the crack velocity can be monitored, but when one tries to impose a velocity V between v_c and v_2, stick-slip motion occurs. According to the simple relaxation model proposed in Refs.[22,29,31] the first branch is followed up to G_c (point A) where the velocity v_2 is too low, so that the velocity jumps to v_1 (point B) where the velocity is too high. The crack slows down to point C where the velocity v_2 is still too high, then jumps to point D where the crack seems arrested. Then, the velocity increases to v_c (too low) and jumps to B, and so on. In this stick-slip motion most of the time is spent on the slow velocity branch DA, and the recorded force is saw-tooth shaped with the higher value giving classically G_i and K_i for crack initiation and the lower value G_a or K_a for crack arrest. However, this simple model does not explain why the amplitude between G_i and G_a decreases when the imposed velocity increases[32,33] and why the recorded force can be sometimes sinusoidal or chaotic. In fact, when the crack accelerates or decelerates, inertial effects cannot be neglected, and the equation

$$G - w = w\varphi(a_T v) - \frac{dU_K}{dA} \qquad (26)$$

must be used, where U_K is the kinetic energy of the system, instead of Eq.(22).

Maugis and Barquins[32.36] have studied stick-slip in peeling using a roller tape or radius R and inertia I unwound at a linear velocity V (up to 20 m/s) by a couplemeter motor allowing the peel force P to be measured. The length L of the peeled band can be chosen (between 0.1 m and 3 m) by changing the distance between the rollers. When peeling is continuous the peel angle is $\pi/2$, and the energy release rate is:

$$G = \frac{P}{b} = \frac{Eh\delta}{L} \qquad (27)$$

where δ is the elongation of the peeled band of width b, thickness h and Young modulus E, the crack velocity v is equal to the imposed linear velocity V, and the angular velocity of the unwound roller is $\Omega = v/R$. When peeling is jerky, the peeling point oscillates along the circumference of the unwinding roller whose angular velocity is not constant, and the peel angle can vary from 0 to π.

Fig.(13) displays the $G(v)$ curve obtained, with its two branches of stable propagation separated by a region where stick-slip occurs. The first branch can be

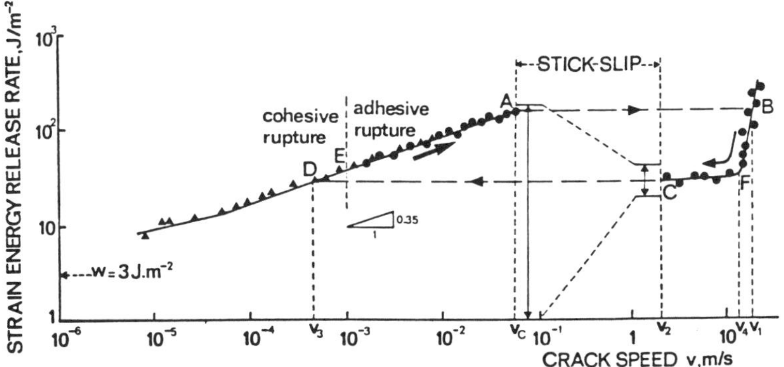

Figure 13 The $G(v)$ vs v curve in peeling: experimental results.

represented by the equation

$$G = w[1 + \alpha(T)v^{n_1}] \tag{28}$$

with $w = 3J/m^2$, $n_1 = 0.35$ and $\alpha(T) = 135$ SI units at $23°C$. At point E a transition from cohesive to adhesive rupture is noted, with no change in the slope of the curve, in agreement with the observations of Aubrey and Sherriff.[33] According to the simple relaxation oscillation model, the period of oscillation can be evaluated from the time spent on the first branch. Neglecting the variation of peel angle around $\theta = \pi/2$, the relation between v and V is:

$$v = V - \frac{d\delta}{dt} \tag{29}$$

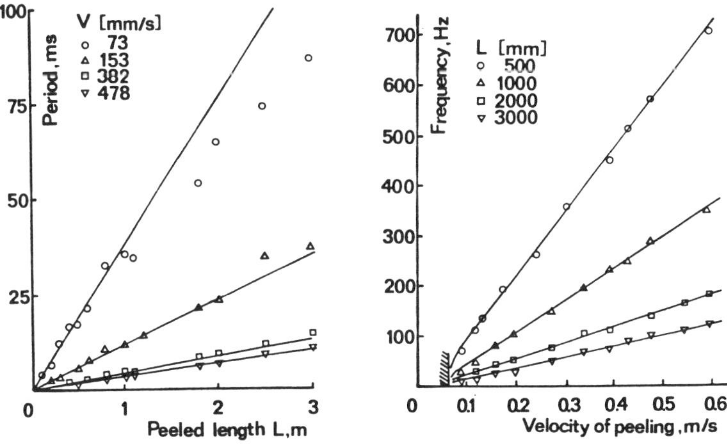

Figure 14 Stick-slip in peeling: comparison between theory and experiment.

Extracting v from Eq.(28), putting it in Eq.(29) and using Eq.(27) gives t:

$$t = \frac{L}{Eh} \int_{G_{min}}^{G_c} \frac{dG}{V - v_c(\frac{G-w}{G_c-w})^{\frac{1}{n_1}}} \tag{30}$$

where G_{min} corresponds to point C. Numerical calculation is shown as full line in Fig.(14) and is in satisfactory agreement with experimental results. This model accounts for the frequencies observed in this range, for the acoustic emission at each velocity jump, for the marks left on the tape at each cohesive-adhesive transition, but does not explain why the amplitude of oscillation decreases with V, why the observed oscillation is sinusoidal, and why the marks disappear at $V > 0.65$m/s.

Inertial effects due to kinetic energy $U_K = (1/2)I\Omega^2$ of the roller can be taken into account in a simple but inexact model where the variations of peel angle are neglected, and the angular velocity of the roller is simply:

$$\Omega = \frac{v}{R} \tag{31}$$

However the resulting equations are most interesting. Letting $m = I/R^2$, Eq.(26) and (31) lead to:

$$\dot{v} = \frac{dv}{dt} = \frac{b}{m}[G - \Phi(v)] \tag{32}$$

where $\Phi(v) = w + w\varphi(v)$, and which shows that the $G(v)$ curve must be carefully distinguished from the $\Phi(v)$ curve when the crack velocity is not constant.

By time derivation of the force elongation relation $P = k\delta$, where $k = Ebh/L$ is the stiffness of the peeled length, and using Eq.(29) one has:

$$\dot{G} = \frac{dG}{dt} = -\frac{k}{b}(v - V) \tag{33}$$

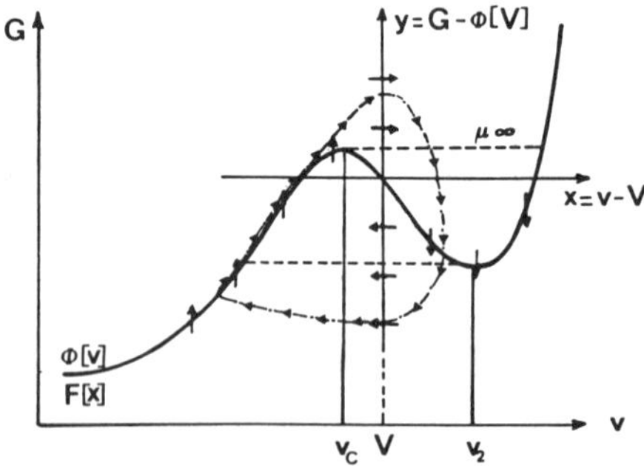

Figure 15 $G(v)$ vs v curve showing the negative slope branch and a cycle of self-sustained oscillations.

Let us take the point $[V, \Phi(V)]$ (stable or unstable) as the origin Fig.(15) and letting:

$$x = v - V$$
$$F(x) = \Phi(v) - \Phi(V)$$
$$f(x) = \frac{dF}{dx} = \frac{d\Phi}{dv}$$
$$\omega = \sqrt{\frac{k}{m}}$$
$$\mu = \frac{b}{\sqrt{km}}$$

Eqs.(32) and (33) lead to:

$$\ddot{x} + \mu\omega f(x)\dot{x} + \omega^2 x = 0$$

which is the classical Lienard equation for self-sustained oscillations, and which is known to have limit cycles when the function $F(x)$ has a branch with a negative slope.[37,38] This equation is equivalent to the autonomous system:

$$\dot{x} = \mu\omega[y - F(x)] \tag{35}$$

$$\dot{y} = -\frac{\omega}{\mu}x \tag{36}$$

with $y = G - \Phi(V)$, and which describes the motion in the space phase (x, y), *i.e.* (v, G) with a translation.

The slope of orbits in this space is:

$$\frac{dy}{dx} = -\frac{x}{\mu^2[y - F(x)]} \tag{37}$$

and one can see in Fig.(15) that the axis $x = 0$ is an isocline $0°$, that the curve $y = F(x)$ (ie the curve $G = \Phi(V)$) is an isocline $90°$, and that the origin is a singular point. When the inertia term tends towards zero $(\mu \to \infty)$, Eq.(37) becomes:

$$dy[y - F(x)] = 0$$

which represents the curve $y = F(x)$ and the horizontal $dy = 0$ on which, from Eq.(32) the acceleration is infinite. It is thus only for $m \to 0$ that the relaxation cycle previously described is valid.

Linearization of the system around the origin $x = y = 0$ (*i.e.* $v = V, G = \Phi(V)$) shows that the behaviour is governed by the single parameter $S = \frac{\mu}{2}(\partial\varphi/\partial v)_{v=V}$, and that the origin can be a node, a centre or a focus. In this last case, orbits are spirals expending from the origin if $-1 < S < 0$, or approaching it if $0 < S < 1$. Points A and C of the $\Phi(v)$ curve are Hopf bifurcation points where stable stationary equilibrium gives way to limit cycles. These limit cycles can grow progressively or appear abruptly when S varies depending on the sign of the third derivative of $\Phi(v)$. They include in their interior either both points A and C or only one of them.

This simple model explains why the oscillations can be sinusoidal or saw-tooth shaped depending on the shape of the limit cycle, and why the amplitude of oscillations changes with the imposed velocity V. It shows that G_i for crack initiation and G_a for crack arrest (the upper and lower points of the $G(v)$ curve) are different from G_c and G_{min} (except for the relaxation cycle) and are not a material property.

The model is more complicated when the variation of peel angle is taken into account, which gives a third degree of freedom and leads to three differential equations, allowing a road to the chaos when the limit cycles are changed into strange attractors.

5. Viscous Drag and Limited Rate of Transport

When a crack propagates in a liquid medium, γ and w can be reduced by the liquid, giving a shift in the $G(v)$ curves as discussed above. However, above a given velocity, the viscous drag due to hydrodynamic phenomena inside the crack may be superior to the drag $w\varphi(v)$ due to the moving stresses. In this case a knee appears on the $G(v)$ curves, and one follows the curve for viscous drag until cavitation occurs. The crack thus propagates without contact with the liquid medium, and the $G(v)$ curve for propagation in air or vacuum is followed, as shown by Carré and Schultz[39] for peeling of elastomers in oils of various viscosities. The curve for viscous drag can even cross the $G(v)$ curve for propagation in air, with a velocity jump when cavitation occurs according to Fig.(16) and as observed by Michalske and Frechettte[40] for fracture of glass in water.

A knee in the $G(v)$ curves (region II) can also be due to a limited rate of transport of active species at the crack tip. When the crack velocity increases, active species present in a gas or diffusing through a liquid are less and less available at the

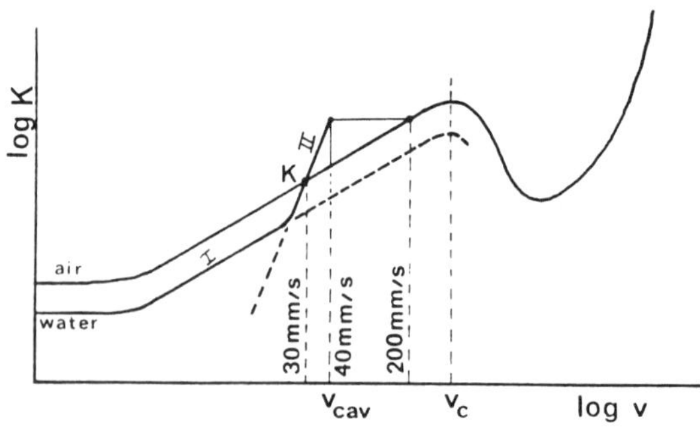

Figure 16 Influence of the viscous drag inside a propagating crack.

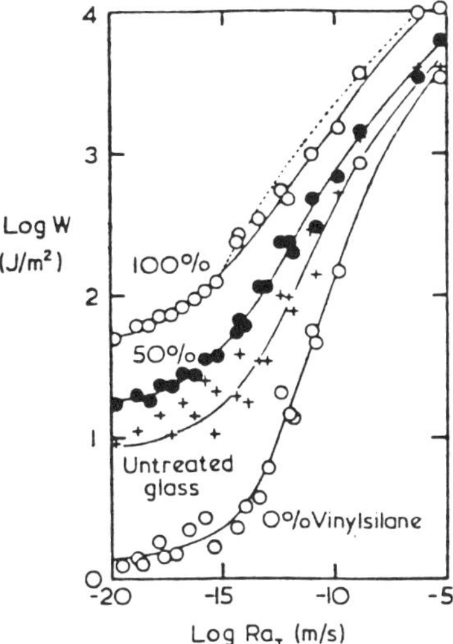

Figure 17 G Increase by surface treatments of the glass substrate.

crack tip, so that the coverage at the crack tip is less than the equilibrium coverage. Although the region II in rupture of ceramics is generally explained by a reduction of stress corrosion due to the limited rate of transport,[41,42] an explanation based on a lower reduction of surface energy works as well.[43]

6. Work of Chains Pull Out in Adherence of Polymers

Hitherto we have only considered dissipations which tend to zero when the crack velocity tends to zero, that is the value G_o of the energy release rate for vanishing crack speed is the Dupré energy of adhesion w. This was the case for glass-polyurethane systems, where Eq.(22) was verified within 2% whatever the geometry and the loading, and where the crack advances for $G > w$, and recedes (crack healing) for $G < w$.[44−46]

However, examination of the literature shows that the threshold values G_0 are generally considerably higher that the thermodynamic values w. (These can exceed 100 J/m^2, whereas w or 2γ cannot exceed 2 J/m^2).

Furthermore, when doing surface treatments to increase interfacial bonding (with silanes for example) the $\log G, \log v$ curves do not remain parallel. Ahagon and Gent[47] have increased the density of covalent bonds (primary bonds) between glass and an elastomer by varying the proportion of vinyl silane in ethylsilane for the glass treatment, so that the density of vinyl groups used for the bonding with the elastomer could vary from 0% to 100%. Fig.(17) shows that G is considerably more increased at low crack velocities than at high velocities, and that the curves do not remain parallel. For 100% vinyl silane, the $G(v)$ curve is identical to that obtained by tearing the elastomer. In another experiment, Chang and Gent[48] have shown that G_0 linearly increases with the density of primary bonds, up to a limit value G_0^* characteristic of the bulk elastomer, Fig.(18) from Ref.[49]. This figure also shows that the less the elastomer is crosslinked the higher the G_0. Similar experiments have been done by Lake and Stevenson[50] and Lake[51] pointing out that $G(v)$ curves do   not remain parallel when interfacial cross-linking is increased at constant bulk properties.

It is well known that if after rupture of a polymer the two lips of the crack are brought in contact under a light pressure a welding occurs by progressive interdiffusion of chains. (This interdiffusion is the mechanism of solvent welding of plastics, where a "good" solvent is used to swell surface layers and increase chain mobilities.[52] Jud, Kausch and Williams[53] have shown that the critical energy release rate G_c to break again the partially welded crack increased with contact time t as $G_c \sim t^{1/2}$ to finally reach the value of the bulk polymer, after some interdiffusion length. For PMMA at 385 K, the maximum of G_c is reached within 10 min after an interdiffusion of 2.5 nm. It is these interdiffused chains which must be pulled out at the moment of re-fracture. Using his reptation theory, de Gennes[54–56] proposed that $G(t)$ varies as

$$G(t) \simeq N f l(t)$$

where N is the number of chains having diffused, f the force to extract them from their tube, and

$$l(t) = 2\sqrt{\frac{DT}{\pi}}$$

the length to the chains penetrated. D is the diffusion coefficient along the tube, and is inversely proportional to the molecular weight M (or to the total chain length

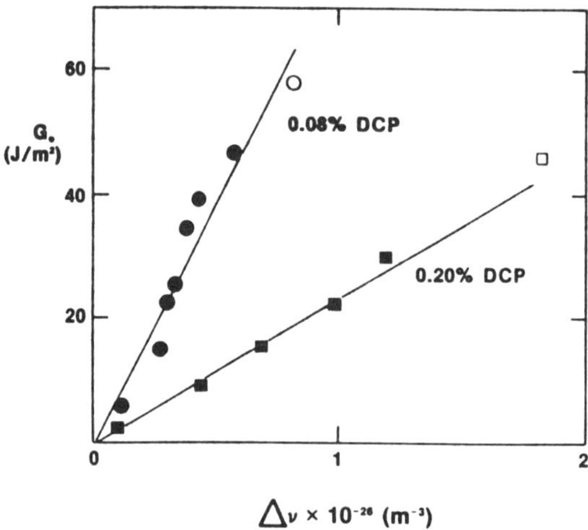

Figure 18 Influence of elastomer crosslink on G_0, from Ref.[49].

L). He thus found the variation proportional to $t^{1/2}$ as observed in experiments. For a healing crack, N is of the same order of magnitude as the number of chains broken by the crack, and thus varies as $M^{-1/2}$. The maximum of $l(t)$ being $L/2$, one has, assuming f to be independent of L

$$G_0(t) \sim t^{\frac{1}{2}} M^{-1} \tag{38}$$

$$G_0(\infty) \sim M^{\frac{1}{2}} \tag{39}$$

Eq.(39) corresponds to Gent and Tobias[57] results relating G_0 to $\sqrt{M}$ for the tearing of various elastomers.

For adhesion experiments, in which the number N of primary bond binding the chains to the interface is increased, these chains having to be pulled out, one can see that G_0^* corresponds to N maximum characteristic of the bulk polymer. Furthermore, at constant N, G_0 increases with l, $i.e.$ the less the polymer is crosslinked, the higher the G_0. As a result of chain extraction a layer of oriented polymer is then left on the surface, as in the Bikerman weak layer theory. Note that experiments on contact of silicon gel spheres or glass have clearly shown that a thin film of silicone was left on the glass.[58]

The work done in this extraction has to be added to viscoelastic losses due to the moving stress, tentatively as

$$G - w = w\varphi(a_T v) + NfL \tag{40}$$

where N is the number of primary bonds firmly holding the chains to the substracts. If one examines more closely the Fig.(17), one can assume that for 0% vinylsilane no chain is chemically bonded and that the curve corresponds to Eq.(22), with $\varphi(a_T v) \sim (a_T v)^{0.43}$ at low velocity. Then, if one substracts from the values of the three higher curves (untreated glass, 50% and 100% vinylsilane) the values of the lower curve, one obtains in this log-log diagram three curves vertically shifted, and corresponding possibly to the term NfL (with a friction coefficient f velocity dependent). There is thus superposition of two dissipative effects, and it is this addition which leads to nonparallelism of curves in a log-log plot. Note that this addition can completely hide the correlation between $\varphi(v)$ and $E''(\omega)$.

In simply adding a supplementary term, we have neglected the fact that the pulled chains apply a stress on the crack lips, introducing a negative K_1 and deforming the crack. (It is the same problem as for the drag by viscous liquid, where the hydrodynamic pressure elastically deforms the crack.[58]) This point has been studied in the rupture of polymers subjected to the phenomenon of crazing, schematically depicted on Fig.(19) from Bevan et al.[60] (For a review on crazes, see Kramer[61].) The crack (where fibrils are broken and retracted) is preceded by a zone (craze) where pulled fibrils (50 to 300 Å diameter) apply a stress σ_0 nearly constant to the crack lips, as in the plastic zone of the Dugdale model.[62] In this model, as in the Barenblatt one, this stress σ_0 leads to a stress intensity factor K_m, which compensates the stress intensity factor K_1 due to external loading. The stress singularity disappears and the length of the "plastic" zone is

$$d = \frac{\pi}{8} \frac{K_1^2}{\sigma_0^2} \tag{41}$$

from which one can deduce the shape of the crack in the plastic zone:

$$\delta = \frac{8}{\pi E} \sigma_0 d \left[\xi - \frac{x}{2d} \ln \frac{1+\xi}{1-\xi} \right] \tag{42}$$

where $x = 0$ is the origin of the plastic zone (at crack tip) and $\xi = (1 - x/d)^{1/2}$. The crack opening displacement (C.O.D) is the value of δ at $x = 0$

$$\delta_t = \frac{8\sigma_0 d}{\pi E} = \frac{K_1^2}{\sigma_0 E} \tag{43}$$

hence

$$G = \sigma_0 \delta_t \tag{44}$$

When an increasing stress is applied to a slit, the length d of the craze, and then C.O.D increase until a critical C.O.D, δ_c, is reached, corresponding to a K_0 or $G_0 = \sigma_0 \delta_c$ applied by the experimenter. The more stretched fibrils are thus broken, and the crack propagate for $G > G_0$. As shown by Ward et al[63–66] the craze profile is correctly given by Eq.(42), and δ_c considerably varies with the molecular weight.

Since Berry[67] in 1964, the influence of molecular weight, chain lengths and entanglement effects on the fracture of polymers is the subject of considerable interest.[68–73] Recently, Prentice[73] has given a model for the rupture of polymers,

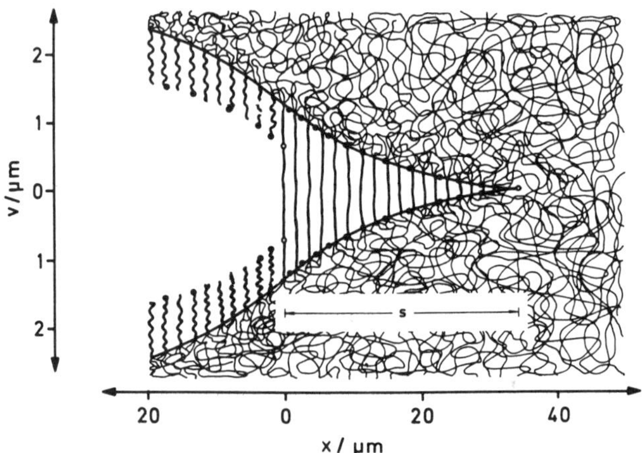

Figure 19 The phenomenon of crazing: a schematic picture from Ref.[60].

in which the force to extract a chain is proportional to its length, and function of the pull out velocity. He obtains $G \sim (M - M_c)^2$ as long as the chain length is not too high (M_c is the critical entanglement weight). Above a critical length, ie critical molecular weight, the chains are broken before being completely pulled out, and G becomes independent of M, as experiments show. The various possibilities of chain extraction, chain extension and chain scission during slow crack propagation, and the various energy losses have been recently studied by Evans.[74] Using the reptation model, and comparing the relaxation times for the various mechanisms he has shown that partial or full chain pull out could exist at very low crack velocity, dissipating energy around $100 J/m^2$, and that chain scission predominates at higher velocity with a lower dissipation, but it is likely that at these last velocities viscoelastic losses due to moving stresses become predominant. This transition between chain pull-out and chain scission when the crack velocity increases must be compared to the cohesive-adhesive transition observed during peeling.

The way to combine these various informations as an equation for crack propagation is not yet very clear. In particular are the viscoelastic losses always proportional to w? The mechanisms of polymers fracture are actively studied at the present time, and one can hope progress in understanding adherence of polymers is the near future.

References

[1] K. Kendall, *The adhesion and surface energy of elastic solids*, J. Phys. D, **4**, 1186, (1971).

[2] K.L. Johnson, K. Kendall and A.D. Roberts, *Surface energy and the contact of elastic solids*, Proc. Roy. Soc., **A324**, 301, (1971).

[3] D. Maugis and M. Barquins, *Fracture mechanics and the adherence of viscoelastic bodies*, J. Phys. D, **11**, 1989, (1978).

[4] C.E. Inglis, *Stresses in a plate due to the presence of cracks and sharp corners*, Trans. Inst. Nat. Archit., **55**, 219, (1913)

[5] K.J. Chang and H.C. Wu, *Angled elliptic notch problem under biaxial loading*, J. Appl. Mech., **47**, 57, (1980).

[6] I.N. Sneddon, *The distribution of stress in the neighbourhood of a crack in an elastic solid*, Proc. Roy. Soc., **A187**, 229, (1946).

[7] H. Kuppers, *Die numerische berechnung der spannungsverteilung in der umgebung einer kerbspitze*, Glastech. Ber., **37**, 185, (1964).

[8] M. Barquins and D. Maugis, *Adhesive contact of axisymmetric punches on an elastic half-space : the modified Hertz-Hubert's stress tensor for contacting spheres*, J. Meca. Theor. Appl., **1**, 331, (1982).

[9] G.I. Barenblatt, *The mathematical theory of equilibrium cracks in brittle fracture*, Adv. Appl. Mech., **7**, 55, (1962).

[10] D. Mc Lean, in *Grain boundaries in metals*, Clarendon, Oxford, 299, (1957).

[11] B.R. Lawn, D.H. Roach and R.M. Thomson, *Thresholds and reversibility in brittle cracks : an atomistic surface force model*, J. Mat. Sci., **22**, 4036, (1987).

[12] H.K. Mueller and W.G. Knauss, *The fracture energy and some mechanical properties of a polyurethane elastomer*, Trans. Soc. Rheol., **15**, 217, (1971).

[13] E.H. Andrews and A.J. Kinloch, *Mechanics of adhesive failure*, Proc. Roy. Soc., **A332**, 385, (1973).

[14] A.N. Gent and J. Schultz, *Effect of wetting liquids on the strength of adhesion of viscoelastic materials*, J. Adhesion, **3**, 281, (1972).

[15] L. Mullins, *Rupture of rubber, Part IX. Role of hysteresis in the tearing of rubber*, Trans. Inst. Rubber Ind., **35**, 213, (1959).

[16] D. Maugis, M. Barquins, *Adhesive contact of sectionally smooth-ended punches on elastic half-spaces: theory and experiments*, J. Phys. D, **16**, 1843, (1983).

[17] M. Barquins, *Kinetics of the spontaneous peeling of elastomers*, J. Appl. Polym. Sci., **29**, 3269, (1984).

[18] M. Barquins, *Influence of the stiffness of testing machine on the adherence of elastomers*, J. Appl. Polymer. Sci., **28**, 2647, (1983).

[19] K. Kendall, *Shrinkage and peel strength of adhesive joints*, J. Phys. D, **6**, 1782, (1973).

[20] A.D. Roberts, *Looking at rubber adhesion*, Rubber Chem. Technol., **52**, 23, (1979).

[21] M. Barquins and D. Maugis, *Tackiness of elastomers*, J. Adhesion., **13**, 53, (1981).

[22] D. Maugis, *Adherence of solids* in *Microscopic aspects of adhesion and lubrication*, J.M. Georges ed., Elsevier, Amsterdam, 221, (1982).

[23] G.P. Marshall, L.H. Coutts and J.G. Williams, *Temperature effects in the fracture of PMMA*, J. Mater. Sci., **9**, 1409, (1974).

[24] R.M. Hill and L.A. Dissado, *Relaxation in elastic and viscoelastic materials*, J. Mater. Sci., **19**, 1576, (1984).

[25] R.S. Rivlin and A.G. Thomas, *Rupture of rubber I. Characteristic energy for tearing*, J. Polym. Sci., **10**, 291, (1953).

[26] R.D. Margolis, R.W. Dunlap and H. Markovitz, *Fracture toughness testing of glassy plastics* in *Cracks and fracture*, ASTM STP 601, Philadelphia, 391, (1976).

[27] D. Maugis, *Subcritical crack growth, surface energy, fracture toughness, stick-slip and embrittlement*, J. Mater. Sci., **20**, 3041, (1985).

[28] T. Kobayashi and J.W. Dally, *A system of modified epoxies for dynamic photoelastic studies of fracture*, Exp. Mech., **17**, 367, (1977).

[29] A.B.J. Clark and G.R. Irwin, *Crack propagation behaviors*, Exp. Mech., **6**, 321, (1966).

[30] J.G. Williams, J.C. Radon and C.E. Turner, *Designing against fracture in brittle plastics*, Polym. Eng. Sci., **4**, 130, (1968).

[31] J.G. Williams, *Fracture mechanics of polymers*, Ellis Horwood, New York (1984).

[32] S. Yamini and R.J. Young, *Crack propagation in and factography of epoxy resins*, J. Mater. Sci., **14**, 1609, (1979).

[33] D. W. Aubrey and M. Sherrif, *Peel adhesion and viscoelasticity of rubber-resin blends*, J. Polym. Sci. Polym. Chem. Ed., **18**, 2597, (1980).

[34] M. Barquins, B. Khandani and D. Maugis, *Propagation saccadée de fissure dans le pelage d'un solide viscoélastique*, C. R. Acad. Sci. (Paris), Ser.II, **303**, 1517, (1986).

[35] D. Maugis, *Propagation saccadée de fissure en pelage, rôle de l'inertie*, C. R. Acad. Sci. (Paris), Ser.II, **304**, 775, (1987).

[36] D. Maugis and M. Barquins, *Stick-slip and peeling of adhesive tapes* in *Adhesion 12*, K.W. Allen ed., Elsevier, London, 205, (1988).

[37] N. Minorsky, *Non linear oscillations*, Van Nostrand, New York (1962).

[38] D.W. Jordan and P. Smith, *Non linear ordinary differential equations*, Clarendon Press, Oxford, (1977).

[39] A. Carre and J. Schultz, *Polymer-Aluminium adhesion IV. Kinetic aspects of the effect of a liquid environment*, J. Adhesion, **18**, 207, (1985).

[40] T.A. Michalske and V.D. Frechette, *Dynamic effect of liquids on crack growth leading to catastrophic failure in glass*, J. Amer. Ceram. Soc., **63**, 603, (1980).

[41] S.M. Wiederhorn, *Influence of water vapor on crack propagation on soda-lime glass*, J. Am. Ceram. Soc., **50**, 407, (1967).

[42] S.M. Wiederhorn, S.M. Freiman, E.R. Fuller and C.J. Simmons, *Effects of water and other dielectrics on crack growth*, J. Mater. Sci., **17**, 3460, (1982).

[43] D. Maugis, *Subcritical crack growth, surface energy and fracture toughness of brittle materials* in *Fracture mechanics of ceramics*, R.C. Bradt, A.G. Evans, D.P.H. Hasselman and FF. Lange eds., Plenum Publ., New York, vol.8, 255, (1986).

[44] D. Maugis, M. Barquins, *Adherence d'une bille de verre sur un massif viscoélastique: étude du recollement*, C.R. Acad. Sci., (Paris), Ser. B, **287**, 49, (1978).

[45] M. Barquins, *Adhesive contact and kinetics of adherence between a rigid sphere and an elastomeric solid*, Int. J. Adhesion and Adhesives., **3**, 71, (1983).

[46] M. Barquins and D. Wehbi, *Study of adherence of elastomers by cyclic unloading experiments*, J. Adhesion, **20**, 55, (1986).

[47] A. Ahagon and A.N. Gent, *Effect of interfacial bonding on the strength of adhesion*, J. Polym. Sci., **13**, 1285, (1975).

[48] R.J. Chang and A.N. Gent, *Effect on interfacial bonding on the strength of elastomers. I. Self-adhesion*, J. Polym. Sci., **19**, 1619, (1981).

[49] A.N. Gent, *The strength of adhesive bonds. An examination of interfacial chemistry, rheology of materials, and fracture mechanics*, Adhesive age, 27, (Feb. 1982).

[50] G.L. Lake and A. Stevenson, *On the mechanics of peeling in Adhesion 6*, K.W. Allen, ed., Applied Science Publ., London, 41, (1982).

[51] G.J. Lake, *Influence of the strength of interfacial bonding on the fracture characteristics of adhesive joints*, Inter. Adhesion Conf. Nottingham 1984, The Plastic and Rubber Institute, 22.1, (1984).

[52] W.V. Titow, *Solvent welding of plastics in Adhesion 2*, K.W. Allen, ed.), Appl. Sc. Publ., London, 181, (1978).

[53] K. Jud, H.H. Kausch and J.G. Williams, *Fracture mechanics studies of crack healing and welding of polymers*, J. Mater. Sci., **16**, 204, (1981).

[54] P.G. de Gennes, *Sur la soudure des polymères amorphes*, C. R. Acad. Sci., (Paris), Ser.B, **291**, 219, (1980).

[55] P.G. de Gennes, *The formation of polymer/polymer junctions in Microscopic aspects of adhesion and lubrication*, J.M. Georges, ed., Elsevier, Amsterdam, 335, (1982).

[56] P.G. de Gennes, *Adhesion: une liste de questions in Adsorption and Adhesion*, 5ème école d'été meditérranéenne ed., Les Editions de Physique, Paris, 1, (1984).

[57] A.N. Gent and R.H. Tobias, *Threshold tear strength of elastomers*, J. Polym. Sci., 20, 2051, (1982).

[58] A.E. Lee, *The role of elastic deformations in the adhesion of solids*, J. Colloid. Interface Sci., **64**, 577, (1978).

[59] S.M. Wiederhorn, S.W. Freiman, E.R. Fuller and C.J. Simmons, *Effect of water and other dielectric on crack growth*, J. Mat. Sci., **17**, 3460, (1982).

[60] L. Bevan, W. Doell and L. Koenczoel, *Micromechanics of a craze zone at the tip of a stationary crack*, J. Polym. Sci. Part B., **24**, 2433, (1986).

[61] E.J. Kramer, *Microscopic and molecular fundamentals of crazing*, Adv. Polym. Sci., **52/53**, 1, (1983).

[62] D.S. Dugdale, *Yielding of steel sheets containing slits*, J. Mech. Phys. Solids, **8**, 100, (1960).

[63] H.R. Brown and I.M. Ward, *Craze shape and fracture in poly(methylmethacrylate)*, Polymer **14**, 469, (1973).

[64] G.P. Morgan and I.M. Ward, *Temperature dependence of craze shape and fracture in poly(methylmethacrylate)*, Polymer, **18**, 87, (1977).

[65] R.A.W. Frazer and I.M. Ward, *Temperature dependance of craze shape and fracture in polycarbonate*, Polymer, **19**, 220, (1978).

[66] G. L. Pitman and I.M. Ward, *Effect of molecular weight on craze shape and fracture toughness in polycarbonate*, Polymer, **20**, 895, (1979).

[67] J.P. Berry, *Fracture processes in polymeric materials.V. Dependence of the ultimate properties of poly(methylmethacrylate) on molecular weight*, J. Polym. Sci. Part A, **2**, 4069, (1964).

[68] A.N. Gent and A.G. Thomas, *Effect of molecular weight on the tensile strength of glassy plastics*, J. Polym. Sci., **10**, 571, (1972).

[69] R.P. Kusy and D.T. Turner, *Influence of molecular weight of poly(methylmethacrylate) on fracture energy in notched tension*, Polym., **17**, 161, (1976).

[70] R.P. Kusy and M.J. Katz, *Effect of molecular weight on the fracture surface energy of poly(methylmethacrylate) in cleavage*, J. Mater. Sci., **11**, 1475, (1976).

[71] R.P. Kusy and M.J. Katz, *Generalized theory of the total fracture surface energy of glassy organic polymers*, Polymer, **19**, 1345, (1978).

[72] P. Prentice, *Influence of molecular weight on the fracture of poly(methylmethacrylate) (PMMA)*, Polymer, **24**, 344, (1983).

[73] P. Prentice, *The influence of molecular weight on the fracture of thermoplastic glassy polymers*, J. Mater. Sci., **20**, 1445, (1985).

[74] K.E. Evans, *A scaling analysis of the fracture mechanisms in glassy polymers*, J. Polym. Sci. Part B, **25**, 353, (1987).

Chapter 12

Damage Evolution, Instability and Fracture in Ductile Solids

Alan Needleman

Division of Engineering, Brown University
Providence, RI 02912, U.S.A.

Selected issues arising in the continuum mechanics analysis of instability and fracture in ductile metals are discussed. Three topics are considered: (i) the role of material rate dependence in setting the character of governing equations, (ii) deformation and failure in porous plastic solids and (iii) the direct calculation of fracture toughness from an analysis of the microscale ductile fracture mechanism.

1. Introduction

My aim is to introduce some topics of current interest (at least to me), in a manner that gives non-specialists an indication of what the issues are. References are given to papers where the topics covered here are dealt with in more detail and in more depth. A common thread through the paper is provided by the tendency for deformation in ductile solids to localize into narrow bands, with the band direction being set by the constitutive relation and the local stress state. Localization of deformation occurs in a wide variety of solids, in rocks and concrete as well as in structural metals. Strain localization plays an important role in limiting ductility and, accordingly, the phenomenon of strain localization has received a great deal of attention over the last decade, see *e.g.* Refs.[1,2].

Mathematically, for non-linear elastic solids or rate independent plastic solids, localization of deformation is associated with a change in type of the governing differential equations. This change of type renders boundary value problems ill-posed and means that, in general, the theoretical framework used to predict the onset of localization is inadequate for determining the post-localization response. Ill posed problems arising from a change in type of the governing equations are not peculiar to solid mechanics, but also arise in rheology[3] and in fluid mechanics.[4] The first topic discussed illustrates how accounting for the inevitable time and rate dependence of inelastic processes acts to regularize the governing equations and provides a framework for a meaningful investigation of post-localization response.

Disorder and Fracture, Edited by J. C. Charmet *et al.*
Plenum Press, New York, 1990

Experimental studies of ductile fracture have shown the central role played by microvoid growth in the ductile fracture of metals, *e.g.* Ref.[5]. The voids nucleate mainly at second phase particles by decohesion of the particle-matrix interface or by particle fracture, and final rupture involves the growth of neighboring voids to coalescence. The weakening due to microvoid nucleation and growth promotes the localization of deformation, which, in turn, plays a major role in setting the ductile fracture mode. Phenomenological constitutive relations for porous plastic solids have been developed[6–8] and analyses based on them have reproduced observed ductile failure modes in remarkable detail, *e.g.* Refs.[9-11]. In these phenomenological constitutive relations the porosity is characterized by a single scalar parameter, the void volume fraction. Rather recently, quantitative predictions of ductility and fracture have been made within such a framework based on detailed analyses of the microstructural failure process.[12,13]

Detailed comparisons of theory and experiment have suggested the need for going beyond the single parameter characterization of the second phase distribution in order to make reliable quantitative predictions. A recent investigation[14] of the stress-strain response of a random array of voids is described where the distribution relative to preferred material directions has a major influence on the aggregate response.

The third, and last, topic concerns the prediction of toughness, *i.e.* a material's resistance to crack growth. Here, one aim is to relate phenomenological measures of fracture toughness to measurable (and controllable) features of the material's microstructure. Another aim is to assess the range of applicability of various phenomenological ductile fracture criteria and to provide a method of failure analysis when such criteria cannot be applied. It turns out that localization in crack tip fields can strongly affect toughness. An analysis[11] that illustrates distribution effects on the ductile rupture mode at a crack tip is described.

2. Material Rate Dependence and Shear Band Instabilities

When ductile solids are deformed sufficiently, it is often observed that a smoothly varying deformation pattern gives way to one involving highly localized deformations. The localized deformation mode consists of one or more narrow bands of intensely deforming material. In metals the deformation within such a band is largely shear. Hence, such bands are termed shear bands. When the solid can be modelled as rate and time independent, a theoretical framework that goes back to Hadamard[15] is available in which shear band formation coincides with a change of type (typically from elliptic to hyperbolic) of the equations governing continuing quasi-static deformations. When inertia is accounted for, the change in the character of the momentum balance equation is from hyperbolic to elliptic, *i.e.* wave speeds become imaginary.

Shear band localizations have a dual significance. Often the large localized strains in a shear band precipitate a shear fracture. In other circumstances shear bands do not lead to fracture but localized shearing becomes an important mecha-

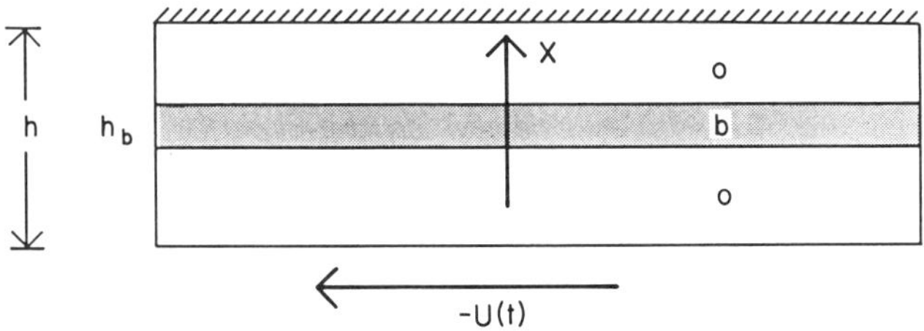

Figure 1 Simple shearing of a planar strip.

nism for subsequent plastic deformation. Thus, shear bands can be a direct precursor to fracture or can act as a mechanism of large strain plastic response. Accordingly, predicting material response after the onset of shear banding is an essential ingredient in a quantitative theory of ductile fracture and, in some circumstances, is necessary for a quantitative description of large strain plastic response.

Unfortunately the change of type of the governing equations leads to ill-posed problems so that solutions to post-localization boundary value problems exhibit pathological behavior. The problem is that the theoretical framework is not sufficient for the task. The pathology can be viewed in terms of the governing equations admitting arbitrarily narrow shear bands, with nothing in the formulation to set the shear band width. Physically then the task is to introduce appropriate characteristic lengths. There are a number of ways to do this, *e.g.* Refs.[16,17]. Perhaps a surprising one is by accounting for the rate dependence of inelastic material response.

Here, a simple one dimensional problem is used to illustrate how accounting for material rate dependence regularizes the problem by implicitly introduces a characteristic length. The discussion is adapted from Ref.[18]. Phenomenology and methods of analysis for localization in more than one dimension are discussed in Refs.[1,2,8,19].

Consider the planar block of height h subject to simple shearing displacement boundary conditions pictured in Fig.(1); t is time and the only relevant spatial dimension is x.

Balance of linear momentum implies

$$\tau_{,x} = \rho u_{,tt} \tag{1}$$

where $u(x,t)$ is the displacement, $\tau(x,t)$ is the shear stress and a comma denotes partial differentiation with respect to the indicated variable.

The boundary conditions for the simple shearing boundary value problem sketched in Fig.(1) are

$$u(0,t) = -U(t) \qquad u(h,t) = 0 \tag{2}$$

with the sign in Eq.(2) chosen so that positive $U(t)$ corresponds to positive shearing.

The shear strain, γ, is introduced via

$$\gamma = u_{,x} \tag{3}$$

Quasi-static deformations are those for which the inertial term in Eq.(1) is negligible. Then Eq.(1) implies that the shear stress is independent of x, *i.e.* $\tau = \tau(t)$.

What remains to be specified is the constitutive response of the material. In the simple example here this reduces to the dependence of τ on γ. An elastic solid is one for which the stress is a direct function of the strain, *i.e.* $\tau = F(\gamma)$; elasticity implies history independence. Plastic deformation is dissipative and in plasticity history matters. In a plastic process, a cycle that returns the stress to its original value does not necessarily return the strain to its initial value. At this point it is convenient to introduce symbols for the velocity, the strain rate and the stress rate,

$$v = u_{,t} \qquad \dot{\gamma} = \gamma_{,t} \qquad \dot{\tau} = \tau_{,t} \tag{4}$$

Now, consider a solid deforming by two distinct mechanisms, one that is reversible and one that is irreversible. For example, in a crystalline metal these can, respectively, correspond to deformation induced by stretching of the atomic lattice and by dislocation motion. The strain rate, $\dot{\gamma}$, is written as

$$\dot{\gamma} = \dot{\gamma}^e + \dot{\gamma}^p \tag{5}$$

with the elastic strain rate given by

$$\dot{\gamma}^e = \frac{1}{G}\dot{\tau} \tag{6}$$

where $G = dF(\gamma)/d\gamma > 0$. In the circumstances of interest here, it is appropriate to regard G as constant.

In the rate independent idealization, the plastic strain rate is then specified by

$$\dot{\gamma}^p = \begin{cases} \dot{\tau}/K, & \text{for plastic loading;} \\ 0, & \text{otherwise} \end{cases} \tag{7}$$

Here, K is the plastic hardening modulus and plastic loading refers to circumstances where $\tau \geq \tau_y$ and $\dot{\gamma}^p$ is positive. According to Eq.(7) the accumulation of plastic strain is independent of the time scale but, because of the loading-unloading behavior, the response is path dependent.

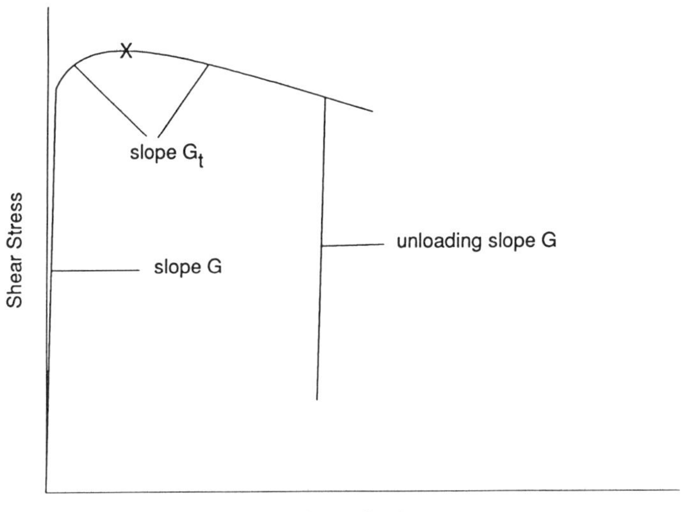

Figure 2 Shear stress versus shear strain response for a hardening-softening solid.

Combining Eqs.(5),(6) and (7), and solving for $\dot{\tau}$ gives

$$\dot{\tau} = \begin{cases} G_t\dot{\gamma}, & \text{for plastic loading;} \\ G\dot{\gamma}, & \text{otherwise} \end{cases} \tag{8}$$

where $1/G_t = 1/G + 1/K$.

Fig.(2) sketches a non-linear τ–γ relation and shows the unloading branch. The shear stress, τ, in this figure attains a maximum, at which point $G_t = 0$, and then decreases. Such softening can be the result of adiabatic heating or of the development of some internal damage process. In any case, the simple strain softening model in Fig.(2) is not an accurate model of the physical process, but it suffices for the present purposes. It is also worth noting that a nonlinear elastic solid is one for which $\dot{\tau} = G_t\dot{\gamma}$ always and that the key feature for localization, the dependence of G_t on strain, does not depend on the material being characterized as inelastic.

For a rate dependent solid, the plastic strain rate is written as

$$\dot{\gamma}^p = \dot{a}f(\tau, g) \tag{9}$$

where $\dot{a}$ is a material parameter of dimension 1/time and g is a hardness parameter that evolves with accumulating plastic strain, *i.e.* ,

$$\dot{g} = H(\gamma^p)\dot{\gamma}^p \qquad \gamma^p = \int_0^t \dot{\gamma}^p dt \tag{10}$$

Fig.(3) shows a function $g(\gamma^p)$ that qualitatively leads to the sort of hardening-softening response sketched in Fig.(2). By writing Eqs.(9) and (10) in terms of $\dot{\gamma}^p$ and τ, instead of absolute values, attention is confined to circumstances where τ is non-negative. Substituting Eqs.(9) and (6) into Eq.(5) and solving for $\dot{\tau}$ gives

$$\dot{\tau} = G[\dot{\gamma} - \dot{a}f(\tau, g)] \tag{11}$$

In Eq.(9) the evolution equation for $\dot{\gamma}^p$ is written as a kinetic relation, that is, given the current stress, τ, and hardness, g, the plastic strain rate can be calculated. This contrasts with the rate independent relation Eq.(7) where the strain rate must be known before it can be decided whether the plastic loading or elastic unloading branch is active. In the present context, this is the fundamental distinction between rate dependent and rate independent plasticity; for rate dependent plasticity the plastic strain rate does not depend on rate quantities whereas for rate independent plasticity it does. This distinction is of importance even when the effect of a change in strain rate on stress level is small.

2.1 Localization in rate independent solids

For a homogeneous solid (one with properties independent of x in Fig.(1)) one solution to the quasi-static boundary value problem is a state of homogeneous shear. At some stage of the deformation history, the possibility of bifurcation into a shear band mode can be considered. Equilibrium requires that the stress state remain homogeneous so that

$$\dot{\tau}_b = \dot{\tau}_o \tag{12}$$

where $()_b$ and $()_o$ denote quantities inside and outside the band in Fig.(1).

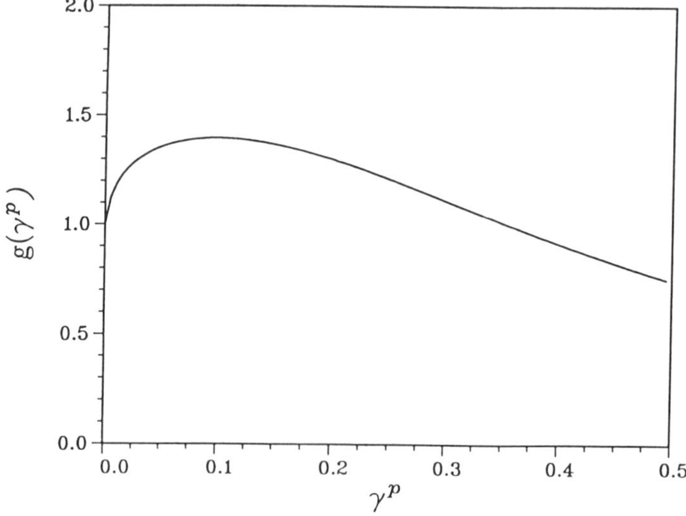

Figure 3 Hardness function $g(\gamma^p)$ for a hardening-softening rate dependent solid.

For a rate independent solid

$$\dot{\tau}_b = G_t \dot{\gamma}_b \qquad \dot{\tau}_o = G_t \dot{\gamma}_o \tag{13}$$

Earliest bifurcation is possible when

$$G_t(\dot{\gamma}_b - \dot{\gamma}_o) = 0 \tag{14}$$

A non-trivial solution to Eq.(14) is possible when G_t vanishes, that is at the peak of the stress-strain curve.

In the postbifurcation regime, the material inside the band continues to undergo plastic loading, while the material outside the band is elastically unloading (if the material outside the band were to continue plastic loading, the shear strain rates in the two regions would be identical and there would be no band). The total shear strain rate is

$$\dot{\gamma} = (1 - \beta)\dot{\gamma}_o + \beta\dot{\gamma}_b \tag{15}$$

where $\beta = h_b/h$.

Rate equilibrium and the constitutive relation give the overall shear stress-shear strain response as

$$\dot{\tau} = \frac{G_t \dot{\gamma}}{[\beta + (1 - \beta)\frac{G_t}{G}]} \tag{16}$$

The value of β is undetermined by the analysis and the solution is inherently non-unique. Furthermore, the presence of an initial imperfection does not resolve the non-uniqueness.[18]

Under dynamic loading conditions, the momentum balance equation (1) and constitutive equation (7) can be combined to give a wave equation for the velocity field

$$\begin{cases} \rho v_{,tt} = G_t v_{,xx}, & \text{for plastic loading;} \\ \rho v_{,tt} = G v_{,xx}, & \text{otherwise} \end{cases} \tag{17}$$

where ρ is the mass density and commas denote partial differentiation. For plastic loading Eq.(17) is a wave equation with wave speed c_t given by

$$c_t = \sqrt{G_t/\rho} \tag{18}$$

A clear discussion of how a vanishing tangent modulus leads to "wave trapping" is given in Ref.[20]. To summarize that discussion, suppose that at some shear strain G_t vanishes. Then the wave speed, c_t in Eq.(18) vanishes. In the particular boundary value problem under discussion here, the maximum strain occurs at the boundary $x = 0$ so that c_t first vanishes at this point. With $c_t = 0$, the boundary itself is a characteristic and it is no longer possible to communicate boundary data to interior points. Additional increases in deformation are not transmitted to interior points and there is a strain discontinuity at $x = 0$.

Hence, under both quasi-static and dynamic loading conditions, the solutions to localization problems for rate independent solids permit arbitrarily narrow bands of intense deformation. There is nothing in the formulation to set a minimum width to such bands. Also note that in both the quasi-static and dynamic cases the condition governing a change of type is a local one, depending only on the value of the material stiffness at the spatial point in question.

2.2 Localization in rate dependent solids

For the rate dependent solid under quasi-static loading the homogeneous solution is unique and bifurcation into a shear band mode is ruled out. Since the pre-bifurcation state is a homogeneous one, in Eq.(9) $f_b = f_0$ at bifurcation. Bifurcation then requires,

$$G(\dot{\gamma}_b - \dot{\gamma}_o) = 0 \tag{19}$$

As long as the elastic shear modulus, G, is positive, the unique solution to Eq.(19) is $\dot{\gamma}_b = \dot{\gamma}_o$.

Next, suppose that there is a band in which the material properties differ from those in the surrounding material. A solution can be constructed in which there is uniform straining inside the band and uniform straining outside the band, but with a jump in strain rate across the band. Within the band

$$(\dot{\gamma}^p)_b = \dot{a}_b f_b(\tau, g_b) \tag{20}$$

and outside the band

$$(\dot{\gamma}^p)_o = \dot{a}_o f_o(\tau, g_o) \tag{21}$$

Combining Eqs.(20) and (21) with Eqs.(11) and (12) gives two equations for the two unknowns, $\dot{\gamma}_o$ and $\dot{\gamma}_b$, namely

$$(1 - \beta)\dot{\gamma}_o + \beta\dot{\gamma}_b = \dot{\gamma} \tag{22}$$

$$-\dot{\gamma}_o + \dot{\gamma}_b = -\dot{a}_o f_o(\tau, g_o) + \dot{a}_b f_b(\tau, g_b) \tag{23}$$

The determinant of coefficients of Eqs.(22) and (23) is unity for all values of β. Hence, the band solution is unique and the imperfection sets the width of the band. Localization can still occur in the sense that $\dot{\gamma}_b/\dot{\gamma}_o$ may become arbitrarily large. However, at each instant of time the solution to Eqs.(22) and (23) is unique.

It is worth noting that material rate dependence does not smooth out the strain discontinuity across the band. The solution is sensitive to initial conditions; for example, in the simple shear problem, the band width is set by the initial inhomogeneity thickness. By rendering the solution unique, material rate dependence makes the initial inhomogeneity width the relevant length scale. The simple shear problem is highly degenerate, but the conclusion holds in a more general context.

Under dynamic loading conditions, the character of the governing equations is revealed by differentiating Eq.(1) with respect to time and using the constitutive relation (9) to obtain

$$\rho v_{,tt} = G v_{,xx} - G \dot{a} f(\tau, g)_{,x} \tag{24}$$

The material parameters in Eq.(21) can be grouped into a characteristic time, T, and a characteristic length, L, via

$$T = \frac{1}{\dot{a}} \qquad L = \frac{\sqrt{G/\rho}}{\dot{a}} \tag{25}$$

The parameter L is a propagation length; it is the distance an elastic wave travels in the characteristic time T.

Introducing the dimensionless variables

$$\bar{t} = \frac{t}{T} \qquad \bar{x} = \frac{x}{L} \qquad \bar{v} = (\frac{T}{L})v \tag{26}$$

into Eq.(21) gives

$$\bar{v}_{,\bar{t}\bar{t}} = \bar{v}_{,\bar{x}\bar{x}} - f(\tau, g)_{,\bar{x}} \tag{27}$$

The last term in Eq.(24) can be rewritten as

$$f(\tau, g)_{,\bar{x}} = f(\tau, g)_{,\tau} \tau_{,\bar{x}} + f(\tau, g)_{,g} g_{,\bar{x}} = G f(\tau, g)_{,\tau} \bar{v}_{,\bar{t}} + f(\tau, g)_{,g} g_{,\bar{x}} \tag{28}$$

Substituting Eq.(28) into Eq.(27) gives

$$\bar{v}_{,\bar{t}\bar{t}} + \lambda \bar{v}_{,\bar{t}} = \bar{v}_{,\bar{x}\bar{x}} + \xi \bar{g}_{,\bar{x}} \tag{29}$$

where

$$\lambda = G f(\tau, g)_{,\tau} \qquad \xi = -G f(\tau, g)_{,g} \qquad \bar{g} = \frac{g}{G} \tag{30}$$

The governing equations are Eq.(29) together with Eq.(11) and the evolution equation for g Eq.(10).

The highest order derivative terms in Eq.(29) correspond to a wave equation with characteristics at 45° in the $\bar{x} - \bar{t}$ plane, which is consistent with the experimental observation that incremental disturbances in plastically deformed solids propagate at the elastic wave speed. The character of Eq.(29), as determined from the highest order derivative terms, is independent of whether the material is hardening or softening, which contrasts with the situation for rate independent solids. The rate independent limit is $\lambda \to \infty$ and $\xi \to \lambda$ in Eq.(29). In this limit only the lower order derivative terms remain. The highest order derivative terms in Eq.(29) correspond to a wave equation. By way of contrast, adding linear viscosity to a rate independent stress strain relation gives a diffusion type equation.[18]

As can be anticipated from the character of Eq.(29), numerical solutions exhibit a boundary layer behavior. In the numerical solutions in Ref.[18], the strain localizes into a continually narrowing boundary layer. For a continually decreasing g, as sketched in Fig.(3), an unhappy ending at a finite time appears inevitable. The possibilities include: (i) the shear stress τ vanishes with $\dot{\gamma}_b/\dot{\gamma}_o$ remaining finite and (ii) $\dot{\gamma}_b/\dot{\gamma}_o$ becomes unbounded before τ vanishes. In the former case, fracture precedes instability while in the latter case instability precedes fracture. The conditions under which one of these is favored over the other are unknown. It should be noted that neither the physics of fracture nor a physical localization limiter, such as heat conduction, are incorporated into the strain softening material description.

In effect, material rate dependence introduces a length scale. In quasi-static problems, the length scale is one characterizing the imperfection or inhomogeneity; in dynamic problems, it is a characteristic length of propagation of elastic waves. In any particular case other length scales may be relevant and may in fact play a dominant role in setting the width of shear bands. Boundary value problems can be formulated incorporating the length scale that sets the shear band width; for example, by accounting for heat conduction, which introduces a material dependent length scale, or by incorporating appropriate geometric or material inhomogeneities into the formulation. In other circumstances the dimensions of some microstructural feature of a size scale smaller than any modelled in the boundary value problem may set the length scale. In such a case a constitutive characterization explicitly incorporating a material dependent length parameter appears appropriate.

3. Deformation and Failure of Porous Plastic Solids

Phenomenological constitutive relations for the plastic flow of porous solids have been based on a one parameter characterization of the porosity, *e.g.* Ref.[6-8]. Within this framework the voids are represented in terms of a single parameter, the void volume fraction, and the voids give rise to dilatancy and pressure sensitivity of the macroscopic plastic deformations. In particular, analyses of the influence of microscopic voids on plastic flow have been based on a constitutive formulation for progressively cavitating solids due to Gurson[6]. Failure, the complete loss of stress carrying capacity, arises as a natural outcome of the deformation history. This contrasts with the usual approach to fracture analysis where the constitutive characterization of the material and a fracture criterion are specified separately. An inherent limitation of the traditional approach stems from the fact that ductile failure occurs progressively and is inherently path dependent, so that achievement of a critical value of some parameter, *e.g.* a critical strain, cannot hold in general.

A detailed discussion of the Gurson constitutive framework for porous plastic solids will not be given here. The basis of the constitutive relation is a flow potential of the form

$$\Phi = \frac{\sigma_e^2}{\bar{\sigma}^2} + 2q_1 f \cosh\left(\frac{3q_2\sigma_h}{2\bar{\sigma}}\right) - 1 - q_1^2 f^2 = 0 \tag{31}$$

Here, σ_e is the Mises effective stress, $\sigma_h = \frac{1}{3}\boldsymbol{\sigma} : \mathbf{I}$ is the hydrostatic stress, f is the void volume fraction and $\bar{\sigma}$ is the strength of the matrix material. The parameters q_1 and q_2 were introduced by Tvergaard[21] to bring shear band bifurcation predictions of the Gurson[6] constitutive relation into closer agreement with corresponding results of full numerical analyses for a periodic array of voids. A further modification that is important for failure predictions involves replacing the void volume fraction f in Eq.(31) by a bi-linear function $f^*(f)$ in order to more accurately represent void coalescence.[9]

Simulations of ductile fracture processes based on Eq.(31) have reproduced observed phenomenologies in remarkable detail. For example, a characteristic feature of ductile fracture in structural metals is the contrast between the shear fracture mode observed in plane strain tensile specimens and the cup-cone mode observed in axisymmetric tensile specimens. In both cases, the deformations remain essentially homogeneous up to the maximum load point after which a diffuse neck develops. In the round bar tension test, diffuse necking is followed by a cup-cone type fracture. On the other hand, in the plane strain tensile test of the same material, the deformation mode usually shifts to one involving localized shearing while the diffuse neck is rather shallow, see *e.g.* Ref.[22].

A finite element analysis of necking and failure in the round bar tensile test has been carried out in Ref.[9] This is a quasi-static analysis, fully accounting for finite strains and rotations. A crack forms in the center of the neck and propagates across the specimen in Fig.(4), where the shaded region corresponds to material that has undergone a complete loss of stress carrying capacity. It can be seen in this figure that there is a tendency for the crack to zig-zag. This is a consequence of shear localization being inhibited by the additional plastic work associated with the hoop strains that accompany shearing in the axisymmetric geometry. As the free surface is approached this axisymmetric constraint is relaxed, permitting the cone of the cup-cone fracture to form. This analysis shows how the interaction of the tendency to localization in a material weakened by void nucleation and growth together with a constraining geometrical effect lead to the cup-cone fracture.

There is no corresponding geometrical constraint in plane strain tension so that, once initiated, a shear band can propagate across the entire specimen as illustrated in Fig.(5). These results clearly demonstrate the fracture mode transition characteristic of ductile metals.

The form of Eq.(31) was arrived at in Ref.[6] through an approximate rigid-plastic limit analysis of a thick walled spherical shell. Koplik and Needleman[23] have carried out finite element cell model analyses for an axisymmetric representation of a uniform array of spherical voids, fully accounting for void interaction effects and for void shape changes. Cell models represent uniform periodic arrays and are widely used to calculate constitutive properties of multi-phase materials. The cell model used in Ref.[23] and shown in Fig.(6) is an axisymmetric approximation to a three dimensional hexagonal array of voids.

Fig.(7) shows a comparison of finite element results for void volume fraction evolution and stress carrying capacity with corresponding predictions of the

Alan Needleman

Gurson[6] constitutive relation at three triaxiality levels. With the macroscopic effective stress denoted by Σ_e, and the macroscopic hydrostatic stress by Σ_h, the triaxiality T is given by

$$T = \frac{\Sigma_h}{\Sigma_e} \tag{32}$$

For a range of material parameters and triaxiality levels, rather good agreement was found in Ref.[23] using $q_1 = 1.25$ and $q_2 = 1.0$ in Eq.(31), with Σ_e identified with σ_e and Σ_h identified with σ_h. The cell model calculations also show a shift in strain state to a mode of uniaxial straining at which point the plastic deformation localizes to the ligament between neighboring voids. This event was associated with the accelerated void growth accompanying coalescence. The stress-strain response, at least for low void volume fractions, was found to be well approximated as a function of void volume fraction, independent of cell aspect ratio, but the strain at the initiation of accelerated void growth was sensitive to the cell aspect ratio.

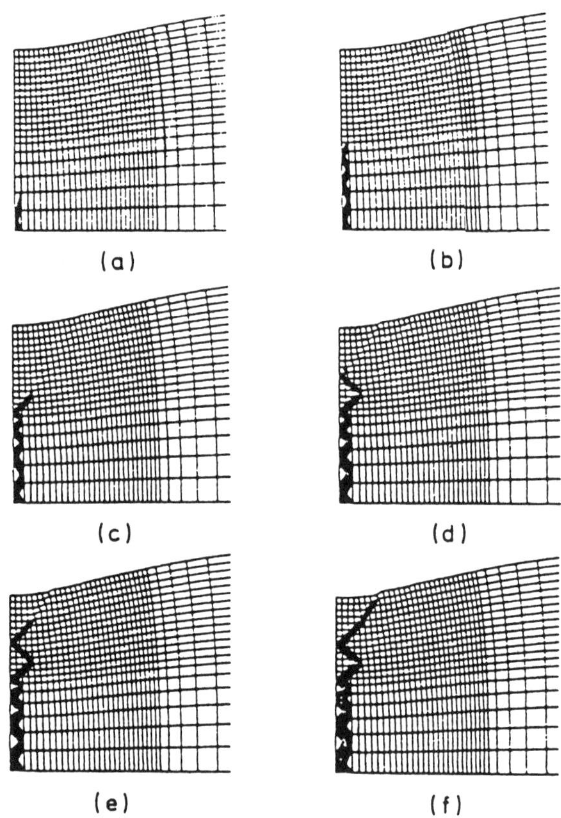

Figure 4 Crack growth in the neck of an axisymmetric tensile specimen. Triangular finite elements which have undergone a complete loss in stress carrying capacity are painted black. From Ref.[9].

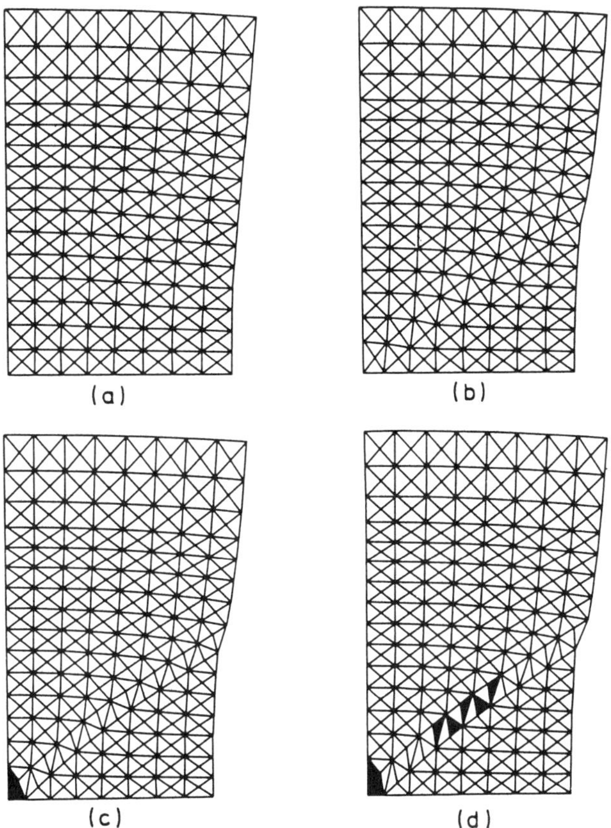

Figure 5 Shear band development in the neck of a plane strain tension specimen. Triangular elements which have undergone a complete loss in stress carrying capacity are painted black. From Ref.[10].

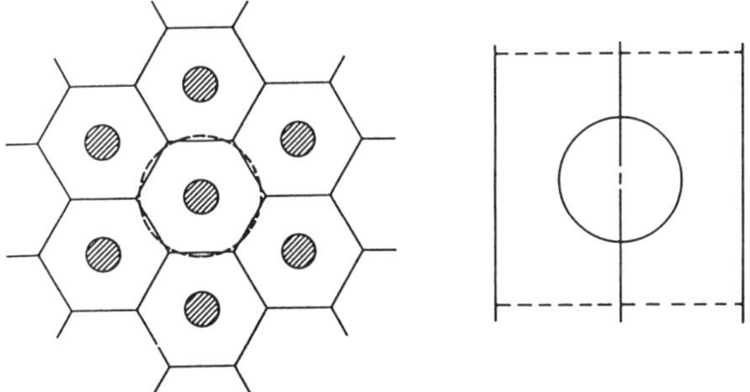

Figure 6 Axisymmetric cell model for a uniform distribution of voids.

Non-uniform void distributions are known to significantly (and adversely) effect ductility, both from experiment[24] and from theoretical analysis.[25] The role of
the non-uniformity of the distribution on the stress-strain response of porous plastic
solids is less well explored. In Ref.[14], the small strain stress-strain response of a
porous plastic solid subject to plane strain tension and containing a doubly periodic
array of clusters of cylindrical voids is analyzed. The void distribution within each
cluster is irregular. Fig.(8) shows various distributions, all having an initial porosity of 9.375 per cent. The predicted macroscopic stress-strain response for these
distributions, along with a cell model calculation based on a planar doubly periodic
array of circular voids, is shown in Fig.(9).

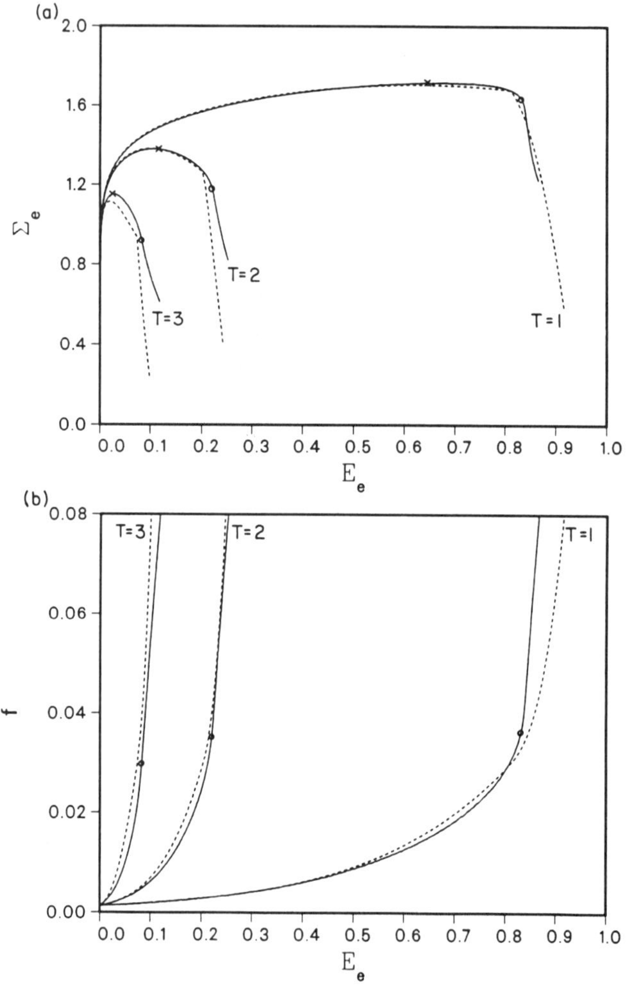

Figure 7 Comparison of (a) Aggregate stress-strain response and (b) evolution of
the void volume fraction for a cell model and the Gurson [6] model with $q_1 = 1.25$,
$q_2 = 1.0$. The initial void volume fraction is 0.0013. From Ref.[23].

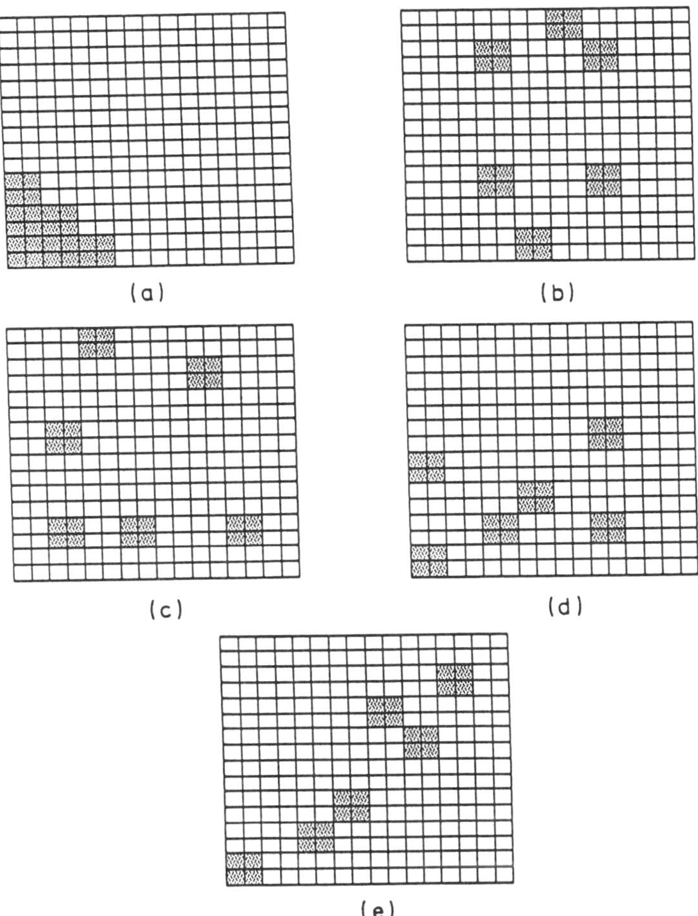

Figure 8 Various void distributions, all having an area fraction of $6/64 = 0.09375$. Symmetry conditions are applied so that a doubly periodic array of cells is analyzed. From Ref.[14].

The largest effect of void distribution is on the aggregate flow strength. Elastic properties exhibit a much lower sensitivity to distribution. Hence, the deviation of the aggregate response from that of the double periodic circular cylindrical array of voids is greater in the plastic range than in the elastic range. In particular, the sensitivity of the aggregate flow strength to distributions of voids aligned at 45° to the principal stress directions arises from the directional sensitivity of plastic yielding. For a rate independent ideally plastic solid in plane strain the spatial equations governing incremental equilibrium are hyperbolic and the 45° directions correspond to the directions of the characteristics. For the strain hardening and strain rate hardening constitutive relation used to characterize the matrix material in Ref.[14], the incremental equilibrium equations remain elliptic. Nevertheless, the preferred directions of the limiting ideally plastic solid seem to govern the directional sensitivity. It is hoped that these sorts of results for clustered periodic arrays will serve as an ingredient in a yet-to-be-developed extension of current approaches to homogenization for inelastic solids; see Ref.[26] for a discussion of the current state of homogenization theory for inelastic solids.

4. Damage Mechanics and Fracture

One aim of fracture analysis is to identify a parameter (if one exists!) that characterizes the intensity of stress and strain acting on the material near the crack tip, where the fracture process occurs. Typically, the actual crack is idealized as being mathematically sharp and the amplitude of the crack tip singular fields serves as the characterizing parameter. Several requirements are necessary for such a characterizing parameter approach to be useful. The first, sometimes termed autonomy, is for the local crack tip stress and strain fields to be universal functions, having the same form regardless of specimen geometry or loading conditions. The second is for the singular stress and strain distributions to dominate, *i.e.* to actually describe conditions, over a region large enough to encompass the fracture process zone. Finally, to be useful, it must be possible to relate the value of the characterizing parameter to the remote loading conditions.

The characterizing parameter approach provides a framework for predicting fracture behavior in structures from measurements on test specimens. When this approach is applicable, identical values of the characterizing parameter in a structure and in a test specimen correspond to identical stressing conditions at the crack

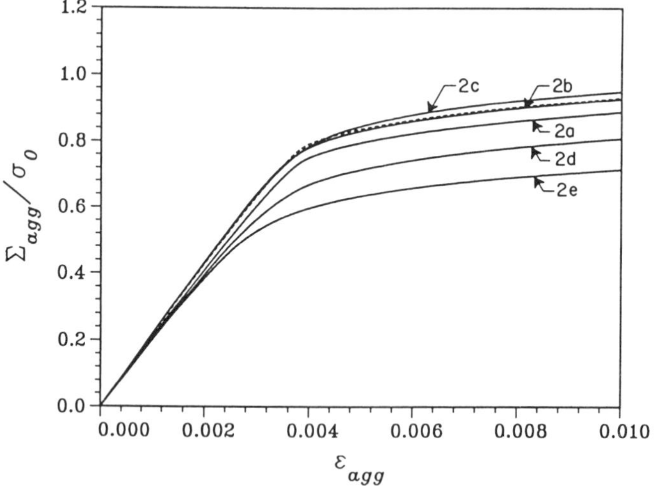

Figure 9 Stress strain curves for the void patterns in Fig.(8) (the letter marked for each curve refers to the corresponding void pattern). The dashed line corresponds to a cell calculation for a uniform distribution of circular voids. From Ref.[14].

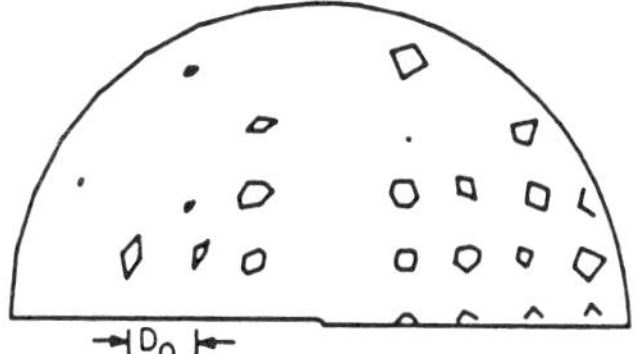
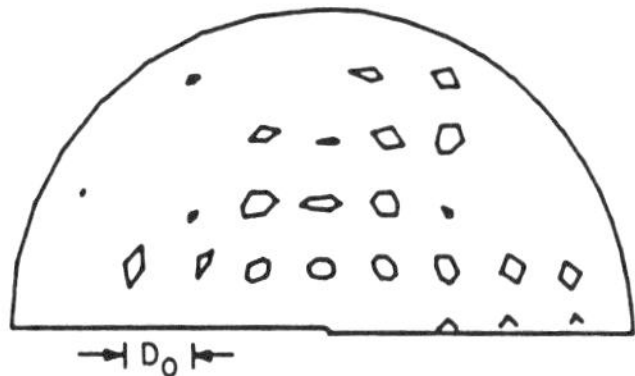

Figure 10 Contours of void nucleation amplitude showing two of the inclusion distributions analyzed in Ref.[11].

tip. Hence, fracture in a test specimen gives, through the characterizing parameter, a direct indication of when fracture will occur in an engineering structure.

A different approach to the analysis of fracture involves the solution of full boundary value problems based on a constitutive relation that has one or more internal variables to characterize the microstructural failure mechanism. When used without recourse to the framework of the characterizing parameter approach, a separate full analysis is required for each specimen geometry and loading condition. One use for such detailed fracture analyses is the quantitative prediction of fracture mechanics characterizing parameters in terms of microstructural attributes, to serve as a basis for designing tougher materials. Another is to ascertain the limits of validity of the characterizing parameter methodology and to provide an alternative when it is not applicable. Perspectives on non-linear fracture mechanics can be found in Refs.[27,28].

A common fracture mechanism in ductile metals involves the interaction of two populations of void nucleating particles with an existing crack; large inclusions that nucleate voids at relatively small strains and small particles that nucleate voids at much larger strains. In Ref.[11] this fracture mechanism was analyzed using the Gurson[6] constitutive framework.

At a crack tip, relevant microstructural size scales are greater than the distance over which large changes in stress and deformation state take place so that a material length scale must enter into the formulation. In Ref.[11], the distribution of the small scale particles was taken to be uniform and nucleation at these particles was governed by a strain controlled criterion. Nucleation at the large inclusions was taken to be stress controlled and their spatial distribution was modelled by a square array of "islands" of increased density of the amplitude of the void nucleation function. The spacing between these "islands" served as the material length scale. Fig.(10) shows two of the distributions analyzed in Ref.[11].

Fig.(11) illustrates the dependence of a qualitative feature of crack growth on the distribution; namely, the tendency of the crack to deviate from the initial crack line for certain inclusion distributions. It is clear from Fig.(11) that shear localization plays a strong role in this process. A further outcome of the analysis in Ref.[11] is the quantitative prediction of macroscopic toughness in terms of measurable and

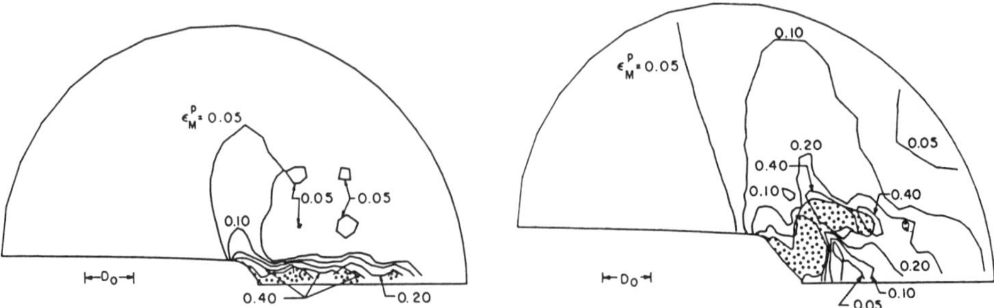

Figure 11 Contours of constant matrix plastic strain (denoted by ϵ^p_M in the figure) in the near tip region for the two inclusion distributions shown in Fig.(10). The dotted region is where the void volume fraction exceeds 0.10 and is shown in order to indicate the current crack location. From Ref.[11].

(hopefully) controllable parameters characterizing the micro-scale fracture process, *i.e.* the distribution and nucleation characteristics of inclusions. Toughness measures, such as the critical crack tip opening displacement and the tearing modulus, which measures the resistance to crack extension, depend on the attributes of the two particle populations as well as on matrix material properties such as strength and strain hardening. Although no attempt was made to model a particular real material, the calculated values of critical crack opening displacement and tearing modulus were typical of those measured in tough steels. The calculations also give a detailed picture of the process of crack growth. One interesting feature, consistent with experimental observation, is the increase in tearing modulus associated with crack branching. Another is the role of mechanical interactions in confining high porosity to very near the fracture surface. A detailed comparison of prediction and experiment for a different microstructure is contained in Ref.[13].

Acknowledgements

The support of the Office of Naval Research through grant N00014-89-J-3054 is gratefully acknowledged.

References

[1] J.R. Rice, *The Localization of Plastic Deformation*, Proc. 14th Int. Congr. Theoret. Appl. Mech., ed. W.T. Koiter, North-Holland, 207, (1977).

[2] A. Needleman and V. Tvergaard, *Finite Element Analysis of Localization in Plasticity*, in *Finite Elements - Special Problems in Solid Mechanics*, ed. J.T. Oden and G.F. Carey, Prentice-Hall, 94, (1983).

[3] D.D. Joseph and J.C. Sault, *Change of Type and Loss of Evolution in the Flow of Viscoelastic Fluids*, J. Non-Newt. Fluid Mech., **20**, 117, (1986).

[4] H. Aref, *Finger, Bubble, Tendril, Spike. An Essay on the Morphology and Dynamics of Interfaces in Fluids*, Polish Acad. Sci. Fluid Dyn. Trans., (1986), also Univ. Cal. San Diego Preprint.

[5] K.E. Puttick, *Ductile Fracture in Metals*, Phil. Mag. Ser. 8, **4**, 964, (1960).

[6] A.L. Gurson, *Plastic Flow and Fracture Behavior of Ductile Materials Incorporating Void Nucleation, Growth and Interaction*, Ph.D Thesis, Brown University, (1975).

[7] T. Guennouni and D. François, *Constitutive Equations for Rigid Plastic or Viscoplastic Materials Containing Voids*, Fatig. Fract. Eng. Mater. Struct., **10**, 399, (1987).

[8] V. Tvergaard, *Material Failure by Void Growth to Coalescence*, Adv. Appl. Mech., in press.

[9] V. Tvergaard and A. Needleman,*Analysis of the Cup-Cone Fracture in a Round Tensile Bar*, Acta Metall., **32**, 157, (1984).

[10] R. Becker and A. Needleman, *Effect of Yield Surface Curvature on Necking and Failure in Porous Plastic Solids*, J. Appl. Mech., **108**, 491, (1986).

[11] A. Needleman. and V. Tvergaard,*An Analysis of Ductile Rupture Modes at a Crack Tip*, J. Mech. Phys. Solids, **35**, 151, (1987).

[12] R. Becker, R. Needleman, O. Richmond and V. Tvergaard, *Void Growth and Failure in Notched Bars*, J. Mech. Phys. Solids, **36**, 317, (1988).

[13] R. Becker,A. Needleman, S. Suresh, V. Tvergaard and V.K. Vasudevan, *An Analysis of Ductile Failure by Grain Boundary Void Growth*, Acta Metall., **37**, 99, (1989).

[14] A. Needleman and A.S. Kushner, *An Analysis of Void Distribution Effects on Plastic Flow in Porous Solids*, Europ. J. Mech., in press.

[15] J.J. Hadamard, *Leçons sur la Propagation des Ondes et les Equations de L'Hydrodynamique*, Librairie Scientifique A. Hermann, Paris, (1903).

[16] Z.P. Bažant, T. Belytschko and T.P. Chang, *Continuum Theory for Strain-Softening*, J. Engng. Mech. ASCE, **110**, 1666, (1984).

[17] N. Triantafyllidis and E.C. Aifantis, *A Gradient Approach to Localization of Deformation I. Hyperelastic Materials*, J. Elast., **16**, 225, (1986).

[18] A. Needleman, *Material Rate Dependence and Mesh Sensitivity in Localization Problems*, Comp. Meths. Appl. Mech. Eng., **67**, 69, (1988).

[19] A. Needleman, *Continuum Mechanics Studies of Plastic Instabilities*, Rev. de Phys. Apliqu., **23**, 585, (1988).

[20] F.H. Wu and L.B. Freund, *Deformation Trapping due to Thermoplastic Instability in One-Dimensional Wave Propagation*, J. Mech. Phys. Solids, **32**, 119, (1984).

[21] V. Tvergaard, *Influence of Voids on Shear Band Instabilities under Plane Strain Conditions*, Int. J. Fract., **17**, 389, (1981).

[22] G.R. Speich and W.A. Spitzig, *Effect of Volume Fraction and Shape of Sulfide Inclusions on Through-Thickness Ductility and Impact Energy of High-Strength 4340 Plate Steels*, , Metall. Trans., **13A**, 2239, (1982).

[23] J. Koplik and A. Needleman, *Void Growth and Coalescence in Porous Plastic Solids*, Int. J. Solids Struct., **24**, 835, (1988).

[24] E.M. Dubensky and D.A. Koss, *Void/Pore Distributions and Ductile Fracture*, Met. Trans., **18A**, 1887, (1987).

[25] R. Becker, *The Effect of Porosity Distribution on Failure*, J. Mech. Phys. Solids, **35**, 577, (1987).

[26] P.M. Suquet, *Elements of Homogenization for Inelastic Solid Mechanics* in *Homogenization Techniques for Composite Media*, ed. E. Sanchez-Palencia and A. Zaoui, Lecture Notes in Physics 272, Springer-Verlag, 193, (1986).

[27] J.R. Rice, *Elastic-Plastic Fracture Mechanics* in *The Mechanics of Fracture*, ed. F. Erdogan, AMD Vol. 19, ASME, 23, (1976).

[28] J.W. Hutchinson, *Nonlinear Fracture Mechanics*, Department of Solid Mechanics, The Technical University of Denmark, (1979).

Chapter 13

Aspects of Nonlinearity and Selforganisation in Plastic Deformation[†]

Elias C. Aifantis

Department of Mechanical Engineering-Engineering Mechanics
Michigan Technological University, Houghton, MI 49931, USA

Even though there is strong evidence of multiplicity, nonlinearity and selforganization in plasticity, there is not a general consensus or systematic effort to utilize recent methods of nonlinear physics for understanding bifurcation and pattering of deformation phenomena. An approach recently advocated by the author and his co-workers is briefly outlined here. Emphasis is put on illustrating the origin of plastic instabilities at various levels of observation. Details are given on the problem of the persistent slip bands noted, for example, during cyclic deformation of Cu-monocrystals, and of the Portevin-Le Chatelier bands noted, for example, during monotonic deformation of Al-5% Mg polycrystals.

1. Pattern Forming Instabilities in Plasticity

It is well known that a dislocation may be viewed as a result of an instability inducing a local disturbance which persists, in the sense of metastability, over uniform lattice states. At somewhat larger or microscopic scales, such pattern forming instabilities result in static or travelling dislocation pile-ups. At even larger or mesoscopic scales, there is a plethora of phenomena related to the inhomogeneity of slip: slip lines, slip bands, slip bundles, and deformation bands. The cases of kink and coarse slip bands observed in monotonic deformation and of the persistent slip bands observed in cyclic deformation are characteristic examples of pattern forming instabilities at the mesoscopic scale. Finally, at macroscopic scales we have the phenomena of shear band formation including the development of adiabatic shear bands. While many of these phenomena have been studied extensively in their own right, there has not been an underlying principle or framework to consider their description from a unifying point of view. In particular, no effort has been made

---

[†] An invited lecture based on similar ideas was also delivered at the International Symposium "Plasticity 89" held in Tsu, Japan, July 31-August 4,1989. A short version of this paper was printed in the proceedings of this symposium entitled *Advances in Plasticity* 1989, pp. 537-540, edited by A. Khan and M. Tokuda, and published by Pergamon Press (1989).

Disorder and Fracture, Edited by J. C. Charmet *et al.*
Plenum Press, New York, 1990

to isolate their common characteristics and describe their occurrence and evolution with partial differential equations of nonlinear physics. It will be shown here that multiplicity or non-monotone equations of state is the basic feature in all these problems and that introduction of higher order gradients for capturing the evolution of the system in the post-instability regime is a feasible approach.

At the mesoscopic scale, in particular, a common framework for the analysis of dislocation patterning phenomena is obtained by employing the method and techniques of "selforganization". As in the case of typical physico-chemical systems which organize themselves into various forms and structures under the influence of an external driving force, the governing equations for the case of dislocation populations turn out to be nonlinear partial differential equations of the reaction-diffusion type. The physical origin of the reaction-diffusion terms is related to the production, annihilation, transport, and interaction of dislocation clusters and point defects.The corresponding mathematical analysis involves the application of stability and bifurcation theory together with the so-called method of "adiabatic elimination" or "slow mode dynamics" for describing the system in the nonlinear domain beyond bifurcation points. At macroscopic scales, partial differential equations for the plastic strain are derived exhibiting a higher order spatial dependence than the standard equations of classical continuum mechanics. The higher order gradients are introduced into the constitutive equations, *e.g.* in the yield condition or the expression for the flow stress. Their presence may be justified on the basis of the "adiabatic elimination" of internal variables with diffusive transport, such as point defects and dislocation ensembles, characterized by short life-times and evolving at scales smaller than the macroscopic scale.The explicit appearance of higher order gradients of the macroscopic variables is necessary in connection with non-monotone equations of state and the occurrence of softening in the stress-strain relationship.

Many of the implications of incorporating non-monotonicity and high-order gradients into the analysis of localization and patterning of deformation at the micro- and macro-level have been discussed by the author and his co-workers in recent publications by Aifantis and co-workers.[1−16] It was shown that the introduction of gradient terms in the equations of dislocation dynamics results in a predictive model for the wavelength of persistent slip bands and the symmetry breaking instabilities associated with these dislocation structures. The introduction of higher-order strain gradients into the constitutive expression for the flow stress results in a predictive model for the width of shear bands and the spacing and velocity of Portevin-Le Chatelier bands.

Some of this work is revisited here from a more general point of view. Moreover, other possible mathematical model equations describing pattern forming instabilities at various levels of plastic deformation are outlined.

2. Non-Monotone Equations of State

In this section we give a brief account of non-monotone equations of state encountered in typical phenomena of plastic deformation.

We begin by considering an atomic lattice chain. Each atom of the chain is subjected to the influence of a body force field f due to the presence of the adjacent atomic layers, commonly known as substrate. In the simplest possible situation the effect of substrate is such that each atom of the chain is embedded in a periodic force potential of the form

$$f = \sin(u) \tag{1}$$

where f is normalized by a molecular constant and u a properly normalized (by $2\pi/b, b =$ period of substrate) relative displacement. The form (1) is the most familiar non-monotone force-displacement relation at the atomic level. The atomic chain tries to maintain its integrity within the periodic potential imposed by the substrate through spring-like forces exerted among its atoms. While the most stable equilibrium configuration corresponds to each atom of the chain situated at each trough of the potential, other possible metastable equilibrium configurations can occur as a result of the competition between the "elastic" potential energy of the springs and the "long-range" potential of the substrate. Each atom may then progressively be displaced with respect to its nearest trough and the chain may eventually expand (contract) with the number of atoms being one less (more) than the number of troughs, thus giving rise to a negative (positive) dislocation. This is essentially the physical basis of the Frenkel-Kontorova model. The corresponding model equation is obtained by incorporating inertia and elastic strain gradients in Eq.(1) as it will be seen in the next section. It results to the well-known Sine-Gordon equation of soliton physics.

At the mesoscopic scale, non-monotonicity is inherent in the description of dislocation populations by nonlinear source terms of the form, for example,

$$g(\rho) = a\rho - b\rho^2 \tag{2}$$

where ρ is the dislocation density, the coefficient a measures the production or generation of dislocations and the coefficient b measures annihilation or crowding effects. In fact, Eq.(2) is the simplest possible form of Verhulst dynamics and leads to the so-called "logistic equation" of animal populations. A cubic nonlinearity representing dislocation-dislocation dipole interactions is the next generalization of Eq.(2). Proper scaling of ρ and g in terms of the quantities ν and f yields for Eq.(2) and its cubic generalization (by shifting variables and neglecting quadratic terms) the following expressions

$$f = \nu(1 - \nu) \text{ or } f = \nu(1 - \nu^2) \tag{3}$$

When different kinds or families of dislocations interact among themselves and point defects, Eq.(3) is replaced by a system of equations containing coupling terms determining species competition and coexistence.

Next, we discuss two examples of non-monotone equations of state related to the stress-strain and stress-strain rate responses. More details can be found in the work of Estrin and Kubin[16], where microscopic arguments pertaining to the concept of positive/negative strain hardening and strain rate sensitivity are outlined. The first example is concerned with the negative portion of the (microscopic) strain hardening ($h \sim \frac{\partial \sigma}{\partial \varepsilon}$) graph for small strains. This "unstable" regime is a result of the competition between forest hardening and dislocation production. It is associated with the formation of slip lines and slip bands. At large strains, another zero of the (macroscopic) hardening coefficient occurs and this is associated with the emergence of macroscopic shear bands. The second example is concerned with the non-monotone dependence of the stress on the strain rate. The negative slope regime (negative strain rate sensitivity $s \sim \frac{\partial \sigma}{\partial \dot{\varepsilon}}$) is consistent with microscopic theory based on the dynamical interaction of mobile dislocations and point defects. The aforementioned two cases are commonly known as strain hardening and strain rate sensitivity instabilities and they are respectively related to the regimes where the parameters $h = \frac{\partial \sigma}{\partial \varepsilon}$ and $s = \frac{\partial \sigma}{\partial \dot{\varepsilon}}$ are negative.

Finally, we mention the case of thermal softening as this is related to the occurrence of adiabatic shear bands at the macroscopic scale. This is explicitly seen upon the elimination of the temperature variable between the constitutive equation for the flow stress of the form

$$\sigma = k\varepsilon^n \dot{\varepsilon}^m (1 - d\theta) \tag{4}$$

(for linear thermal softening) or

$$\sigma = k(1 - re^{-s\varepsilon})\dot{\varepsilon}^m \theta^\nu \tag{5}$$

(for power-law thermal softening) and the "homogeneous" part of the energy equation

$$\dot{\theta} = B\sigma\dot{\varepsilon} \tag{6}$$

where σ, ε denote equivalent stress and strain and the rest of the quantities are material parameters.

3. Nonlinear Model Equations

By incorporating time- and space-dependent terms in non-monotone equations of state of the type discussed in the previous section, we derive classes of nonlinear partial differential equations which, in principle, can describe order, chaos, and pattern formation in plasticity. The time-dependent terms are second or first time derivatives for systems without or with friction respectively. The space-dependent terms are usually second or fourth order gradient terms for systems without or with diffusive instabilities. Due to space limitations we do not provide details of the derivation here but only give a few examples pertaining to the form of the relevant differential equation(s) and its basic properties.

In relation to the form (1) the following Sine-Gordon equation modelling the dynamics of a one-dimensional dislocation can be derived

$$u_{,tt} - u_{,xx} + \sin(u) = 0 \tag{7}$$

It is well-known that Eq.(7) possesses travelling wave and pulse-like solutions which may physically be identified with single dislocations and dislocation dipoles. It turns out that under certain conditions (for negligible damping and absence of external force) Eq.(7) also models the structure and motion of a dislocation viewed as a vibrating string in a Peierls potential.

In relation to the form (3) the following Fisher-Kolmogorov equation modelling the propagation of slip can be derived

$$u_{,t} = u_{,xx} + u(1 - u) \tag{8}$$

It is well known that this equation admits travelling wave solutions $u(x - Vt)$ for $V \geq 2$. Analogous results but with different stability properties hold if the last term of Eq.(8) is replaced by $u(1 - u^2)$. If, in addition, a fourth-order spatial term of the form $-\gamma u_{,xxxx}$ is introduced in Eq.(8), then the fronts with oscillatory structure are possible.

In the case of interacting dislocation populations, as for example in the case with mobile and immobile dislocations during cyclic deformation, equations of the type (8) give their place to the system of differential equations.

$$\dot{\rho}_i = \nabla_i (D^{(0)}_{ij} - D^{(1)}_{ijk}\nabla^2_k)\nabla_j\rho_i + f(\rho_i) - b\rho_i + \sum_n c_n\rho_i^n(\rho_m^+ + \rho_m^-)$$

$$\dot{\rho}_m^+ = -\nu\nabla_x\rho_m^+ + \frac{b}{2}\rho_i - \sum_n c_n\rho_i^n\rho_m^+ \tag{9}$$

$$\dot{\rho}_m^- = +\nu\nabla_x\rho_m^- + \frac{b}{2}\rho_i - \sum_n c_n\rho_i^n\rho_m^-$$

which is shown to lead to the prediction and symmetry breaking instabilities associated with persistent slip bands. The origin of this set of equations, the related physical considerations, and their mathematical implications are discussed in Section 4.

The early stages of monotonic deformation can also be modelled by a system of equations of the form

$$\dot{\rho}_m^+ = D\nabla^2\rho^+ - E\nabla^4\rho^+ - \tilde{\nu}\cdot\nabla\rho^+ - NL^+(\rho^+,\rho^-)$$

$$\dot{\rho}_m^- = D\nabla^2\rho^- - E\nabla^4\rho^- - \tilde{\nu}\cdot\nabla\rho^- - NL^+(\rho^+,\rho^-) \tag{10}$$

where D may be negative due to attractive character of dislocations. It can be shown that the plastic flow in the form of a current or drift-like motion of dislocation ν has a stabilizing effect on perturbations with wave vectors parallel to the glide direction, while it does not affect the development of patterns with wave vector orthogonal to the glide velocity with a preferred wavelength $\lambda_c = 2\pi\sqrt{E/|D|}$. The resulting patterns will then correspond to layered structures parallel to the slip planes.

Next, we discuss pattern forming instabilities at the macroscale by considering the solutions of appropriate differential equations for the plastic strain or strain rate. For the problem of stationary macroscopic shear bands it turns out that the "adiabatic elimination" of internal variables with diffusive transport leads to the following nonlinear differential equation for the shear strain distribution across the shear band

$$a(\gamma)\gamma_{,xx} + b(\gamma)\gamma_{,x}^2 = \kappa(\gamma) - \kappa_o \tag{11}$$

where $a(\gamma) > 0$ and $b(\gamma)$ are phenomenological gradient coefficients and the flow stress $\kappa(\gamma)$ is non-monotone. It turns out that solution of Eq.(11) leads to the prediction of shear band widths in accordance with experiments.

For the case of viscoplastic materials with a gradient-dependent flow stress σ of the form $\sigma = h\varepsilon + f(\dot\varepsilon) + c\varepsilon_{,xx}$, with f being non-monotone and c constant, the differential equation of quasistatic equilibrium yields Lienard's equation

$$z_{,\eta\eta} + \mu f'(z)z_{,\eta} + (z - z_s) = 0 \tag{12}$$

for the strain rate $z = \dot\varepsilon(x - Vt)$, with η being the travelling wave variable properly normalized. This equation possesses periodic solutions which can physically be identified with travelling Portevin-Le Chatelier bands. More details on the problem of Portevin-Le Chatelier effect are given below in Section 5.

We conclude by suggesting a model equation for a (unit density) viscous material with a higher order gradient dependent stress of the form $\sigma = \mu v_{,x} - cv_{,xx}$. Then, the one-dimensional momentum equation $\sigma_{,x} = \dot v$ implies

$$v_{,t} + v\,v_{,x} + cv_{,xxx} = \mu v_{,xx} \tag{13}$$

which is recognized as the Korteweg-de Vries-Burgers equation.

4. A Dislocation Reaction-Transport Model for PSB's

Here we provide more details on a generalized dynamical system of coupled nonlinear differential equations describing the collective behavior of dislocation populations during fatigue experiments and discuss the related patterning instabilities leading to the formation of PSB structures. This dynamical model correctly predicts the wavelength of the ladder-like structure. Furthermore, the breaking of translational and rotational symmetries at the bifurcation naturally leads to layer-splitting and imperfection development in accordance with experimental observations. Below we discuss briefly the physical basis of the proposed model.

Consider a monocrystal oriented for single slip in a direction parallel to the x axis and submitted to cyclic loading. After an initial period where hardening occurs, the forest of immobile dislocations is already well developed and this motivates the distinction between two types of dislocations: "trapped" or nearly immobile ones of density ρ_i, and "free" or mobile ones gliding on the primary slip plane of density ρ_m. In this regime, the dislocation density is sufficiently high ($10^{13} - 10^{14}m^{-2}$) and

can be represented by a continuous concentration field on a space scale larger than
a few lattice spacings. For each type of dislocation a balance equation is set up.
The source and flux terms are determined by the following basic processes:

• The creation of dislocations under an applied stress is described by means of internal sources. In the case of PSB formation, a large number of dislocation clusters
is already present in the crystal. Hence, the majority of the newly created dislocations will almost be immediately pinned by the forest of nearly immobile ones.
Such mechanisms induce "sourc" terms $f(\rho_i)$ in the kinetic equation for the immobile dislocation density ρ_i,

• When the applied stress reaches certain thresholds such that thermal activation
(along with internal stress — or secondary dislocation — induced activation) dominates local energy barriers, dislocations may break free and move rapidly with a
stress-dependent velocity in their glide planes. The freeing processes of trapped
dislocations lead to a linear source term $b\rho_m$ in the kinetic equation for the mobile
dislocation density ρ_m and to a corresponding sink term in the immobile dislocation
kinetics. The parameter b, specifying the freeing or mobilization rate, is vanishingly
small below threshold and suddenly reaches a finite value beyond the critical stress
level. This parameter will be taken as the bifurcation parameter of the system and
its variation may depend on the strain rate, the temperature, and other mechanical,
crystallographic, and internal structure parameters.

• The pinning of mobile dislocations by immobile dipoles or multipoles induces a
sink term in their evolution equation and a corresponding source term in the equation describing the dynamics of the latter. This term is generally nonlinear and
depends on the elementary interactions between the two types of dislocation populations. It is of the form $\sum_{n>1} c_n \rho_i^n \rho_m$ where the index n models the order of the
cluster participating in the pinning process. The dipole formation corresponding
to quadratic couplings between mobile dislocations is mainly efficient in the early
stages of the deformation process. It will not be considered in the present discussion which is concerned with situations where the forest of multipoles is already
well-developed.

• The motion of free or mobile dislocations is of a drift type giving rise to plastic
flow. It is described by a "complete balance law" including both a material derivative and a divergence term. For the trapped or immobile dislocations, the motion
in the slip and cross-slip directions is diffusive, their mobility being mostly due to
thermal effects or to the coupling with vacancies and other defects.

By taking into account the aforementioned dynamical processes, the balance
equations for the two dislocation populations may be written in the form given by
Eq.(9)

$$\dot{\rho}_i = \nabla_i (D_{ij}^{(0)} - D_{ijk}^{(1)} \nabla_k^2) \nabla_j \rho_i + f(\rho_i) - b\rho_i + \sum_n c_n \rho_i^n (\rho_m^+ + \rho_m^-)$$

$$\dot{\rho}_m^+ = -\nu \nabla_x \rho_m^+ + \frac{b}{2}\rho_i - \sum_n c_n \rho_i^n \rho_m^+ \tag{14}$$

$$\dot{\rho}_m^- = +\nu \nabla_x \rho_m^- + \frac{b}{2}\rho_i - \sum_n c_n \rho_i^n \rho_m^-$$

which are also listed here for convenience. We remark that a further distinction be-

Elias C. Aifantis

tween positive and negative mobile dislocation densities ρ_m^+ and ρ_m^-, of a constant average glide velocity $\nu = \nu^+ = -\nu^- = \bar{\nu}\sin(\omega t)$ (with $\bar{\nu}$ denoting the amplitude and ω the frequency of the cyclic process), has been made. A mechanical justification for the origin of the various terms included in Eq.(14) on the basis of effective mass and momentum balances for dislocation populations has been given by Aifantis .[5,6,15] By expressing the dynamical equations (14) in terms of the sum $\rho_m = \rho_m^+ + \rho_m^-$ and the difference $\delta_m = \rho_m^+ - \rho_m^-$ we obtain

$$
\begin{aligned}
\dot{\rho}_i &= \nabla_i(D_{ij}^{(0)} - D_{ijk}^{(1)}\nabla_k^2)\nabla_j\rho_i + f(\rho_i) - b\rho_i + \sum_n c_n^n \rho_i\rho_m \\
\dot{\rho}_m &= -\nu\nabla_x\delta_m + b\rho_i - \sum_n c_n\rho_i^n\rho_m \\
\dot{\delta}_m &= -\nu\nabla_x\rho_m - \sum_n c_n\rho_i^n\delta_m
\end{aligned}
\tag{15}
$$

The terms $f(\rho_i)$ and $-b\rho_i$ in the above equations represent respectively creation, annihilation and mobilization of trapped dislocations. Due to the attractive character of the elastic interactions between dislocations, the diffusion coefficient tensor $D_{ij}^{(0)}$ may be negative definite in the high density regime while $D_{ij}^{(1)}$ remains positive definite. Hence, a diffusive instability may occur for high creation rates or high stress levels, similar to Holt's theory of dislocation cell formation. The anisotropy of the underlying crystal structure is reflected here in the diffusion tensors.

On time scales larger than the period of the fatigue process and on space scales larger than the mean distance between obstacles, the effective motion of mobile dislocations is diffusive. It turns out (*e.g.* Aifantis[15]) that this is a consequence of the combined effect of the back and forward motion of dislocations during each fatigue cycle and to the presence of a large number of pinning clusters. The effective diffusion coefficient D_M deduced from Eq.(15) is then proportional to the square of the mean velocity amplitude $\bar{\nu}$ [which may be related to the stress intensity τ via phenomenological laws of the type, for example, $\bar{\nu} = \bar{\nu}_o\exp-(\frac{\tau}{\tau_o})^m$] and inversely proportional to the total pinning rate. Specifically, it turns out that $D_M = \bar{\nu}^2/2\sum_n c_n\rho_i^{on}$ and that its value is much larger than the values of the diffusion coefficients of the trapped dislocations. Our working version of the kinetic equation for ρ_m will then be given by

$$
\dot{\rho}_m = D_M\nabla_x^2\rho_m + b\rho_i - \sum_n c_n\rho_i^n\rho_m
\tag{16}
$$

less general version of (14) and (15) has been adopted and analyzed earlier by Walgraef and Aifantis.[8-9] In that version the diffusivity $D_{ij}^{(0)}$, was assumed isotropic, the diffusivity $D_{ij}^{(1)}$ was assumed to vanish and the index n was taken equal to 2 (thus accounting only for the trapping of mobile dislocations by immobile dipoles). It turns out that the generalizations introduced in this paper do not alter the basic features and properties of the elementary model discussed earlier.

To see this let us consider the pattern forming instabilities associated with the dynamical system (15) and (16). The linear stability analysis of the uniform steady state ($f(\rho_i^0) = 0$, $\rho_m^0 = b\rho_i^0 / \sum_n c_n \rho_i^{on}$) gives, in Fourier space, the following linear evolution matrix,

$$L = \begin{pmatrix} \beta - a - q_i q_j \left(D_{ij}^{(0)} + q_k^2 D_{ijk}^{(1)} \right) & \gamma \\ -\beta & -\gamma - q_x^2 D_M \end{pmatrix} \tag{17}$$

where $a = -f'(\rho_i^0)$ is positive as a requirement of stability of the uniform steady state at low stress intensities. The coefficients γ and β are defined by the relations $\gamma = \sum_n c_n \rho_i^{on}$ and $\beta = b(\rho_i^0 \frac{\partial \ln \gamma}{\partial \rho_i^0} - 1)$. The computation of the eigenvalues of this evolution matrix shows that, in the absence of diffusional instability ($D_{ij}^{(0)}$ is positive definite), by increasing the stress intensity (and since the motion of mobile dislocations in the primary slip direction is much faster than in any other), the uniform steady state becomes unstable versus density modulations first along the x direction at a bifurcation point given by

$$\beta = \beta_c = \left(\sqrt{a} + \sqrt{\gamma D_I / D_M} \right)^2$$

$$\tilde{q} = q_c \tilde{e}_x \ , \ q_c = \frac{2\pi}{\lambda_c} = \left[\frac{a\gamma}{D_I D_M} \right]^{1/4} \tag{18}$$

where $D_I \equiv D_{xx}^0$ and $\tilde{e}_x$ is the unit vector in the x direction.

It follows that the expressions given in Eq.(18) are exactly the same with those obtained from the initial elementary model provided that the mobilization and pinning coefficients β and γ are properly defined.

Hence, beyond this pattern forming instability, ladder-like structures are expected to develop with a wavelength λ_c which is a material property depending on the dislocation mobilities, as well as their multiplication and pinning rates. In fact, the wavelength selection results from a compromise between dislocation motion and interactions: the diffusion tries to increase it, while the interactions try to decrease it. By relating the parameters of the model to experimental quantities, it may be shown that the wavelength satisfies standard empirical or phenomenological relations such as $q_c \sim \sqrt{\rho_i^0} \sim \sqrt{\tau_s}$ where τ_s is the resolved shear stress. When $D_{ij}^{(0)}$ is not positive definite, a diffusional instability occurs, leading first to the formation of cellular structures which may be associated with the vein structure of the matrix. By increasing the stress intensity, the freeing of trapped dislocations becomes more and more efficient. As a result of the highly anisotropic motion of free dislocations, this cellular structure is destabilized versus ladder-like structures with wave vectors parallel to the primary slip direction (Walgraef and Aifantis[8−10]).

Having found a pattern forming instability in the dynamical model, we must now consider the selection and stability properties of the pattern evolving beyond the instability. The nonlinear character of the dislocation dynamics does not allow, in general, analytic solutions for ρ_i and ρ_m . However, near the instability (or the weakly nonlinear regime around the bifurcation point) the dynamics may be reduced to much simpler forms, as a result of the time and space scale separations which occur between stable and unstable modes. This approximation procedure, justified by physical (slow mode dynamics) or mathematical (center manifold theorem) arguments leads in our problem to the following asymptotic dynamics for the dislocation microstructure, when attention is restricted to the primary x - y slip plane

$$\tau_o \partial_t \sigma = \{\varepsilon - d_x(q_c^2 + \nabla^2)^2 + d_y \nabla_y^2\}\sigma - \nu\sigma^2 - u\sigma^3 \tag{19}$$

where $\varepsilon = (\beta - \beta_c)/\beta_c$, $d_x = D_I/\beta_c q_c^2$, $d_y = D_I/\beta_c(1 + \gamma/qc^2 D_M)$, $\nabla^2 = \nabla_x^2 + \nabla_y^2$. The order parameter-like variable σ is a linear combination of the two dislocation densities, while u and ν may be obtained explicitly in terms of the phenomenological coefficients of the model. [The "characteristic time" like constant τ_o is also directly related to the original parameters of the model and should not be confused with the reference yield-like or threshold stress τ_o used in other places of the text.]

In isotropic cases, $(d_y = 0)$, the only stable steady states beyond threshold $(\varepsilon > 0)$ correspond to hexagonal or layered patterns, while in highly anisotropic media the cellular structures become unstable. It thus turns out again that layered structures with walls perpendicular to the primary slip direction will be the only possible ones. Their amplitude equation is given by

$$\tau_o \partial_t A = \varepsilon A + (\xi_x^2 \nabla_x^2 + \xi_x^2 \nabla_x^2)A - 3u|A|^2 A \tag{20}$$

where $\sigma(\tilde{r}, t) = A(\tilde{r}, t)\exp[iq_c x] + A^*(\tilde{r}, t)\exp[-iq_c x]$ with * denoting complex conjugate, $|\ |$ magnitude, and $\xi_x^2 = 4q_c^2 d_x$, $\xi_y^2 = d_y$. This equation describes the slow amplitude modulations on th critical length scale of the pattern. The preferred solution (with respect to the associated Lyapounov function) for steady state modulations of uniform amplitude in a medium with natural boundary conditions is then given by

$$\sigma_o = 2R_o \cos(q_c x + \phi_o), R_o = \sqrt{\frac{\varepsilon}{3u}} \tag{21}$$

$$\phi_o = \text{constant}$$

5. A Gradient Model for PLC Deformation Bands

Here we give a brief account of a higher-order strain gradient model pertaining to the PLC phenomenon and discuss its suitability in predicting shear band widths and spacings. First we show that higher order gradients in strain can be generated by "adiabatically eliminating" the concentration of defects obeying diffusing dynamics on scales much smaller than the macroscopic scale of interest. Having illustrated that second order strain gradients can indeed appear in the expression for the flow stress, we consider in detail the structure and spacing of propagating shear bands in a viscoplastic material with strain rate softening (PLC bands).

The underlying goal is to develop a nonlinear differential equation for the plastic strain rate by utilizing the equations of continuum mechanics. This is possible only if the standard constitutive equations for rigidly viscoplastic solids are properly modified to account for the strongly heterogeneous and strongly nonlinear character of deformation. In other words,the inhomogeneous (diffusive) evolution of the underlying microstructure (*e.g.* point defects or dislocations) enter implicitly into the problem in the form of higher order strain gradients which modify the standard constitutive equations of (homogeneous or nearly homogeneous) continuum mechanics and alter the qualitative character of the governing differential equations.

To render this argument more explicit we consider the generation of second order strain gradients into the one-dimensional constitutive equation describing a viscoplastic material $\sigma = h\varepsilon + f(\dot\varepsilon)$ (where σ is the stress,ε is the strain, and f a nonconvex function of the strain rate $\dot\varepsilon$). Under PLC conditions, the dislocations (which are accounted here for through Orowan's relation for the strain rate $\dot\varepsilon \sim b\rho_m v$) interact strongly with the solute population ρ which, as usual, obeys a diffusive dynamics. We can thus expect that, in general, the system is described by the equations

$$\sigma = h\varepsilon + f(\dot\varepsilon) + g(\varepsilon, \rho)$$
$$\dot\rho = D\rho_{,xx} + r(\rho, \varepsilon) \tag{22}$$

where the subscript x denotes partial differentiation, the term $g(\varepsilon, \rho)$ represents the solute-dislocation interaction, and $r(\rho, \varepsilon)$ is a measure of the trapping/freeing mechanism of solute atmospheres in the dislocation cores. However, the solute concentration ρ is the fast variable of the system and can thus be "adiabatically" eliminated. Specifically, by assuming that g and r are linear forms, taking the Fourier transform of Eq.(22) and then solving for ρ_q ($\dot\rho_q \simeq 0$) , we can show that for "sufficiently large space scales" Eq.(22) is reduced to

$$\sigma = h\varepsilon + f(\dot\varepsilon) + c\varepsilon_{,xx} \tag{23}$$

where c is a positive constant and the strain hardening modulus h is renormalized.

We note that (23) is a direct gradient-dependent generalization of the constitutive equation for viscoplastic materials discussed in detail by Kubin, Estrin and co-workers (see references quoted in Aifantis [15]). Specifically, they proposed the following constitutive equation, in one dimension, for materials exhibiting the PLC effect

$$\sigma = h\varepsilon + f(\dot\varepsilon) \tag{24}$$

where σ and ε are the axial stress and strain and h is the strain hardening modulus. The function f is the viscous part of the flow stress and is assumed to be a single loop or nonconvex (negative slope regime *i.e.* $s = \partial\sigma/\partial\varepsilon < 0$ for $\dot\varepsilon_1 < \dot\varepsilon_s < \dot\varepsilon_2$). When the applied stress rate $\dot\sigma_o$ is such that the corresponding homogeneous steady state solution $\dot\varepsilon$ ($= \dot\sigma_o/h$) lies in the negative slope region $\dot\varepsilon_1 < \dot\varepsilon_s < \dot\varepsilon_2$, the homogeneous solution becomes unstable and a periodic succession of shear bands cross the specimen with constant velocity. To quantify the above discussion we assume a gradient dependent flow stress of the form given by Eq.(23).

With this simple modification to Eq.(23) and on assuming a travelling wave solution of the form $\dot{\varepsilon} = Z(x - Vt)$, Eq.(23) can be written as

$$Z_{,\eta\eta} + \mu f'(Z)Z_{,\eta} + (Z - Z_s) = 0 \tag{25}$$

where $\eta = -\sqrt{h/c}, \mu = V/\sqrt{ch}$. Equation (24) is the well-known Lienard's equation, a classical example of relaxation oscillations. According to LaSalle's theorem (see references quoted in Zbib and Aifantis [12–14]) a stable periodic solution exists for $\dot{\varepsilon}_1 < \dot{\varepsilon}_s < \dot{\varepsilon}_2$. Moreover, the natural speed of the travelling wave may approximately be obtained through the relation $V = 2\sqrt{ch}/|f'(Z_s)|$. More details on this problem together with appropriate numerical results illustrating the periodicity of the PLC bands, their velocity, as well as their direct influence on obtaining serrated and staircase stress-strain curves can be found in Refs.[12-15].

Acknowledgements

NSF support under grant CES-88 00459 is acknowledged.

References

[1] E.C. Aifantis, *Dislocation Kinetics and the Formation of Deformation Bands, Defects* in *Fracture and Fatigue*, G.C. Sih and J.W.Provan eds., Martinus-Nijhoff, 75, (1983).

[2] E.C. Aifantis,*On the microstructural origin of certain inelastic models*, J. Mat. Engng. Tech. **106**, 326, (1984).

[3] E.C. Aifantis, *On the Mechanics of Modulated Structures* in *Modulated Structure Materials*,T. Tsakalakos ed., Martinus-Nijhoff, 357, (1984).

[4] E.C. Aifantis, *Continuum Models for Dislocated States and Media with Microstructures. Mechanics of Dislocations*, E.C. Aifantis and J.P. Hirth eds., Metals Park, 127, (1985).

[5] E.C. Aifantis, *Mechanics of microstructures I, II, III. Mechanical Properties and Behaviour of Solids: Plastic Instabilities*, V. Balakrishnan and C.E. Bottani eds., World Scientific, Singapore, 314, (1986).

[6] E.C. Aifantis, *On the dynamical origin of dislocation patterns*, Mat. Sci. Eng. **81**, 563, (1986).

[7] D. Walgraef and E.C. Aifantis, *On the formation and stability of dislocation patterns - I, II, III*, Int. J. Eng. Sci. **23**,1351,1359 and 1365, (1985).

[8] D. Walgraef and E.C. Aifantis, *Dislocation Patterning in fatigued metals as a result of dynamical instabilities*, J.Appl.Phys. **58**, 688, (1985).

[9] D. Walgraef and E.C. Aifantis, *Dislocation patterning in fatigued metals: Labyrinth structures and rotational effects*, Int. J. Eng. Sci., 1789, (1986).

[10] D. Walgraef and E.C. Aifantis, *Plastic instabilities, dislocation patterns and nonequilibrium phenomena*, Res Mechanica **23**,161,(1988).

[11] M. Triantafyllidis and E.C. Aifantis, *A gradient approach to localization of deformation - I. Hyperelastic materials*, J.of Elasticity **16**, 225, (1986).

[12] H.M. Zbib and E.C. Aifantis,*On the localization and postlocalization behaviour of plastic deformation - I: On the initiation of shear bands; II: On the evolution and thickness of shear bands; III: On the structure and velocity of the Portevin-Le Chatelier Bands*, Res Mechanica **23**, 261, 279 and 293, (1988).

[13] H.M. Zbib and E.C. Aifantis, *On the structure and width of shear bands*, Scripta Met. **22**, 703,(1988).

[14] H.M. Zbib and E.C. Aifantis, *A gradient-dependent model for the Portevin-Le Chatelier effect*, Scripta Met. **22**, 1331, (1988).

[15] E.C. Aifantis, *The physics of plastic deformation*, Int. J. Plasticity **3**, 211, (1987).

[16] Y. Estrin and L.P. Kubin, *Plastic instabilities: Classification and physical mechanisms* In *Material Instabilities, Special Issue*, E.C. Aifantis et al. eds., Res Mechanica **23**, 197, (1988).

V Materials and Applications

Granular media constitute a privileged class of heterogeneous materials due to the disorder in the contacts even independently of that in the distribution of the properties of the individual grains. The transport properties of packings of grains are often described in terms of homogeneization treatments. These methods apply best when the disorder is small and when the heterogeneities are weak as would be the case if the void space was filled with a material having similar properties to those of the grains. This tends to smooth out the effect in the heterogeneities of contacts. However for *dry* packings, the distribution of contacts is a *dominant factor* and strongly limits the applicability of the above approaches. The variation of the contact distribution with applied stress has a direct impact on the non linear elastic behavior. The first lecture, by Bideau *et al.*, reviews the techniques of determination of the "coordination number" and the general characteristics of central force percolation and contact non-linearity which have a broad range of applicability in non linear mechanical effects related to disorder.

In all, including reinforced or prestressed, concrete structures, concrete is cracked, in its normal state. This cracking is due both to physical processes (thermal and/or hygral shrink) and mechanical loading. These two types of cracking develop at a large range of scale. Because of the different scales for the description of cracks in concrete, there exist various theoretical approaches to the modelization of this phenomenon. In this second lecture, Acker *et al.* distinguish three main types: (i) linear and non linear fracture mechanics, (ii) macroscopic approach (no discrete crack) and (iii) stochastic approach (showing that the observed scale effect of the structural behavior can be accounted for by a scaling law of the materials which constitute the structure). As a lot of technological problems and innovative impediments could be reduced with a predictive modelling of cracking (including explicitly cracks opening) and each type of modelization corresponds to a certain scale, the authors emphasize their range of application and their limits.

The last two lectures, by Knopoff, deal with a peculiar application of statistical modelling in the field of geophysics, as recent developments in nonlinear science give impetus to an attack on the problems of developing physical models for seismicity. One of the principal characteristics of earthquake occurrence is clustering in the five dimensional space of time, space and size. The first lecture is intended to be a primer of statistical earthquake phenomenology and describes not only the space-and-time-average properties, but also some of the clustering phenomenology. Unfortunately, in most cases, we have information about the events themselves but relatively little information about the deformation in the intervals between the earthquakes. Few concrete results are available to date from the theoretical study of the state of deformation as a stochastic process; so the second lecture focuses on the problems of a theory of earthquake occurrence with the hope that it can simulate satisfactorily this part of the total history of deformation.

Chapter 14

Granular Media: Effects of Disorder

Daniel Bideau
Etienne Guyon and Luc Oger[†]

Groupe de Physique Cristalline, URA CNRS 04804
Université de Rennes I, F-35042 Rennes Cédex, France

In practically all fields of the sciences, one finds granular media where disorders, making different physical properties, coexist at different scales. Emphasis has been put on recent developments based on ill condensed matter.

1. Introduction

The first main characteristics of a granular medium comes from the fact that it presents two phases:

- The solid phase, also called grain space, whose structure is generally complex. Its behaviour is somewhat intermediate between that of compressible liquid and of a solid (it flows and keeps the shape of the container).
- The porous phase, also called void space, generally well connected.

One can think that because these two phases are geometrically dual, their structural characterisation should be done in terms of the same geometrical parameters and by the same experimental and theoretical tools. But in fact the porosity (void over total volume ratio), which is the simpler geometrical characteristic of the pore space, and the mean coordination number (mean number of real contacts by grains), the simpler one for the grain space, are not correlated even if some empirical rules have been proposed.[1] In this article, we will concern ourselves by the solid phase, and we shall describe it as a disordered packing of grains. Such a system has been often studied in the literature from different points of view. Generally, it presents simultaneously different types of disorder:

- a *geometrical* disorder, which is originated from the size distribution of the grains, their shape, *etc* ..., and can greatly be modified by the effects of the walls of the

---

[†] Laboratoire de Physique et Mécanique de la Matière Hétérogène, URA CNRS 857, Ecole Supérieure de Physique et Chimie Industrielles de Paris, 10 Rue Vauquelin, F-75231 Paris Cédex 05, France

Disorder and Fracture, Edited by J. C. Charmet *et al.*
Plenum Press, New York, 1990

container. A limit case of this geometrical disorder is that of segregation which leads to a more or less spatially extended phase separation essentially due to grain size distribution.

• a disorder of *contacts*, occurring because there are some fluctuations in the distances between grains, due to geometrical disorder or to grain size distribution: then, for example even under a weak compression, the number of real contacts varies. This number of contacts is essential for transport or mechanical properties in the grain space, because currents and stresses are only transmitted through these contacts.

One can consider also a substitutional disorder, when two or more phases coexist in the material. But we shall study here only the geometrical disorder and the disorder of contacts (part 2), and their effects on the properties of granular media (part 4). Experimentally, it is very difficult to determine the mean coordination number z (mean number of contacts by grain) which is yet one of the most important geometrical characteristics of such a medium in its grain space. Part 3 is devoted to this problem.

2. Disorder in Granular Media

Structural order —or disorder— has been well described in condensed matter physics. The different aspects of the cristalline state, in particular, has been well studied. It is not the case for the structure of granular media. Two reasons can be invoked to explain this difference:

• The size of the grains in such media can vary from some micrometers to several centimeters. So the usual experimental tools of solid state physics such as scattering techniques which provide directly space analysis are not adapted to that *mesoscale*.
• The structure of these materials is extremely complex owing to the above described disorders: the effects of the grain size distribution or of the shape of the grains are not easy to grasp. So, some structural models are necessary to study what is the most important parameter from a geometrical point of view: the effects of sterical exclusions. These effects have been much studied using packings of spheres or disks: such an approach is very fruitful because these packings are also used as structural models in physics of dense liquids[2] or metallic glasses.[3] The work has been very intensive during these ten last years on the later system[4]: metallic glasses are characterised by very weak bonding, and then interactions between atoms can be considered as *hard sphere interactions.*

Now, let us try to understand why a packing of spheres (3D) or disks (2D) is disordered? From a general point of view,[5] one can consider that order or disorder in condensed matter is the result of the competition between the most stable unit cell involving short range interactions, and long-range order. Figure 1 illustrates this idea. The most stable unit cell for equal disk packings with *hard sphere interaction* is one in which the centers of three disks form an equilateral triangle.

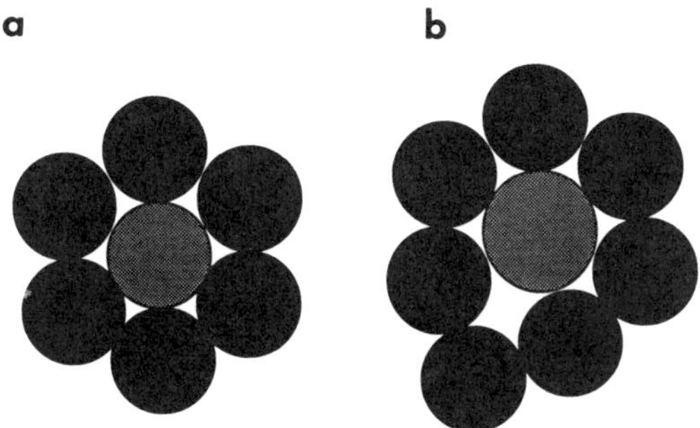

Figure 1 Sterical arrangement of identical disks around a central one; (left): dense seed of an ordered packing of equal disks; (right): dense seed of a disordered packing of equal disks due to a large central one. Long range disorder is induced by the stacking fault at the perimeter of the central impurity.

This unit cell is obviously compatible with cristalline order as seen in figure 1a: six equilateral triangles fill exactly the space around the central disk. If the central disk has a slightly larger diameter than the six other ones, as in figure 1b, the six surrounding disks can not fill exactly the space: a gap, or stacking fault, appears. The same situation occurs if the system of figure 1a would be replaced by spheres glued at the surface of a balloon: the curvature of the space or the grain space distribution leads to the same stacking defects as in two dimensional packings. In the limit where the central sphere has the same diameter as the surrounding one, a regular tetrahedron is the basic subunit which corresponds to the equilateral triangle in the 2D case. However in such a case, it is not possible to completely fill the 3D space by these units: so, dense packings of equal spheres are intrinsically disordered. But an *order seed*, like a plane wall, can lead to more or less ordered structure. Such an order propagation from a seed is generally blocked up in the case of grain size distribution, at 2D or 3D. So, we can consider that granular media, which generally present grain size distribution, can only be short-range ordered.

Grain size distribution not only does create disorder in packings, but also greatly modifies the structure of the system. In the case of binary mixtures for example, the porosity is very sensitive to the composition of the packing, as shown by figure 2 for spheres.

The "V" shape of the curves is quite generally observed in packings of grains, but the observed minimum value of the porosity is very dependent on the *building rule* of the sample, due to the fact that segregation, at a more or less large scale, always occurs in this case.[7] Segregation can lead to complete phase separation (small grains at the bottom, large ones at the top) for large diameter ratio, and for composition corresponding to the left part of the "V": small grains percolate through the packing of large grains. This more or less extended segregation, and the heterogencities which it creates, can greatly modify the properties of the material, mainly in its porous space.[8,9]

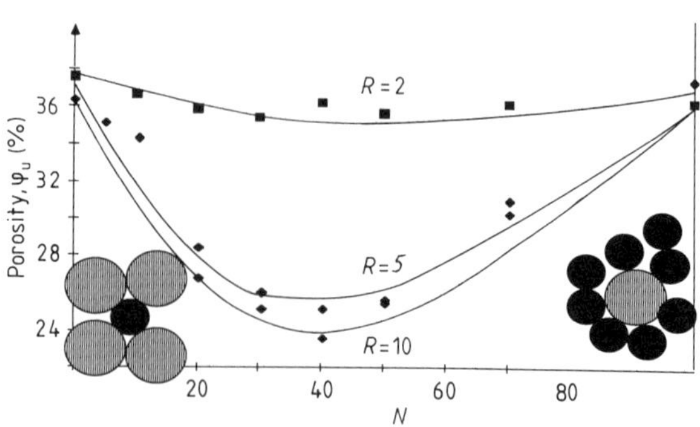

Figure 2 Variations of the porosity of binary mixtures of spheres, versus composition, expressed by the volume percentage of small spheres.[6] R is the diameter ratio.

But even in the case of quasi-ordered packings of equal spheres (3D) or disks (2D), a granular media can appear disordered from a physical point of view: all the possible contacts are not present. If one wants to study the properties of the material in its grain space, the simple geometrical statistical model proposed by Dodds[10] is very useful, because it permits to take into account the disorder of contacts. The base of the model is a representation of a packing of spheres (3D) or disks (2D) by a network in which vertices are the centers of the grains, and bonds join the nearest neighbours: such a network is a completely triangulated one in 2D, and is the dual of the Voronoi tesselation built from the packing. One can show, from the corresponding Euler equation, that the mean coordination number of this network is just 6 in 2D; it is nearly 13.4 in 3D. Starting from this representation, it is easy to recover the mean coordination of the real packing we try to model by a random cutting of the wanted number of bonds. Figure 3 gives a 2D illustration of the model.

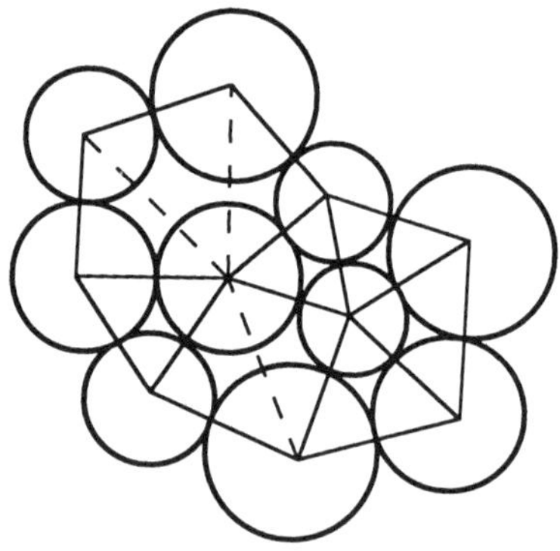

Figure 3 2D packing of non-equal disks and its representation by a network according to Dodds model. The solid lines pass through the real contacts, and the dotted ones correspond to gaps (cut bonds).

In this model, the final diluted network can be considered as random, and one can think to apply percolation ideas to such a system. If one is interested in the transport properties of granular media, an active bond can be replaced by a resistance, whose value depends essentially on the resistance of contact between the two considered grains.[11] A first approach of the mechanical properties of the material can be made by replacing existing bonds by springs or elastic bars,[12] as discussed below.

In fact, real packings can never be really represented by a randomly diluted network. A first explanation comes from the fact that real packings are built under gravity. Then the minimum number of contacts for each grain is imposed from stability criterion: a sphere in a packing of sphere is stabilized only if it has at least three (3D) or two (2D) contacts with its neighbours. The building procedure can lead to more complicated correlations.[13] But, generally, the corrections to randomness are not very important if one wants to understand phenomenologically the physical behaviour of a granular media. So, the determination of the mean coordination number of packings appears to be the first step in the modelling of any granular medium.

3. Determination of the Mean Coordination Number

For an ordered array of equal spheres, the coordination number Z is exactly the number of spheres in contact with a given one. This number is an integer and has a constant value for a given ordered packing ($Z = 6$ for SC; $Z = 8$ for BCC; $Z = 12$ for FCC or HCP). In disordered packings, it is difficult to define precisely the number of real contacts. Here, we define the mean coordination number as the mean number of contacts around one sphere: it may not be an integer. But even this definition leads to different values of the coordination number depending on the way it has been evaluated.

3.1 Review of different studies of the mean coordination number

Historical examples can be found in the early work of Sir d'Arcy Thompson[14] or in the review of Cargill.[15] For example, Smith et al.[16] used a packing of mono-size lead spheres and filled the pore space with acetic acid at first, then pulled out the liquid and counted the number of clean zones around each sphere. A match method has been used by Bernal and Mason[17] and Wadsworth.[18] Several authors[2,19,20] have made consolidated packings of spheres obtained by just assembling hard spheres and then filling the pore space by non viscous epoxy; after polymerization, they extracted sphere after sphere and measured the real position of each one in the space. By using a very old study made by Hale with a compressed sample of deformable peas, Bernal[21] set up a new technique in counting the total number of faces obtained in a packing of compressed spheres. A last approach used the computation of disordered packings with different algorithms (with a central force, inside a periodic box, with relaxation process).[22,23,24,25,26,27]

These different studies provided different results depending upon how the number of contacts per sphere was estimated. The values fluctuate from 6 up to 13.56[2] for a random disordered packing of equal spheres. Values between 6 and 7 correspond, in many cases, to the real number of contacts that we can define as $L = 2R$ with L the distance between the two centers of the spheres of radius R. Gotoh and Finney[28] found a value, equal to 6 at a packing fraction of 0.6366, as the limit of the number of contacts at a distance from the center of a sphere equal to the diameter for the maximal possible disordered packing. A close contact defined as: $2R < L < 2.1R$ gives a coordination number between 7 and 8.5. A local contact $2R < L < 2.2R$ gives a value between 7.7 and 9.3. The upper limit of the contact can be defined by $L \leq 2\sqrt{2}\,R$ and the coordination number obtained in this case goes up to 13.54 by using a Voronoi tesselation.[29,2] Another value (13.4) has been expressed by Dodds.[30] In the general cases of the numerical packings of spheres the value is governed by the packing process; thus in 3D a mean value of 6 was generally found[25] even in packings with grain size distribution.[27]

Some progress has been made using stereological methods[31] combined with image analysis[32,33]. This, and also some assumptions of the structure, in addition to giving the metric properties such as porosity and interfacial area per unit volume make it also possible to give a quantitative account of the connectivity number of the grain space or of the pore space.

3.2 Stereological studies of the mean coordination number

First, Pomeau and Serra[34] derived the average number n_c of real contacts per unit volume from the limit behaviour of the distribution function of the closest distances between the discs cuts in a plane section. Their result was checked by Chermant et al.[35] on experimental FCC regular dense packings. E. Gardner[36] proposed a one dimensional approach which appears to be simpler: she calculated n_c from the distribution of the distances per unit length between two consecutive intercepts (called separators in the following) of spheres by a random line.

3.2.1 2D Studies

Planar cuts of assemblies of spheres are made of assemblies of (non touching) disks of various radii whose centers are the projections of the centers of the spheres. If we use packings of identical spheres, the information obtained in a random metallographic plane can be related to two possible positions of the centers of each sphere which intercept the plane, but by combining all the couples of disks we can eliminate these uncertainties (Fig 4).[37]

The other possibility is to study the line defined as the smallest distance between adjacent disk perimeters ρ which joins the centers of the disks (Fig.(5)).

The calculation of the average number of real contacts then relies on the remark that, when the distance between two circular sections is small, it is likely to corre-

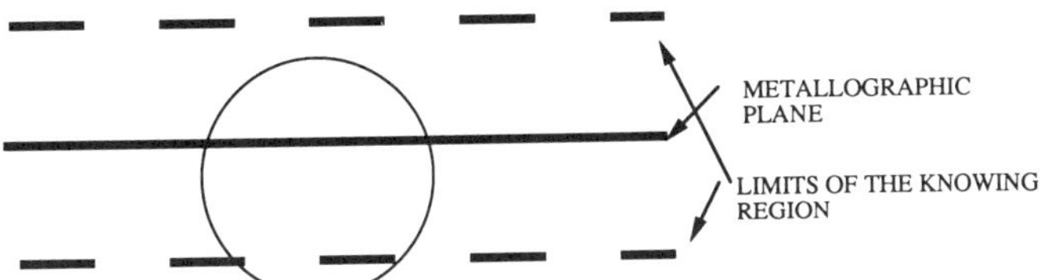

Figure 4 A cut inside 3D spheres packings can give information within a space limited by two planes at a distance equal to the radius of the sphere of the 2D cut.

spond to two spheres in contact. This is rigorously true for ordered packings, as the distances between spheres are exactly known. It is only a limit law for disordered stackings as the distances between spheres varies continuously.

For monosize spheres of radius R, the average number per unit area of neighbouring disks with closest distance smaller than ρ behaves as

$$F(\rho) \cong n_c \sqrt{\pi \rho R}\ \Gamma(7/4)\ \Gamma(9/4) \tag{1}$$

for small ρ, $(\rho \to 0)$. It is then relatively easy to get function $F(\rho)$ for any ρ through image analysis and extrapolate it to the origin. Coordination number Z is related to n_c through

$$\frac{Z}{2} = \frac{n_c}{n_v} \tag{2}$$

where n_v is the average number of spheres per unit volume. Formula (1) has been checked on a FCC array[35] giving, with a good accuracy, the expected value $Z = 12.0 \pm 0.3$. This argument may be extended to binary mixtures, to assemblies of spheres with size distribution, and to *soft* contacts, which is a first step towards slightly sintered packings.

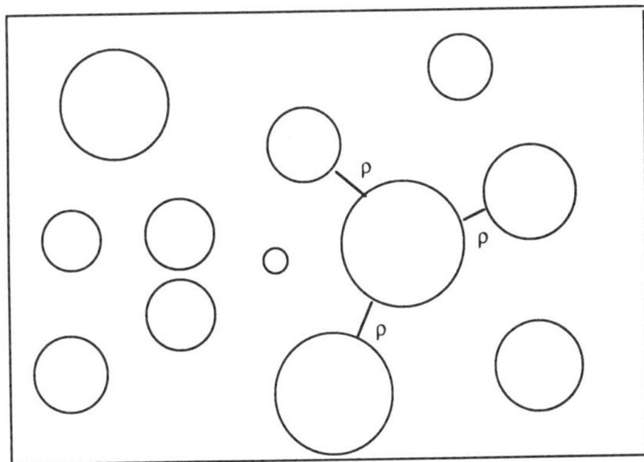

Figure 5 A metallographic cut inside 3D packings of identical spheres ρ is the measurement of the smallest distance between two neighbouring disks.

3.2.2 1D Studies

Oger et al.[38] have revised E. Gardner's calculation for monosize assemblies of spheres. The idea is to consider the intersection of random lines with the spheres of the packing. A line which crosses one sphere determines inside the sphere a chord which is called an *intercept*. The distance between two intercepts on the line is called a *separator*; when the spheres are in contact, we have a *true separator*. This approach may give further information. In particular, it is possible to calculate the complete distribution function for true separators (i.e., belonging to touching spheres) which is a universal function; we have begun to investigate an extension to non touching particles, which amounts to evaluating of the pair correlation function, in the scope of a geometrical description of the intersticial phase. Results are validated on numerical packings of equal spheres and disks.

It is possible to get the number of separators smaller than ω per unit line by the same kind of analysis as before. Let $N_L(\omega)$ (L stands for linear) be this function. For small ω, it coincides with the number $N_L^*(\omega)$ of true separators smaller than ω as, like above, small separators arise from touching spheres. In contrast to the former situation, the complete calculation of $N_L^*(\omega)$ has been performed; it is given by Oger et al.[38] Its main feature is its universality: the reduced function $N_L^*(\omega)/n_c R^2$ is a unique function of the reduced variable $u = \omega/2R$. The behaviour is shown in Fig.(7) for $0 < u < 1$ (full curve).

For small ω, $N_L^*(\omega)$ is linear and

$$N^*L(w) \cong \frac{2\pi}{3} n_c(R\omega) \quad \text{when } \omega \to 0 \tag{3}$$

and n_c (whence, the coordination number Z) is obtained from the slope at the origin. For $w = 2R$, $N_L^*(\omega)$ saturates to its maximum value

$$N_L^*(\omega) = \pi n_c R^2 (1 - \frac{8}{3\pi}) \qquad \omega \geq 2R \tag{4}$$

Incidentally, let us notice that the ratio $h = \frac{N_L^*(2R)}{N_L(2R)}$ where $N_L(2R) = \pi n R^2$ (n is the density number), is smaller than 1; this gives an upper bound for the coordination number: $Z_{max} = 13.23$ which is not far from the similar limits obtained by Coxeter[29] ($Z_{max} = 13.56$) and Dodds[10] ($Z_{max} = 13.40$).

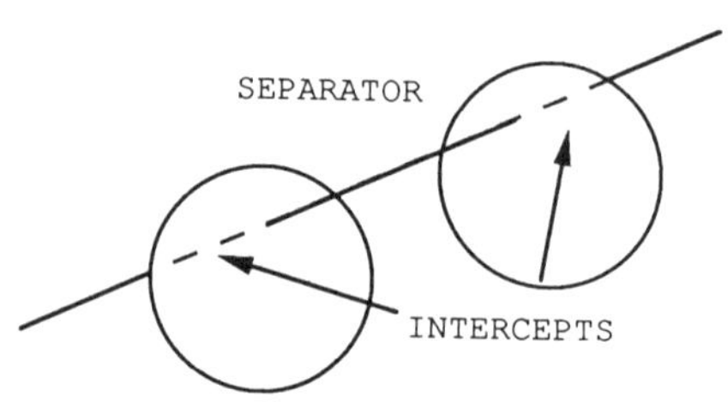

Figure 6 Intersection of a biparticle by a line.

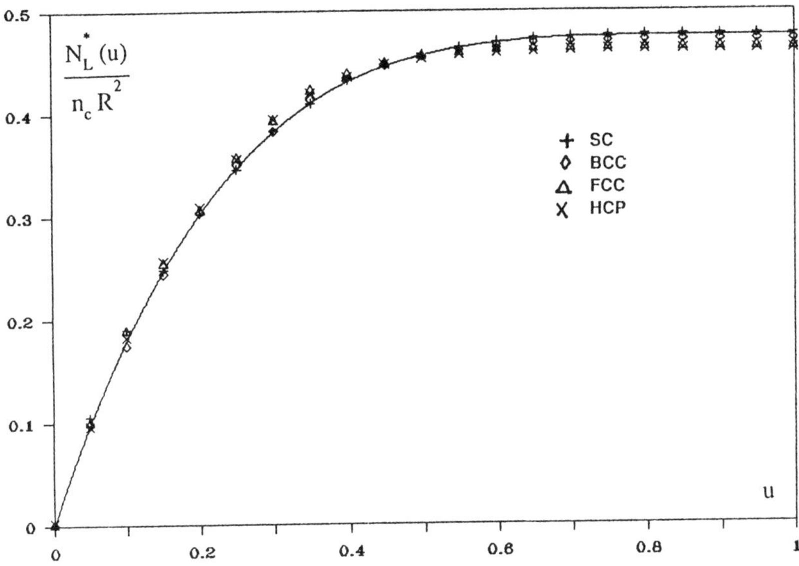

Figure 7 Plot of the reduced function $N_L^*(\omega)/n_c R^2$ vs u. The continuous line represents the theoretical curve and the four kinds of symbols the normalized distribution functions for the four ordered 3D packings (SC,BCC, FCC, HCP).

4. Physical Consequences of Disorder

The transport properties of packings of grains are often described in terms of homogenesation treatments. These methods apply best when the disorder is small and when the heterogeneities are weak as would be the case if the void space was filled with a material having similar properties with grains. This tends to smooth out the effect in the heterogeneities of contacts. The work of Batchelor[39] as well as the recent effective medium treatment of the thermal conductivity of composites by Acrivos[40] are of this type. However for *dry* packings considered in this article, the distribution of contacts is a dominant factor and strongly limits the applicability of the above approaches. In particular, we will see that the non linear mechanical response of a granular array is stronger than what would be given by simply using the non linear response of the local contacts (Hertz's law).

4.1 Central force percolation

The lattice model of Dodds above corresponds to an off-lattice percolation problem where a fraction $q(= 1 - p)$ is removed at random ($q = 1/3$ if the real coordination number of a 2D packing $Z = 4$). Quite generally the fraction q can be varied from zero to one and there is a percolation threshold q_c above which long range connectivity is lost. At this value, the electrical conductance of a packing of conducting grains would vanish. There is however a second threshold for a small values q_r which relates to the mechanical rigidity of the lattice. Let us replace the contact between two grains by an elastic bar. Two grains in contact with a third one may roll on it and, consequently, change the angle between them. Thus the packing can be modeled by bars freely articulated at their nodes, and continuity is not sufficient to insure rigidity of the structure. For equal length bars (corresponding to packings of disks of equal diameters) the continuity threshold is $q_c = 0.65$ whereas the rigidity one is $q_r = 0.34$. The properties of percolation near rigidity threshold (the so-called central force percolation problem) have been well studied recently and turned out, somewhat surprisingly, to correspond to the classical percolation universal ones obtained near q_c. The behavior of a lattice of articulated bars for values of q such as $(q_r < q < q_c)$ has been much less studied. The lattice can undergo large deformations under streching under infinitesimal forces. The relative force elongation δ varies from zero to infinity when q varies in this range. The elongation δ is upper bounded by the length of the minimal continuous path on the lattice whose critical behavior near q_c is known. However additional steric constraints on the lattice will give rigidity for a smaller extension. This behavior has been approached from a monte-Carlo simulation[41] as well as from an analog simulation[42] but the non linear response is mostly to be studied. The description as such should apply best to stretched lattices such as unwoven textiles under tension but we can imagine the opposite problem of freely suspended (buoyant) grains in a liquid and consider the effect of applying a pressure: continuity would be achieved for a lower pressure (also lower packing fraction) than rigidity.

4.2 *Diode* like effect

The description above has indeed omitted an essential factor: packing of grains resists compression not extension (symetrically an unwoven textile resists stretching not compression). This non linearity can be included in the geometry of the active lattice as the mechanical non linearity of the Hertz contact. The problem has been approached on electrical analog simulations using full lattices of Zener diodes with a random distribution of thresholds.[42] It has also been modeled on mechanical models of disks.[43] The essential message of the different approaches can be schematized on the former example: if the individual threshold of a given bond i is a random variable v_i no current will flow across the lattice as long as the minimum of the sum of v_i taken over all possible lattices joining opposite electrodes does not exceed the applied voltage V. This defines a lower critical voltage V_c; the non linear characteristics $I(V)$ for $V \geq V_c$ is controlled by the progressive enrichment of the lattice in active bonds and is thus of geometrical origin. This description is in qualitative agreement with the photoelastic observation of Dantu and Josselin de Jong of parallel plexiglass cylinders under increasing pressure whose recent film[44] presents a striking demonstration. The lattice of stressed cylinders progressively enriches as the number of active contacts increases with decreasing distances. The mechanical experiments in Rennes[45] on arrays of cylinders also support the description and lead to a force-displacement law $F \propto d^m$ with an exponent $m \approx 3$ to 4 – much larger than the value of the Hertz contact ($m = 3/2$). The correspondences between this problem and classical percolation have been stressed[12] and indicate a much broader class of applications for percolation than from the limited examples of weakly connected lattices.

4.3 Rupture of grains

Breaking grains in order to obtain a finely divided powder is an important industrial and agricultural process. It is also energetically a very inefficient one. One can consider the breaking of individual grains submitted to a standard pressure test (compression of a spherical grain between parallel planes gives a maximum tensile stress in the equatorial plane at the center of the grain). Heterogeneities within the grains have been considered in the spirit of the statistical approaches of Roux and Herrmann (this volume) by Van den Born.[46] The pressure test would correspond to the impact of a sphere in a ball mill. When a full assembly of spheres is considered in such a process, the evolution of the population of particles with time as a function of their size has received descriptions[47] in the spirit of the recent treatment by Redner (this volume). A more efficient and more recent class of breakers are roll mills. The packing of grains is fed between two rapidly counter rotating cylinders of parallel horizontal axes and the resulting agglomeration of finely broken particles is easily redispersed. This problem suggests a correspondence with the previous paragraph. Due to the heterogeneities in the packing, the stress are distributed along a subset of lines of stresses along which grains are broken. It is a subject of present interest to study how the distribution of grains sizes, the conditions of feeding the grains as they are compressed is a way to redistribute the lines of stresses among all grains and may lead to a greater efficiency of the process. An additional parameter not yet considered is the fact that the breaking process takes place rapidly in a non equilibrium situation and that comminution can occur by a shock action.

5. Conclusion

We have seen how granular media constitute a priviledged class of heterogeneous materials due to the disorder in the contacts even independently of the distribution of properties of individual grains. The geometrical characteristics of these media, including the contacts can be characterized at a statistical level. The variation of the contact distribution with applied stress has a direct impact on the non linear elastic behavior. It is probably of crucial importance to study the breaking characteristics of unconsolidated packings. Meanwhile, this short review has led us to introduce general characteristics of central force percolation and diode like ones which have a broad range of applicability in non linear mechanical effects related to disorder.

Annexe 1

As we have explained in this article, it is very difficult to find a good homogeneous, controlled and reproducible granular medium. So we have decided to create with a CNRS (PIRMAT) grant a special tool to perform samples made with glass beads in which we can control exactly the size distribution of the grains, the history of the packing, the final compactness (we can sinter the unconsolidated packing), and so on. This apparatus works in Laboratoire Central des Ponts et Chaussées at Orly.

In our case we can work with one or two sizes of grains, thus if we want to obtain one particular porosity we can use several different procedures. We can choose either the ratio of sphere sizes, the percentage of spheres 2 mixed with spheres 1 or the degree of sintering (see Fig.(2)). By combining these three variables, we can increase the number of possible ways of creating packings with the same porosity. Therefore, we have the procedures at hand to fabricate a set of samples with a wide distribution of properties of the packings (pore size distribution, mean coordination number, and so on...).

The apparatus for creating these packings is composed of several pieces: two electromagnetic vibrating distributors have been used to establish a constant flow of each species, by adjusting the ratio we obtain the correct percentage of spheres. The mould, which receives the beads, rotates at a speed fast enough to pack less than one large grain size thickness per revolution. The last step of the process is

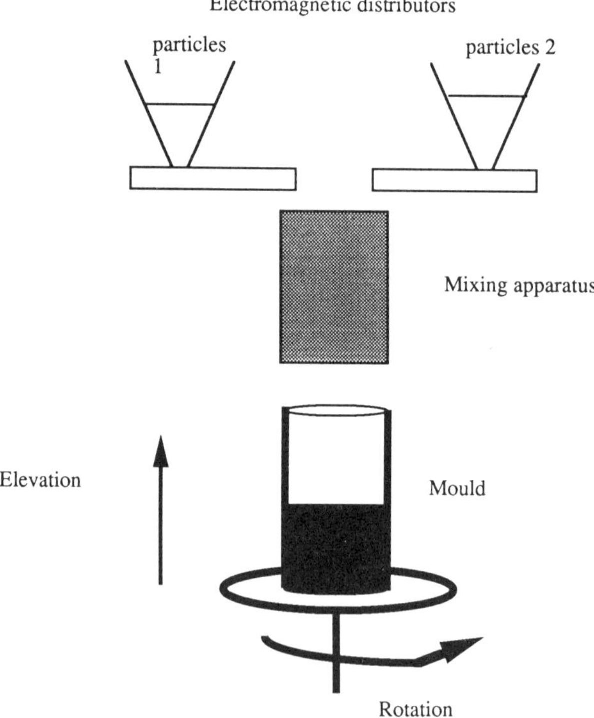

Figure 8 Schematic view of the packing process.

a consolidation of the packing by a heat sintering with particular temperature and time to produce the required consolidated porosity. A typical sintering temperature is 650°C.

Samples for further studies are cored from the sintered consolidated material with a diameter up to 7 cm and a height up to 10 cm. We can perform the measurements of the porosity, the conductivity and the permeability of the pore space on the one hand, and the mechanical strength, the acoustical properties, the coordination number on the other.

Particular samples are available for particular studies: ask Dr C. Gauthier in LCPC Orly.

References

[1] R. Ben Aim, Ph. D. Thesis, Nancy, (1970).

[2] J.D. Bernal, Proc. Roy. Soc. London A **280**, 299, (1964).

[3] P. Chaudhari and D. Turnbull, Science **199**, 11, (1978).

[4] D.R. Nelson and F. Spaepen, Solid State Physics, **42**, 1, (1989).

[5] J.F. Sadoc and R. Mosseri, Pour la science, Janvier, 10, (1985).

[6] E. Guyon, L. Oger and T.J. Plona (1987), J. Phys D, **20**, 1637.
     L. Oger (1987) Ph. D. Thesis , Université de Rennes.

[7] D.E.G. Williams, Private communication, (1988).

[8] J.P. Hulin and D. Salin, in *Disorder and mixing*, E. Guyon, Y. Pomeau and J.P. Nadal eds., Martinus Nijhoff, 89, (1988).

[9] J.C. Bacri , J.P. Bouchaud, A. Georges, E. Guyon, J.P. Hulin, N. Rakotomalala and D. Salin (1989), Proceedings of the 4th EPS conference on the Hydrodynamics of dispersed media, Arcachon, E. Guyon, F. Carmona, A.M. Cazabat and J.P. Hulin eds., (North Holland Publisher).

[10] J.A. Dodds, Nature, **256**, 5514, 187, (1975).

[11] J.P. Troadec and D. Bideau, J. Physique, **42**, 113, (1980).

[12] E. Guyon , S. Roux, A. Hansen, D. Bideau, J.P. Troadec, and H. Crapo, Rep. Progr. Phys. to appear (1989).

[13] S. Roux, D. Stauffer and H.J. Herrmann, J. Physique, **48**, 341, (1987).

[14] D. Thompson, *On Growth and Form*, (J.T. Bonner ed.), Cambridge University Press, (1961).

[15] C.S. Cargill, in *Physics and Chemistry of Porous Media*, D.L.Johnson and P.N. Sen eds., American Institute of Physics, New York, (1984).

[16] W.O. Smith, P.D. Foote and P.F. Busang, Phys. Rev., **36**, 524-530, (1930).

[17] J.D. Bernal and J. Mason, Nature, **188**, 910, (1960).

[18] J. Wadworth, National Res. Lab. Ottawa, Rep MT 41, (1960).

[19] G.D. Scott, Nature, **194**, 936, (1962).

[20] G. Mason and N. Clark, Nature, **207**, 512, (1965).

[21] J.D. Bernal, Nature **183**, 141, (1959).

[22] M.M. Levine and J. Chernick, Nature **208**, 68, (1965).

[23] B.E. Clancy, Brookhaven Nat. Lab. Rep. 997 (T424) Physics TID 4300, (1966).

[24] J.L. Finney, Proc. Roy. Soc. A319, 495, (1970).

[25] C.H. Bennett, J. Appl. Phys. 43, 2727, (1972).

[26] M.J. Powell, Powder Tech., **25**, 45, (1980).

[27] L. Oger, J.P. Troadec, D. Bideau, J.A. Dodds and M.J. Powell, Powd. Techn. **46**, 121 and 133, (1986).

[28] K. Gotoh and J.L. Finney, Nature, **252**, 202, (1974).

[29] H.S.M. Coxeter, Illinois J. Math., **2**, 746, (1958).

[30] J.A. Dodds, J. Colloid and Interf. Sc., **77**, 317, (1980).

[31] E.E. Underwood, *Quantitative stereology*, Addison-Wesley, (1970).

[32] J. Serra, *Image Analysis and Mathematical Morphology*, Academic Press, London, (1982).

[33] M. Coster and J.L. Chermant (1985), *Précis d'analyse d'images*, Ed. du CNRS, Paris.

[34] Y. Pomeau and J. Serra (1985) J. Microsc., **138**, 179.

[35] J.L. Chermant, M. Coster, G. Giraud, J.P. Jernot and H. Robine, J. Physique, **47**, 829, (1986).

[36] E. Gardner, (1985) J. Physique, **10**, 1655.

[37] J.N. Roberts and L.M. Schwartz (1985) Phys Rev B **31**, 5990.

[38] L. Oger, M. Lichtenberg, A. Gervois and E. Guyon (1989) J. of Microscopy,**156**, 65 oct./I;
A. Gervois, M. Lichtenberg, L. Oger and E. Guyon (1989) J. Phys A, **22**, 2119.

[39] G.K. Batchelor (1974) Ann. Rev. Fl. Mech., **6**, 227.

[40] A. Acrivos (1988) Proceedings of the 4th EPS conference on the Hydrodynamics of dispersed media, Arcachon, edited by E. Guyon, F. Carmona, A.M. Cazabat and J.P. Hulin (North Holland Publisher).

[41] B.K. Chakrabati, D. Chowdhury and D. Stauffer, Z. Phys B **62**, 343, (1986).

[42] A. Gilabert, S. Roux and E. Guyon, (1987), J. Physique **48**, 1609.

[43] D. Stauffer, H.J. Herrmann and S. Roux (1987) J. Physique, **48**, 347.

[44] *Au coeur des milieux granulaires: la photoélasticité* (1987) film available at LCPC, 58 Bd Lefebvre, Paris.

[45] T. Travers, M. Ammi, D. Bideau, A. Gervois, J.C. Messager and J.P. Troadec (1987) Europhys. Lett, **4**, 329.
T. Travers, D. Bideau, A. Gervois, J.P. Troadec and J.C. Messager (1986) J. Phys A, **19**, L1033.

[46] L.C. Van den Born (1988) preprint.

[47] A. Austin (1976) Ind. Eng. Chem. Proc. 15,187.

Chapter 15

Cracking and Fracture of Concrete:
Size Effect or Scaling Law?

Paul Acker, Pierre Rossi, and Jean-Michel Torrenti

Laboratoire Central des Ponts et Chaussées
58 Bd Lefebvre, F-75015 Paris, France

1. Introduction: normal state concrete is cracked

In all concrete, reinforced concrete (RC), or even prestressed concrete (PC) structures, concrete is cracked, in its normal state. This (normal) cracking is due both to physical processes (thermal and/or hygral shrink) and mechanical loading.

These two types (physical and mechanical) of cracking develop *at a large range of scale*. Without coating, concrete surface is always cracked more or less deeply. The depth of cracks due to drying is generally low (never more than 15 cm, except for cases of concreting joint) and the crack pattern is very dense, and therefore less open and less visible. In certain cases, however, cracks spacing becomes very large and the crack opening is catastrophic (concreting joint, when shrinkage is restrained by the underlying concrete thick layer). In concrete roads, crack spacing can reach 20 or even 30 m!

In a reinforced concrete beam, RC-bars only contribute to the mechanical strength *after concrete cracking*. In both cases, civil engineers try to limit the crack opening for two reasons: first to preserve the continuity of the structure, and second to avoid the corrosion of the reinforcing steels (it is well known that, below a certain crack opening value, which depends on the environment, RC-bars remain protected against corrosion).

A lot of technological problems and innovative impediments could be reduced with a predictive modelling of cracking, including *explicitly cracks opening*.

2. Cracking of concrete: a multi-scale mechanism

Cracking generally appears simultaneously or successively in several steps depending on the physical (drying, thermal gradient) or the mechanical (flexure, concentrated force) origin of cracking.

Disorder and Fracture, Edited by J. C. Charmet *et al.*
Plenum Press, New York, 1990

Just after demoulding, drying induces intense tensile stresses at the surface, exactly as for a thermal shock. The estimated stresses are larger than the tensile strength and it is clear that concrete skin is cracked. These cracks are generally not visible because of their small depth (a few millimeters) and therefore their small opening. As far as the drying process progresses into the concrete mass, all cracks cannot propagate at the same rate: only a selected number of cracks propagate[1] and the other cracks do not (completely) remain open.[2]

We obtain the self-similar type pattern of cracking described on fig.1. For this reason, crack opening does not progress accordingly to the drying process, but the sum of the openings does (as well as the ratio opening/spacing).

In a concrete beam, RC-bars that support the flexure stresses are placed at the lower part of the beam, at 2 or 3 cm from the surface. Even with relatively low stresses (dead loads, for example), cracks appear, in vertical planes, beginning from the lower face. The spacing between cracks is determined by the transverse bars (*"stirrups"*) or by the "bond-length" of the bars. This bond-length, which is characteristic of the couple concrete-bar, is defined as the distance influenced by the crack. The bond mechanism involves cone-shaped centimetric cracks around the steel bar (see figure 2, from Ref.[3]).

In concrete structures, both mechanisms are generally simultaneous, thus a complete prediction of the crack opening is actually not possible.

3. Modelization of the cracking process of concrete

Because of the different scales for the description of cracks in concrete, there exist various theoretical approaches of the modelization of this phenomenon. We can distinguish three main types :
- linear and non linear fracture mechanics
- macroscopic approach (no discrete crack)
- stochastic approach

As each type of modelization corresponds to a certain scale we will emphasize their range of application and their limits.

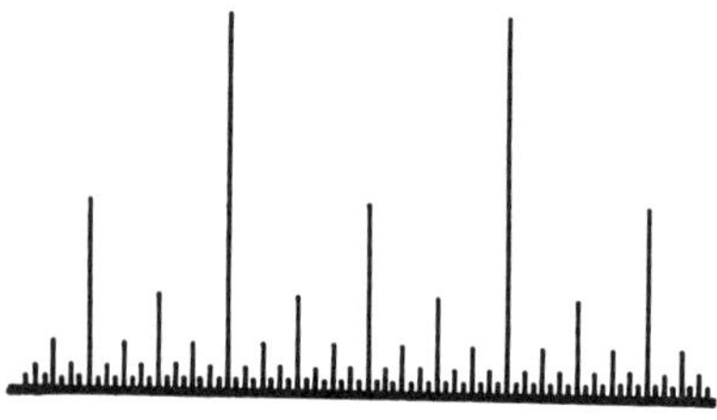

Figure 1 Self-similar pattern of skin concrete cracking.

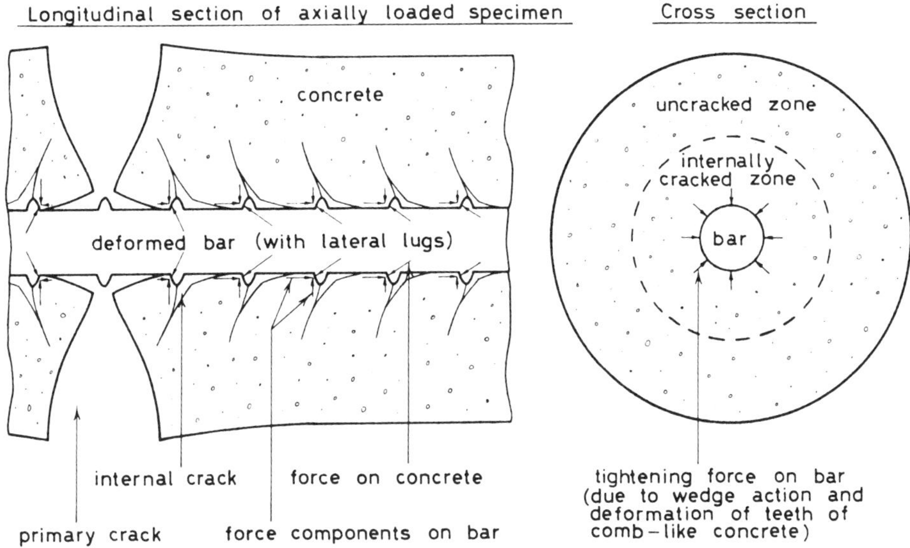

Figure 2 Cracking pattern around a RC-bar proposed by Goto.[3]

3.1 Linear and non-linear fracture mechanics

Fracture mechanics models permit to describe the propagation of one crack in a continuum. The main difference between linear and non linear fracture mechanics is the fact that, for the first one, the medium is always linear elastic when the second considers non linearities at the crack tip. Let us consider now the concepts involved by these models.

3.1.1 Linear elastic fracture mechanics (LEFM)

Consider a material that is linear elastic in infinite transformation and in a plane situation, in which a crack is propagating at a velocity $\dot{\ell}$, in a straight line, along an axis Ox (figure 3).

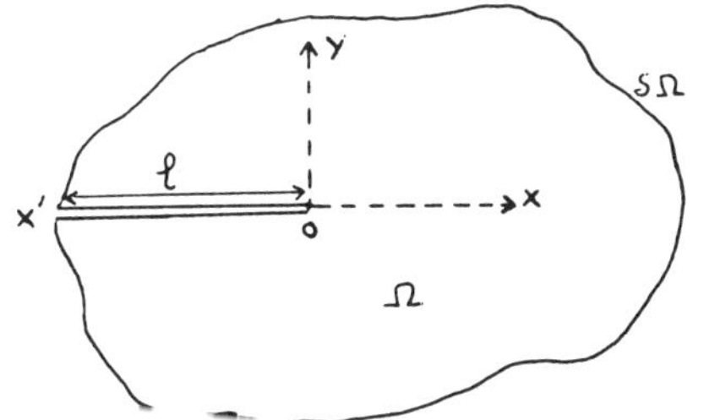

Figure 3 Crack in a linear elastic medium.

Let us assume that we are in a quasi-static situation. In this case, Nguyen[4] has demonstrated that the thermodynamical equilibrium of the system gives the following relation:

$$D_\ell = -(\delta U / \delta \ell)\dot{\ell} \tag{1}$$

in which: U is the total potential energy at time t, D_ℓ is the surface energy dissipated, required for separation of the material into two parts, immediately before cracking. The energy ratio G is introduced:

$$G = (\delta U / \delta \ell) \tag{2}$$

so that $D_\ell = G\dot{\ell}$. This latter expression shows that G is a thermodynamical force which governs the crack propagation. It should be noted that Eq.(1) is based on the assumption that $\dot{\ell}$ is an overall normal-state variable (this a key point for LEFM).

For linear elastic materials the form of the stress field around the crack tip is classical and well known (see figure 1):

$$\sigma_{yy}(x, y = 0) + i\sigma_{xy}(x, y = 0) = \frac{1}{\sqrt{2\pi x}}(K_I + iK_{II}) \tag{3}$$

in which: K_I and K_{II} are intensity factors respectively in tension and shear.

Using Eqs.(2) and (3), we can demonstrate Irwing's formula in plane strain state:

$$G = \frac{(1 - \nu)}{E}(K_I^2 + K_{II}^2) \tag{4}$$

Note that G, K_I and K_{II} depend of the crack length, the geometry of the structure and the boundary conditions.

Griffith's criterion for brittle fracture can be expressed as:

$$\begin{aligned} \text{if} \quad & G < 2\gamma = G_c \quad \text{no propagation} \\ \text{if} \quad & G = 2\gamma = G_c \quad \text{propagation occurs} \end{aligned} \tag{5}$$

in which γ is the critical energy dissipated per unit area. The toughness can be obtained for the tensile mode K_I, using Eqs.(4) and (5):

$$K_{Ic} = \sqrt{\frac{(2\gamma E)}{(1 - \nu)}} \tag{6}$$

In fact the main limit of this theory is that linear elasticity does not describe perfectly the behavior of usual materials (concrete, steel, ceramics) which always present some non-elastic dissipation at the crack tip (dislocations for steel or microcracks for concrete and ceramics).

For these real materials Rice[5] has demonstrated that if the volume dissipation zone is confined (that means very small in front of the linear elastic part) we are allowed to use LEFM even in the case of non linear materials. So when we use LEFM the non elastic area is completely ignored.

For concrete, the non elastic area depends very much of the size of the aggregates.[6] For an ordinary concrete formulation with a maximum size of the aggregate of 20 mm, the length of this non elastic zone is about 10 cm. Thus, as shown by Rossi,[7] the hypothesis of LEFM application is never valid for usual civil engineering structures (due to a *structural scale effect*).

So LEFM is not really useful to describe macrocrack propagation in concrete structures when the length of the crack exceeds a few aggregate size. However, to study microcrack (length shorter than the aggregate size) evolution[8] the answer is not so obvious. If we were able to determine the G_c values for cracks propagation in the cement paste, in the aggregates and in the interfaces between aggregates and cement paste and if the non elastic zones of each microcracks were decoupled, then it would be theoretically possible to describe microcracks propagation by LEFM. Let us however note that it is very difficult to determine experimentally the values of G_c, and that the localization of microcracks cannot be described because of the coupling of the non elastic zones.

To conclude, we can say that LEFM, *due to scale effect*, is not an accurate approach to study the cracking process of concrete.

3.1.2 Non linear fracture mechanics (NLFM)

In that case, at the opposite of LEFM, the non elastic area is taken into account in the mechanical analysis during the crack propagation. Different models exist based on this approach; the Hillergorg's fictitious crack model[9] is the most famous. In this type of models, one considers that what happens in the non elastic area could be compared with the mechanical behavior of a volume of concrete loaded in direct tension. So four main material parameters are introduced in order to describe the behavior of the non elastic area: the Young's modulus E, the tensile strength f_t, the energy dissipation G_f and the slope of descending branch of the force-displacement curve in a direct tensile test (figure 4).

Although this approach is more sophisticated than LEFM, we find again the same limitations *due to the scale effect*: indeed, as it was pointed out before, this model is based on the tensile behavior of concrete and it is now clear that this behavior is strongly *scale dependent*.

As a conclusion, this approach, like LEFM, cannot describe correctly the cracking process of concrete whatever scale one is interested in.

3.2 Macroscopic approach (no real crack described)

Unlike Fracture Mechanics models, the macroscopic approach considers concrete at a scale where cracks do not appear as discontinuities but are taken into account by means of the constitutive relations between stress and strain. The assumption of homogeneity included in this approach implies that only two scales could be described: cement paste where aggregates are considered as macroscopic inclusions like steel reinforcement (numerical concrete of Roelfstra[10]) and concrete including a lot of aggregates (in order to obtain a statistical homogeneity).

Note that in this family of macroscopic models, by introducing some hypothesis concerning the orientation of the microcracks as a function of the maximum principal tensile stress, we are led to an orthotropic damage model, which is more sophisticated than the other macroscopic models (the *band model* of Bažant[11]).

There exist many macroscopic models to describe the behavior of concrete. Since their abilities and their limits are comparable, we only present one of them: the continuum damage model (Mazars[12]).

In this model, damage is described by the decrease of the elastic characteristics:

$$\sigma_{ij} = (1 - D).M_{ijkl}.\varepsilon_{kl} \tag{7}$$

where M is the elastic stiffness tensor and D the damage parameter which grows from 0 (no damage) to 1 (full damage).

In this relation, valid for compression and tension, the strain localization, which exists experimentally, is not explicitly described, but results from a softening behavior in the stress-strain relation. This implies some limitations to the model:

• uniqueness of the solution is not proved

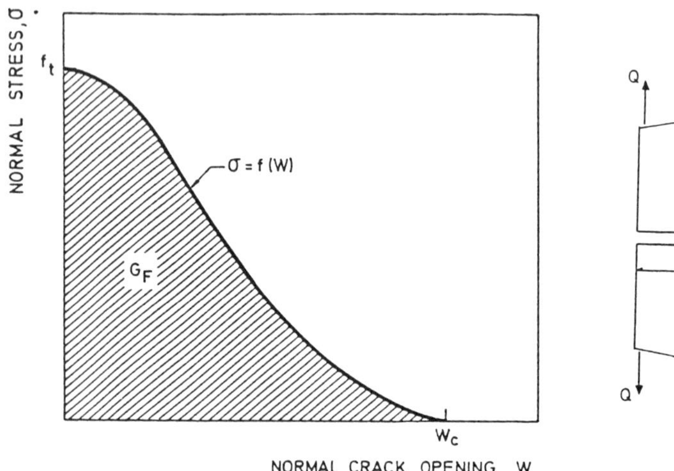
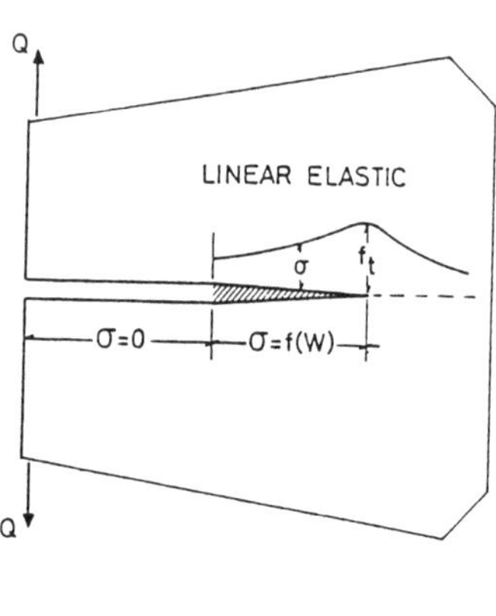

Figure 4 Classical force-displacement curve used in NLFM.

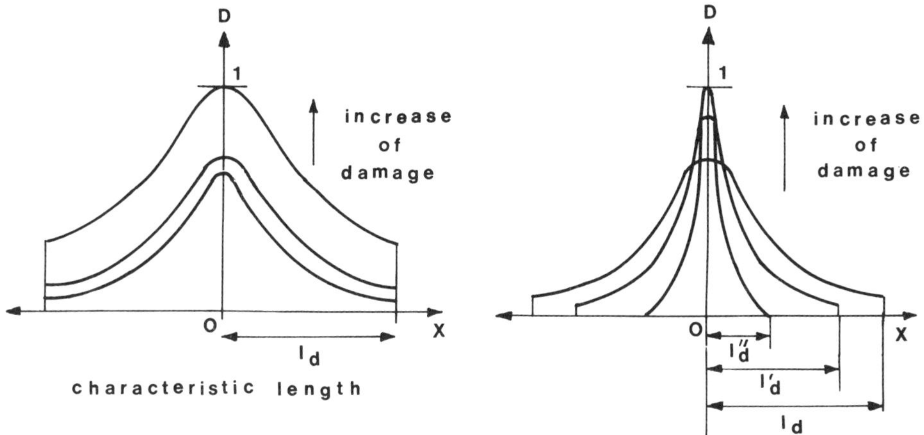

Figure 5 Characteristic length: a) case of the non local damage; b) our physical view.

- the stress-strain relation is scale dependent
- in calculation of structures, using finite element codes, one observes a strong mesh size dependence of the obtained results.

To remedy this last point a non local damage theory was developed. In it, damage is no more a local value but a mean on a certain volume with a characteristic length. In this case mesh dependence is drastically reduced.[13]

In spite of these results, the physical point of view about the concept of characteristic length has to be discussed. If it clearly appears that the damage at a certain point in a structure depends on its surrounding volume, it is not evident that the characteristic length is an intrinsic and constant material property. Indeed, for us, crack formation in a structure is the result of a succession of localizations at different scales which implies a decrease of the characteristic length (figure 5).

To conclude on this damage model, Saouridis[13] shows that it gives good information on the mechanical behavior of a structure and on the cracking process only when the geometry of this structure and the loading conditions lead to a very localized non elastic zone. As an example, to describe the cracks distribution and evolution during the drying shrinkage of a given structure, the damage model is not effective to give the value of the opening of the (main) cracks.

3.3 Stochastic approach

All the approaches presented before are deterministic. Another way to study the cracking process of concrete is to consider that this phenomenon is due mainly to the heterogeneity of the material and to introduce it with a stochastic approach.

Some researchers have proposed different models whose common point is to introduce a distribution function of some mechanical properties of the material. We can classify them in two families:

• those which are only analytical models, giving informations on failure probability of the material[14] or the fracture load of a structure and the moment it occurs[15] or the strength properties which are influenced by loading, temperature, size and age of the specimen.[16]
• those which are implemented in a numerical code and permit to give more informations than the first one family of models, like the different stages of the cracking process of concrete (from diffuse microcracking to localized macrocracks). Among these type of models, we can distinguish those which use a large number of trusts to simulate the heterogeneous material,[17,18] and those which use the finite element method.[19,20]

The main interests of this second family of stochastic models are first to give a realistic information on the complete cracking process of concrete (diffuse microcracking stage, localization of these microcracks to create some macrocracks) and the propagation of these macrocracks and secondly to show[7] that the observed scale effect of the structural behavior can be accounted for by a scaling law of the materials which constitute the structure.

4. Discussion

From the critical analysis of the models presented before two approaches seem to be more efficient than the others: macroscopic models, with a non local approach to take into account the scale effect, and stochastic models. But their respective advantages are related to the industrial problem one needs to solve.

Indeed, civil engineers are very often interested in estimating precisely the safety of their structures compared to the rupture load as well as in an explicit description of the crack pattern. In the two following examples (figures 6a and 6b), performances of the approaches are different.

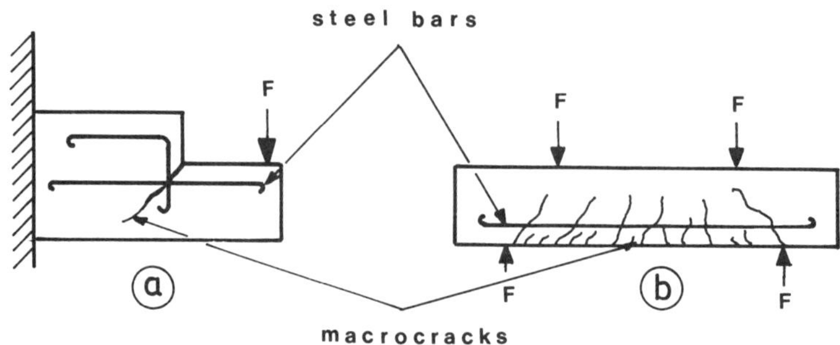

Figure 6 a: Short cantilever beam; b: Four points bending beam.

In the first case, a macroscopic model can give a good prediction of the crack pattern (propagation path and opening of the crack by integration of the local non elastic strains) with a lower cost (computer time) than a stochastic approach. On the other hand, in the second example, the same macroscopic model gives a uniform damaged zone from which we cannot deduce any precise information on the cracks distribution and openings where a stochastic model can give such an information.

5. Conclusion

We have shown that civil engineers need accurate quantitative models for concrete behavior, including cracking, for three types of problems:

- design of complex structures
- durability
- pathology.

Concerning these points, there exist specific models able to take into account cracking. We however believe that these models have to be validated, especially on their ability to reproduce the size effect and the crack distribution (spacing and opening) parameters.

References

[1] Z.P. Bažant, H. Ohtsubo and K. Aoh, *Stability and post-critical growth of a system of cooling or shrinkage cracks*, International Journal of Fracture, **15**, pp. 443, (1979).

[2] P. Acker, *Comportement mécanique du béton: apports de l'approche physico-chimique*, Rapport de Recherche, LCPC, n°152, (1988).

[3] Y. Goto, *Cracks formed in concrete around deformed tension bars*, ACI Journal, **68**, 244, (1971).

[4] Q.S. Nguyen, *Méthodes énergétiques en mécanique de la rupture*, J.M.T.A., **19**, 363, (1980).

[5] J.R. Rice, *Mathematical analysis in the mechanics of fracture*, Liebowitz ed., **2**, 191, (1968).

[6] V.M. Entov and V.I. Yagust, *Experimental investigation of laws governing quasi-static development of macrocracks in concrete*, I 2V AN SSSR Mekhanika, Tverdogo Tela, **10**, 93, (1975).

[7] P. Rossi, *Fissuration du béton: du matériau à la structure. Application de la mécanique linéaire de la rupture*, Thèse de l'E.N.P.C., Rapport de recherche LPC **150**, (1988).

[8] Y.V. Zaitsev and F.H. Wittmann, *Simulation of crack propagation and failure of concrete*, Materials and Structures, **14**, 357, (1981).

[9] A. Hillerborg, M. Modeer and P.E. Petersson, *Analysis of crack formation and crack growth in concrete by means of fracture mechanics and finite elements*, Cement and Concrete Research, **6**, 773, (1976).

[10] P.E. Roelfstra and H. Sadouki, *Total fracture energy of composite materials*, Vorträge der 17. Sitzung des Arbeitskreises Bruchvorgänge, Ed. Deustcher Verband für Materialprüfung, Berlin, 357, (1985).

[11] Z.P. Bažant and B.H. Oh, *Crack band theory for fracture of concrete*, Materials and Structures, **16**, 155, (1983).

[12] J. Mazars, *Application de la mécanique de l'endommagement au comportement non linéaire et à la rupture du béton de structure*, Ph. D. Thesis , Université Paris 6, (1984).

[13] C. Saouridis, *Identification et numérisation objectives des comportements adoucissants: une approche multiéchelle de l'endommagement du béton*, Ph. D. Thesis, Université Paris 6, (1989).

[14] A.S. Jayatilaka, *Fracture of engineering brittle material*, Applied Science Publisher, London, (1979).

[15] L. Sentler, *A stochastic model of concrete strength*, Mémoire AIPC, IABSE Proceedings, 125, (1985).

[16] H. Mihashi and M. Izumi, *A stochastic theory of concrete fracture*, Cement and Concrete Research, **7**, 411, (1977).

[17] N.J. Burt and J.W. Dougill, *Progressive failure in a heterogeneous medium*, J. of Eng. Mech. Div., 365, (1977).

[18] H. Schorn, *Numerical simulation of composite material as concrete*, in *Fracture Toughness and Fracture Energy of Concrete*, F.H. Wittmann ed., Elsevier, (1986).

[19] P.Rossi and S. Richer, *Numerical modelling of concrete cracking based on a stochastic approach*, Materials and Structures, **20**, 334, (1987).

[20] D. Breysse, *A Probabilistic Model for Damage of Concrete Structures*, Brittle Matrix Composite 2, Cedzyna Poland, Elsevier publ., 20, (1988).

Chapter 16.1

The Phenomenology of Earthquake Occurrence

Leon Knopoff

Physics Department and
Institute of Geophysics and Planetary Physics
University of California, Los Angeles, USA

The seismicity of an active earthquake area, that is its space-time-size earthquake history represents an attractive target for the practitioners of modeling of sequences of fracturing events. One of the principal characteristics of earthquake occurrence is clustering in the 5-dimensional space of time, space and size. Before attacking the problems of modeling, which must perforce invoke the physics of deformation of solids, we first outline the properties of the natural target. This paper is intended to be a primer of statistical earthquake phenomenology. We describe not only the space-and-time-average properties, but also some of the clustering phenomenology.

Earthquake occurrence is not uniformly distributed over the surface of the earth. Most of the earthquakes of the world occur in relatively narrow belts or zones located mainly around the margin of the Pacific Ocean, with a secondary zone of frequent occurrence along a broader belt across the southern part of the Eurasian continent. There is also a well-defined narrow belt of earthquakes under the oceans, coinciding more-or-less with the mid-oceanic ridges. Taken together, these zones of major activity delineate the edges of the great tectonic plates and the earthquakes themselves are the result of the rapid release of stored deformational energy at the mutual boundaries between the plates. In general, rapid sliding at plate margins takes place when the plate driving stresses exceed the static friction at the boundary. Thus the seismicity takes the form of slip events which punctuate long intervals of sticking at these boundaries.

Geological information about these earthquake zones indicates that in most cases earthquakes have been occurring on the same zones for millions of years. The very long-term average relative rates of motion of the major plate boundaries are of the order of a few cm/yr, with most of the displacement taking place in the largest earthquakes. Since the displacements in the largest earthquakes are of the order of several meters, the largest earthquakes in any given region only occur once every few centuries. Earthquakes occur on fracture surfaces called faults; the intersection of the faults with the surface of the earth for the largest earthquakes have been

Disorder and Fracture, Edited by J. C. Charmet *et al.*
Plenum Press, New York, 1990

well mapped — at least where they occur on dry land. The two-dimensional geometry of the surface expression of the fault zones has two distinct characteristics. The major earthquake faults, such as those that support the largest earthquakes, have a dendritic character with a number of laterally branching secondary faults; the largest of these branches can themselves support relatively large earthquakes, although these are not as strong as the greatest of the region. On a much smaller scale, the greatest faults themselves appear to have a braided character with many bifurcations and reconnections; the cross-over between these two geometries is at a scale length of about 10 to 15 km[1], a critical number about which we shall have more to say below. Overall, the plate boundaries give the impression of having been highly damaged over the centuries. Clearly in some sense the fractures must heal in a short time in order that deformational stress can be built up once again in the interval between earthquakes. Thus, on the simplest model, the region has the aspect of a complex relaxation oscillator. But the fault zone is a zone of weakness; earthquakes recur on essentially the same geometry, and have been doing so over millions of years. But the healing or suturing or cementing after fracture is only partially complete. Earthquakes recur on the same fault zones over and over. The fracture of new material does not seem to be involved in the process, except insofar as old faults grow and new faults develop.

With regard to the vertical or radial extent of fracturing, most earthquakes are "shallow focus", and occupy only the thinnest of surficial zones. In California, no earthquakes have hypocenters greater than 15 to 20 km, a zone that is a very small fraction of the earth's total radius of approximately 6400 km. Earthquakes in zones where surficial matter is being subducted, that is where the mobile plate material is returned or recycled to the earth's, interior can originate at much greater depths: the greatest depths of earthquakes known, are about 700 km, but intermediate and deep earthquakes are relatively infrequent in comparison with shallow earthquakes. Even the mobile plate thicknesses are greater than the vertical extent of the seismogenic zone for shallow focus earthquakes; plate thicknesses under the oceans are about 100 km or less, and under the continents they may be very much larger in some regions.[2]

Before discussing further aspects of the 4-dimensional space-time distributions of earthquakes, we describe a fifth dimension, namely the sizes of earthquakes. Without going into extensive detail, we remark that the sizes of earthquakes span such a broad range that they are assigned a magnitude which is a logarithmic measure of size, in analogy with stellar magnitudes. Literally, earthquake magnitudes are a logarithmic measure of peak motion on a standard seismograph at a standard distance from an earthquake. Other magnitude scales have been grafted on. Details are given by Richter.[3] Efforts have been made to relate magnitudes to energies of seismic radiation and, by certain more difficult arguments, to fracture energy. A standard formula in common use is[3]

$$E = 10^{\alpha + \beta M} \quad \text{with} \quad \beta \approx 1.5 \tag{1}$$

but there are very large uncertainties in the two coefficients[4], with β ranging from about 1.4 to 2.1. In a given seismic region of the earth, the cumulative number rate-magnitude distribution of earthquakes is

$$N = 10^{a-bM} \tag{2}$$

The exponent b is around 1.0 and, remarkably to lowest order, shows very little variation from seismic region to seismic region. The overwhelming number of estimates of b for tectonic earthquakes fall between about 0.5 and 1.2.[3,5] (There are exceptions for aftershock series, for volcanic earthquakes and other special sequences, but our main concern will be for the overwhelming number of cases which are "normal"). The relative activity of a region is given by the coefficient a, which varies widely from region to region; evidently 10^a is proportional to the size of the region, for a given homogeneous region. If the exponent b does not depend significantly on local geology, *i.e.* on the chemical composition of the near-surface rocks of the earth, whether they be of a sedimentary, igneous or metamorphic nature, then there must be some universal property that determines this exponent. It has been argued that the exponent must be due to the geometry of fracture. Efforts at simulations of fractures can produce the energy-frequency law $N \sim E^{-b/\beta}$ relatively easily, starting from a variety of assumptions. The observational determination of the exponent b/β evidently has large uncertainties.

At the upper end of the energy/magnitude scale, there is considerable uncertainty as to the validity of the logarithmic magnitude-frequency law. Large earthquakes occur so rarely in the time scale of seismic instrumental history, namely over the last 50 to 75 years, that we do not have good estimates of the rates of occurrence of the largest events. Certainly the distribution must be terminated at some upper magnitude threshold. The upper magnitude threshold is in the neighborhood of $M = 9$, using Kanamori's revision of the magnitude scale.

We can argue on geometrical grounds that it is not necessarily true that the standard magnitude-frequency law should apply equally at small and at large magnitudes up to the maximum. The energy released is greater for larger fracture surfaces, simply because a larger fracture surface will release deformational energy stored in a larger volume. But in general a fracture of dimensions less than let us say 10 km will not intersect either boundary to the seismogenic zone, and has, some sort of quasi-circular or oval shape. On the other hand, a larger fracture that grows vertically to intersect these two surfaces, cannot grow in any other direction than horizontally, achieving a more-or-less rectangular aspect. Thus earthquakes smaller than some critical size, have a statistics governed by local inhomogeneities while larger earthquakes are regulated by the large-scale geometrical properties of a broad region. An earthquake with fracture dimensions of about 10 km has a magnitude of 6.0. Thus we may expect the familiar magnitude-frequency law to apply at magnitudes up to $M = 6.0$, if it is to hold in any range. Above $M = 6.0$ in a region such as Southern California, for example, there are insufficient data to justify the application of any statistical formula.

Seismic activity on the fault segment of the San Andreas Fault in Southern California that ruptured in the great earthquake of 1857 ($M > 8$) has been zero at least over the last 55 years down to $M \approx 2$, and may have been quiet for much longer. Similarly the segment of the San Andreas Fault that tore in the 1906 San Francisco earthquake ($M > 8$) has been quiet; it is not certain that the low level activity that has been present in the vicinity has been located on the San Andreas Fault itself, although it is certainly in the neighborhood. There is no doubt that in the latter case, there is a dramatic reduction in seismicity in comparison with the region to the south of the 1906 fault break. Some seismologists postulate that earthquake fault segments will support only a "characteristic earthquake" whose size is determined by the inhomogeneities or barriers at the edges of the fracture surface that stop the fracture itself; no earthquakes that are any larger or any smaller than the characteristic earthquake will occur on a given fault segment and the 450 km long segment of the 1906 fault break will only tear in earthquakes of the size of the 1906 event. On the characteristic earthquake model, small earthquakes do not occur on the faults that tear in great earthquakes. Thus, the magnitude-frequency law in fact describes a series of ruptures on a myriad of faults or faultlets, rather than on a single fault surface, and thereby gives credence to the notion that the neighborhood of a plate boundary is a melange of highly damaged material, although the macroseismicity is dominated by one very large through-going fracture – the plate boundary itself. For reference, the largest earthquakes of the world have fracture sizes of the order of 750 km, with large variance.

In a statistical sense, the magnitude-frequency law, is a long term average in time, and a large scale average in space of the seismicity of a given region. Thus it is the zeroth order moment in space and in time of the distribution. We can construct the first-order temporal moment by calculating the distribution of the time intervals between all possible pairs of earthquakes $(t_i - t_j)$ without regard to their location, in a given seismic region. This distribution is fit very well over the range 2 hours to 10 years by the function $1/t$.[6] This is the Omori[7] law of aftershocks and is the oldest single piece of statistical information about earthquakes in the literature. After any earthquake, often numerous secondary earthquakes take place called aftershocks; observable aftershocks may persist for as long as a number of years after a main shock. Aftershocks will have their own aftershocks *ad infinitum*, and the self-similar character of Omori's law should come as no surprise. The failure of the distribution to be exponential is testimony to the fact that few earthquakes in the catalogues are Poisson independent events. Although the statistical evidence[6] refers to the complete catalog, there are so many more aftershocks than foreshocks, that we may as well assume that foreshocks do not play an important part in *this* portion of our statistical summary. The fact that aftershocks can persist for many years, implies that a viscous rheology for the crust must be an important ingredient to provide the time delays in the stress transfer that triggers the aftershocks. The largest aftershocks occur at the earliest times in the aftershock series. There are some speculations about the precise value of the exponent $1/t^p$, but p lies mainly between 1.0 and 1.4.[8] Thus we may take the probability density of occurrence of an aftershock of a large earthquake to be proportional to $1/t$. Aftershocks play a special role in our phenomenology; not only are they evidence of triggering of one earthquake by another but also, on a broader scale, represent the most obvious manifestation of clustering of seismicity.

Large earthquakes are occasionally preceded by measurable foreshocks. The statistics of foreshocks are difficult to document. But the work of Kagan and Knopoff[6] and Jones and Molnar[9] has shown that an inverse Omori law holds for foreshocks alone, that is the rate of occurrence of foreshocks increases as $(t_0 - t)^{-1}$ where t_0 is the time of the main shock, and the largest foreshocks occur closest in time to the main shock. This law is only a statistical average and cannot be used as a predictor of earthquake times in individual cases because of the extraordinary variability of foreshock occurrence, if foreshocks occur at all.

With regard to the spatial moments, Kagan and Knopoff[10] have shown that the second order moment is self-similar and varies as $1/x$. Although there are some uncertainties in the exponent, it is unity to lowest order; this is also a universal, since the same exponent is found in all the local catalogs of hypocenters studied thus far. Local catalogs are dominated by earthquakes with magnitudes less than, let us say, 6.0. Thus the $1/x$ distribution is descriptive of earthquakes that do not break the surface. Let us suppose that there are two kinds of faults, the one being the geologist's faults, defined as mappable features that represent the places where severe damage has occurred in repeated strong earthquakes over thousands or even millions of years. On the other hand the seismologist observes that small earthquakes today in the crustal interior also cluster spatially around the mapped faults, and calls these features faults as well. Does the seismologist's fault coincide with the geologist's fault? The $1/x$ distribution is not incompatible with a distribution in a plane and hence cannot resolve this problem. To give a solution, Kagan[11] calculated the fourth order moment for Central California, which has perhaps the best recent catalog compilation, and finds the function $1/V$ for the distribution of volumes of tetrahedra formed by all quadruples of hypocenters, even if mislocations of hypocenters and edge effects are taken into account. The significance of the $1/V$ distribution is that it is not $\delta(V)$ and hence that these (small) earthquakes do not occur on planes, nor do they occur on a small number of parallel or gently curving surfaces. It is more likely that the geological and seismological faults have separate populations. It is my own belief that the small events occur on the extension to smaller scales of the dendritic surface faulting associated with large-scale fracturing. Further it is suggestive that the braided character of the large faults on a scale less than 10 km is characteristic of those faults that extend from the surface to the bottom of the seismogenic zone, namely the large and mappable geologic faults. Thus the fault systems represent locations, as remarked, of complex boundaries that appear as a melange of highly damaged material, with the smaller-scale damage strongly influenced by the fractures of large scale. The magnitude-frequency law probably describes well the fractures on the small-scale microfeatures of the dendritic complex, and does not have an obvious bearing on the problems of rupture on the larger faults; the latter appear to have a different geometrical description.

We turn to the observations of space-time coupling of earthquakes. Only a few such relationships have thus far been confirmed statistically. The simplest space-time correlations are those of the relation of aftershocks to the main shocks. Aftershocks of large earthquakes usually cluster around the outer edge of a fault break. This is undoubtedly due to the stress concentration around the edge of a crack subjected to an external shear stress. These stresses relax by the action of the aftershocks. Although the time scale of aftershocks may be long (indeed it is probably scaleless in view of the $1/t$ distribution), and measurable aftershocks may endure for months or even years after the main earthquake, the distance scale is relatively short; even in the largest earthquakes, the distance range of aftershocks from the fault break of the main event appears not to be much more than the thickness of the seismogenic zone in the greatest earthquakes, and is probably much less than this in smaller earthquakes. There are some reports of the temporal growth of the aftershock zone with time after a large earthquake.[12] These reports must be considered to be significant case histories.

Kagan and Knopoff[13] have performed a two-point space-time correlation for large shallow-focus earthquakes worldwide ($M > 8$) and find significant correlations between these extremely large events along a broad range of "velocities" of roughly 250 km/yr. Correlated pairs are found to distances of the order of 2500 km and time intervals of the order of 10 years. There is even a significant number of magnitude 8 "aftershock" of magnitude 8 earthquakes, defined as pairs of events occurring within one year and 250 km of one another.

Keilis-Borok et al.[14] have shown that those earthquakes of intermediate size that also have large numbers of aftershocks, are frequently followed by stronger earthquakes within two to three years, and at distances of the order of 300 km or less. This distance/time ratio is of the order of the scale for $M \geq 8 - M \geq 8$ correlations. In the case of Southern California earthquakes, this correlation seems to have been a predictor of future large earthquakes, since 7 out of 8 strong earthquakes with $M > 6.4$ were preceded by events ($M < 6.3$) with large numbers of aftershocks; there were no false alarms; the number of such local events is too small to draw statistical conclusions, but the success ratio for worldwide earthquakes is remarkably great and is statistically significant.[15]

If aftershocks are a short range stress relaxation in the crustal seismogenic zone, then how is it possible to transmit triggering stresses over distances of the order of several thousand or several hundreds of km in the two cases? How is it possible to transmit stresses over great distances that trigger large earthquakes without triggering other earthquakes, and especially small ones, in the intervening space? Probably the stress transmission does not take place in the seismogenic zone at all but in a low-viscosity layer below this zone; in the case of Southern California earthquakes, this zone would seem to be located in the lower crust between the seismogenic zone in the upper crust and the high viscosity mantle at greater depth. The triggered earthquakes are thus only those that have fractures extending to the bottom of the seismogenic zone, and hence are the largest earthquakes. The mechanism of stress transfer is probably not one of stress diffusion nor is it likely to be fluid migration through interstices in the granular lower crust, but instead is probably one of stress relaxation under the fracture zone of the first of the pair.[16] If this is the case, the apparent "velocity" of coupling between pairs is illusory; instead, in those cases where the second region is already "prepared" for fracture, the later earthquake is triggered contemporaneously by a "slow earthquake" directly beneath the earlier earthquake. The distance scale of coupling is correlated to the size of the triggering event, amplified by the stress wave guide in the lower crust; the time scale corresponds to the relaxation time for (slow earthquake) displacements in the lower crust, which is of the order of the ratio of the viscosity to the shear modulus and is perhaps from 3 to 10 years. If this model is correct, then the correlations are not due to diffusion but instead are due to local stress relaxations which perturb the distant stress field. The reason why some moderate earthquakes serve as triggers for distant events and others do not, may be due to the locations of the moderate events near the bottom of the seismogenic zone where the material is at its strongest[17], but this proposal has not been tested. Kagan[18] has pointed out that there are significant space-time correlations between earthquake pairs in the M=6 range in local catalogs with similar space-time relationships.

Other space-time correlations are not as well supported statistically, but seem frequent enough to form a reasonably large set of case histories. I have already mentioned the observation that aftershock zones seem to grow in space and time. In many cases an interlude of seismic quiescence precedes a future main shock; the region of quiescence often spans the aftershock zone of the future great earthquake; a summary and critique of such observations is given by Habermann[19], Wyss and Habermann[20], Matthews and Reasenberg[21] and Reasenberg and Matthews.[22] Occasionally these interludes of quiescence are themselves preceded by a swarm of earthquakes of intermediate size;[23,24] a swarm is a series of earthquakes that may last for several years, which may be widespread, and which spans the zone of fracture of the later great earthquake. Evison[25] proposes that the sequence of swarm, quiescence and mainshock are causally connected. Occasionally the swarm and the zone of quiescence form a "doughnut" pattern, which is a circular or semicircular pattern of seismicity with the quiescent zone and the future large earthquake at the center of the halo.[26]

In summary, the seismicity of the seismically active parts of the world has a set of codifiable, and occasionally complex space and time patterns. To model this complex phenomenology, we will have to introduce at least 5 major constraints. These are

- Plate tectonics to restore the energy dissipated in earthquakes.
- A wide range of earthquake sizes, with a maximum to the size for any given region. It is likely that there is another cutoff in the frequency distribution below the maximum, with a gap in frequency of occurrence between the two.
- A brittle-ductile rheology to provide for the time delays between clustering events and the abrupt occurrence of fractures.
- Stress redistribution upon fracture to provide the mechanical coupling between earthquake events.
- A complex geometry, to take into account the distribution of earthquakes as well as the fact that the fracture surfaces are of finite size. Because of some of the geometrical features we have outlined, the characteristic earthquake model is probably appropriate: each earthquake fault segment probably supports only an earthquake of a characteristic size. The distribution of earthquakes in space and in time are reflections of the three-dimensional configuration of fault structure.

References

[1] C. Scholtz, *Private communication.*

[2] L. Knopoff, *The thickness of the lithosphere from the dispersion of surface waves,* Geophys. J. Roy. Astron. Soc., **74**, 55, (1983).

[3] C.F. Richter, *Elementary Seismology* , W.H. Freeman and Co., San Francisco, (1958).

[4] C.Y. King and L. Knopoff, *A magnitude-energy relation for large earthquakes,* Bull. Seismol. Soc. Amer., **59**, 269, (1969).

[5] V. Karnik, *Seismicity of the European Area, Vol. 1* , D. Reidel Publ. Co., Dordrecht, (1969).

[6] Y.Y. Kagan and L. Knopoff, *Statistical study of the occurrence of shallow earthquakes,* Geophys. J. Roy. Astron. Soc., **55**, 67, (1978).

[7] F. Omori, *On the aftershocks of earthquake,* J. Coll. Sci. Imper. Univ. Tokyo, **7**, 111, (1894).

[8] T. Utsu, *Aftershocks and earthquake statistics I,* J. Fac. Sci. Hokkaido Univ., Ser. 7, **3**, 129, (1971).

[9] L.M. Jones and P. Molnar, *Some characteristics of foreshocks and their possible relationship to earthquake prediction and premonitory slip on faults,* J. Geophys. Res., **84**, 3596, (1979).

[10] Y.Y. Kagan and L. Knopoff, *Spatial distribution of earthquakes: the two-point correlation function,* Geophys. J. Roy. Astron. Soc., **62**, 303, (1980).
Y.Y. Kagan and L. Knopoff, *Dependence of seismicity on depth,* Bull. Seismol. Soc. Amer., **70**, 1811, (1980).

[11] Y.Y. Kagan, *Spatial distribution of earthquakes: the four-point moment function,* Geophys. J. Roy. Astron. Soc., **67**, 719, (1981).

[12] F. Tajima and H. Kanamori, *Global survey of aftershock area expansion patterns,* Phys. Earth and Planet. Interiors, **40**, 77, (1985).

[13] Y.Y. Kagan and L. Knopoff, *Statistical search for non-random features of the seismicity of strong earthquakes,* Phys. Earth and Planet. Interiors, **12**, 291, (1976).

[14] V.I. Keilis-Borok, L. Knopoff and I.M. Rotvain, *Bursts of aftershocks, long-term precursors of strong earthquakes,* Nature, **283**, 259, (1980).

[15] G.M. Molchan, *Private communication.*

[16] V.C. Li and J.R. Rice, *Crustal deformation in great California earthquake cycles,* J. Geophys. Res., **92**, 11533, (1987).

[17] R.H. Sibson, *Fault zone models, heat flow, and the depth distribution of earthquakes in the continental crust of the United States,* Bull. Seismol. Soc. Amer., **72**, 151, (1982).
R.H. Sibson, *Continental fault structure and the shallow earthquake source,* J. Geol. Soc. Lond., **140**, 741, (1983).

[18] Y.Y. Kagan, *Private communication.*

[19] R.E. Habermann, *Precursory seismic quiescence: Past, present and future,* Pageoph, **126**, 279, (1988).

[20] M. Wyss and R.E. Habermann, *Precursory seismic quiescence,* Pageoph, **126**, 373, (1988).

[21] M.V. Matthews and P.A. Reasenberg, *Statistical methods for investigating quiescence and other temporal seismicity patterns,* Pageoph, **126**, 319, (1988).

[22] P.A. Reasenberg and M.V. Matthews, *Precursory seismic quiescence: a preliminary assessment of the hypothesis*, Pageoph, **126**, 357, (1988).

[23] M. Ohtake, *Search for precursors of the 1974 Izu-Hanto-oki earthquake, Japan*, Pageoph, **114**, 1083, (1976).

[24] F.F. Evison, *The precursory earthquake swarm*, Phys. Earth and Planet. Interiors, **15** , 19, (1977).

[25] F.F. Evison, *Fluctuations in seismicity before major earthquakes*, Nature, **266**, 710, (1977).

[26] K. Mogi, *Earthquake Prediction*, Academic Press, Tokyo, (1985).

Chapter 16.2

The Modeling of Earthquake Occurrence

Leon Knopoff

Physics Department and
Institute of Geophysics and Planetary Physics
University of California, Los Angeles, USA

1. Introduction

Recent developments in nonlinear science give impetus to an attack on the problems of developing physical models for seismicity. The irreversibility of earthquake events is simple enough testimony to support the proposal that the problems of earthquake occurrence are imbedded in the broader range of problems of nonlinear science. The nonlinearity is more profound than this: the deformation of the earth in the time and space between earthquake rupture events is a complex process, with strong accelerated deformation prior to ultimate fracture; the earthquake events themselves are rapidly running changes of state taking place on a much shorter time scale than that of the interludes; and the occurrence of each earthquake influences the times and locations of the others. The simulation targets of modeling endeavors are the rich phenomenology of statistical and case history seismicity which I have described in the preceding chapter; in most cases, we have information about the events themselves but relatively little information about the deformation in the intervals between the earthquakes. This is unfortunate for the purposes of mathematics, since it would seem, at least superficially, that one should focus on the state of stress as a continuous variable, rather than on the episodes of rapidly running changes of state that are the earthquake events themselves. A start on a theoretical study of the state of deformation as a stochastic process has been made by Knopoff[1] and Lomnitz-Adler[2], but few concrete results in this direction are available to date. So we turn our attention to the problems of a theory of earthquake occurrence with the hope that we can simulate well this part of the total history of deformation.

The nonlinear theory of seismicity takes its cue from already well–developed and well–known progress in nonlinear science which has appeared with other applications in mind. Nonlinear models of seismicity can be divided into the two usual classifications, namely those with small numbers and those with many degrees of freedom. The consequences of the analysis of models of seismicity with small numbers of d.o.f.'s are those that have also arisen in other fields, namely several scenarios

Disorder and Fracture, Edited by J. C. Charmet *et al.*
Plenum Press, New York, 1990

leading to chaos including bifurcations, as well as the occurrence of deterministic chaos with its own features of fractality, strange attractors, intermittency, etc. The many d.o.f. problems have solutions that yield zeroth order space–time average moments, *i.e.* the equilibrium equations of state, phase transitions and their associated properties such as fractal geometry at/near critical points, and a variety of nonuniform clustering phenomena. I shall only be able to sketch some examples of simulations of both types from a rather fuller catalog that we and others have compiled.

2. Constraints

I have already indicated the geophysical constraints on the problem of modeling in an earlier chapter, but I summarize them here as well as the results of a small number of relevant laboratory experiments on the rheology of rocks under conditions appropriate to earthquake occurrence. First, the models must include a constraint drawn from plate tectonics. The surface of the earth is covered by a small number of very large, relatively rigid plates that are in motion relative to each other at rates of the order of cm/year. Earthquakes are non–uniformly distributed over the surface of the earth and are found near the boundaries between these plates. The plates themselves form the upper boundary layer of a large scale system of thermally driven convection. The largest earthquakes have taken place repeatedly on the same plate boundaries for millions of years in many cases. Thus earthquakes are suddenly occurring fractures in the prestressed material of the plate boundary region; the latter region has the attributes of an incompletely sutured, already damaged material. The energy dissipated in seismic events is restored from the plate tectonics reservoir.

Second, the energies released in earthquakes cover a wide range of scales. A logarithmic measure is invoked to accommodate the wide range of energies; each magnitude unit corresponds to about a 60-fold ratio of energies. The largest earthquakes of the world all seem to have about the same size — colloquially described as magnitude 8 events. But even earthquakes with magnitudes 6 to 6-1/2 are potentially destructive and can cause significant loss of life, as for example the Friuli, Italy earthquake ($M = 6.0$, May 6, 1976). In some regions of the world, the largest earthquakes that can occur are no larger than magnitude 6 to 6-1/2 events. In any case, there is a maximum magnitude (or energy) for earthquakes in a given region.

The horizontal dimension of the fracture surface of a magnitude 8 earthquake is roughly 500 km, more or less; that of a magnitude 6 earthquake is about 10 km. Because of the logarithmic scaling and other arguments, the distributions of almost all properties of earthquakes are self–similar — call them fractals if you wish — and thus gives credence to the statistical scaling arguments. The distributions are truncated at the large–size end. The symmetry breaking by the largest events influences the stresses on the soon–to–occur smaller events and influences the times of occurrence of these events; in turn the stress redistribution by the abundant smaller events produces significant fluctuations in the times and perhaps the sizes of the largest events as well. I have called the distribution of magnitudes (or energies or dimensions) the zeroth order space–time statistical moment of the distribution, notably because it represents both a long–time and a large–space average; it is probably not justifiable to localize the distribution for a real seismic region too much, because of the danger of paucity of data in a reduced space–time region, but also because we run the risk of studying spatial domains that are smaller than the sizes of the largest fractures. There are also unanswered questions of stationarity of earthquake time series even on the time scale of a half–century, which is the time scale of our longest and best catalogs.

It is by no means certain that the frequency distribution of earthquake magnitudes follows the same simple rule that spans the full range from the largest earthquakes down to all smaller scales. There is some support for the point of view that there may be another cutoff in size below the maximum, with a gap in the frequency of occurrence between the two. Thus there may be at least two size scales; the lower one is probably of the order of 10 km and the other one much larger than this.

Third, there are at least three time scales. The longest is that of plate tectonics which is the time scale for recurrence of the largest earthquake considered as a simple relaxation oscillator. If the only earthquakes in a given region were just the largest ones, then the energy dissipated would be regenerated after a time of the order of 100-500 years in active zones. The process may not be periodic because of interactions of the stress fields on all distance scales, even from seismic region to seismic region. In fact, the solution to the problem of the periodicity or otherwise of the largest earthquakes is one of the goals for modeling studies, among others. At the other extreme of time scale, the fracture process itself can be considered to be an instability, *i.e.* a rapidly running transition from one state of deformation to another. This time scale is of the order of the linear dimension of a crack divided by the elastic wave velocity and is of the order of a few minutes for the largest earthquakes. The decrease in elastic deformational energy in the neighborhood of an earthquake rupture goes into the energies of radiated seismic wave motions and of frictional heating.

It is all too easy to find coupling between earthquake events. Aftershocks — which are themselves genuine earthquakes — are events triggered by a larger predecessor earthquake, and occur over intervals spanning days to months, and even longer, after the triggering event. The parent earthquake generates a redistribution of the stress field which takes place elastically, *i.e.* on the short time scale. Elastic redistribution of stresses could, in principle, trigger earthquakes at remote locations, but that would imply that coupling should take place on an unacceptably short time scale. Thus aftershocks as well as other clustering phenomena imply the existence of a non–elastic coupling for stress redistribution, which gives the intermediate time scale; we ascribe the cause of the intermediate time scale as being due to creep or viscosity.

The experimental properties of the creep are as follows: During the later stages of the deformation of a rock subjected to a large shear stress, the rock will undergo accelerated creep that terminates in rupture, if the confining pressure and temperature are not too large. The reaction rates are much influenced by the presence of interstitial water. The time to fracture depends on $\exp(-\sigma V/kT)$ as shown by Griggs[3] and many others, where σ is the difference between the load shear stress and the frictional shear stress. This observation is consistent with that notion that deformation is an activation process and suggests that chemical kinetics is a relevant process for earthquake modeling. Indeed, microscopic examination of the rock fabric near rupture, shows evidence for crystallographic alteration in the presence of the shear stress. Solution, transport and alteration are all processes dominated by chemical kinetics.

If a preexisting sharp crack is subjected to a large external stress, the crack grows with a velocity governed by $\exp(\frac{\sigma\sqrt{L}}{kT} \times \text{const})$,[4] where L is the length of the crack. This relationship is once again evidence for the importance of chemical processes, especially so, since the strengths of rocks near crack tips, and elsewhere, are much influenced by the presence of water according to Griggs and Blacic[5] and many others. Sufficient water is present in the earth's crust not only to make accelerated creep a significant factor in the modeling, but also to influence the (coefficient of) friction on a soon–to–fracture fault. Accelerated, subsonic creep of this type in metals is called stress corrosion in the engineering literature.

A fourth constraint involves the elastic redistribution of stress. The introduction of a fracture into a medium lowers the stress on and near the walls of the crack and raises it dramatically over a short distance beyond the edges of the crack. This reconfiguration of the stress field of course represents a transition to a lower energy state; the difference of the energies in the two states is the energy released in the event. The significant readjustment of the stress field effectively inhomogenizes the stress field. It also has the potential of changing the stress field on nearby regions such that they are either taken closer to or farther from their own condition of criticality for fracture. The time at which fracture occurs on a region that has been elevated to criticality by elastic processes, will of course be delayed by the influence of creep as described above. Thus the system of fractures and elastic redistributions gives the impression of an interactive system that has the imprint of developing a self–organization because of the non–linear interaction among the inhomogeneous elements. We can thus propose that the primary cause of intermediate–time scale clustering are the elastic redistributions while the actual time–scale is set by the creep processes.

A final constraint on our modeling is the influence of geometry. A common perception of earthquake faults is that they are planar surfaces. But since nothing in the physical world is a plane, even in the world of classical physics, there must be some influence of non–planar geometry on the kinds of models that we hope to construct. It is my belief that an appropriate geometry is one of a highly complex system of blocks of crustal matter distributed over a wide range of scales. It is the continual shifting of blocks of already–fractured material that we must describe. On a distance scale less than that of the largest fractures, Kagan and Knopoff[6] and Kagan[7] have found that the geometry of earthquake fault zones is three-dimensional with a fractal distribution of sizes. The distribution of plate boundary features on a scale larger than the scale size of the largest earthquakes will not concern us directly in the present discussion. What the fractality implies is that a fault zone is not a plane, but is rather a melange of barriers to extended rupture for cracks, or potential cracks, of a variety of scale sizes. It is probably the case that small earthquakes, which we assume take place on the smaller cracks of the fractal melange, do not occur on what geologists have identified as the fault strand that will ultimately break in a large earthquake. One way to model this is to assume that the main fault is stronger than the surrounding region for a reason that invokes a process other than those we have discussed to now. Another way to do it is to assume that the mean strengths of all faults are about the same but that the statistical dispersion of these strengths is very much smaller on the main fault than on the region surrounding it. There are seismological reasons why the former model is untenable; I shall describe below the consequences of a simulation of the latter model.

3. Bifurcations

One model of seismicity with small numbers of degrees of freedom, drawn from population dynamics has been considered by Newman and Knopoff.[8-10] Most of the constraints listed above have been incorporated, with the exception of the spatial constraint. In this model, no constraint is imposed that specifies the thickness of the seismogenic zone or the properties of the viscous layer below it; the earth is considered to be homogeneous without internal structure. The details of fractality of structure in the (computer) region are also not taken into account. Of course it is the latter aspect of the spatial constraints that gives the full problem its apparent large–degrees–of–freedom character. We focus on the problems of the interaction of different scale sizes of cracks; each crack can undergo fusions with other cracks of any size after an appropriate time–delay due to creep. The fused cracks now appear after the aforementioned time delay as cracks of larger size. A hierarchical structure can be set up so that small cracks disappear from the system and ultimately, after many episodes of fusion, appear in the system as the largest cracks permissible. The smallest cracks are restored to the system through the influence of the driving force of plate tectonics. We find that the solutions either tend to stability in the sense of a uniform production rate of earthquake events, or to a periodicity via a simple Hopf bifurcation. In the latter case, the rates of production of large and small earthquakes show a periodic fluctuation but the two populations do not crest at the same time. The results in many respects are similar to those of predator–prey models and indeed the details of the models, in which small cracks fuse to form large cracks have strong similarities. The bifurcation arises if the rate of loading, *i.e.* the injection of new cracks into the system, is allowed to depend on the state of strength of the fault region; if the rate of injection is independent of the state of strength, the system is always stable in the above sense. The presence of a bifurcation that depends on the state of strength, *i.e.* whether large earthquakes have recently taken place or not, shows that the time scales are strongly influenced by the rate of suturing which is itself the rate at which strength is restored to the fractured region after a macrorupture has taken place. The bifurcation threshold is a curve that involves a tradeoff between the time delays for fusion and the rate of injection of small cracks in the strength–dependent case.

4. Quiescence

I turn to the synthesis of the above ingredients into an attempt at simulation of a selected part of the earthquake phenomenology, using a model with high–order dimensionality. There are not infrequent observations of extended periods of quiescence before an impending large earthquake; an example,[11] is given by the regional seismic quiescence near the Izu Peninsula (Japan) for several years before a large $M = 7.8$ earthquake in 1978, Fig.(1a). To model this, Yamashita and Knopoff[12] assume that a suite of pre–existing cracks exists having a fractal size distribution. For mathematical simplicity, the cracks are distributed along multiple parallel planes imbedded in an elastically deformable medium. The cracks are actually two–dimensional strips imbedded in a three–dimensional homogeneous elastic medium, or in terms of their projections on an orthogonal plane, the cracks form a series of lines in two–dimensional space. The spaces between the lines are selected from the same fractal population as the gaps on each line. At $t = 0$ a sudden anti–plane stress σ_0 is imposed on the region. The cracks begin to grow under the influence of the experimentally determined velocity law $v = v_0 \exp(K/K_0)$ (we actually use $v = v_0(K/K_0)^n$ with n a large number) where K is the stress intensity factor, and is the coefficient in the rate of fall–off of stress σ with distance x from the edge of a crack; in the neighborhood of a crack tip, $\sigma \propto \sigma_0(L/x)^{1/2}$ where

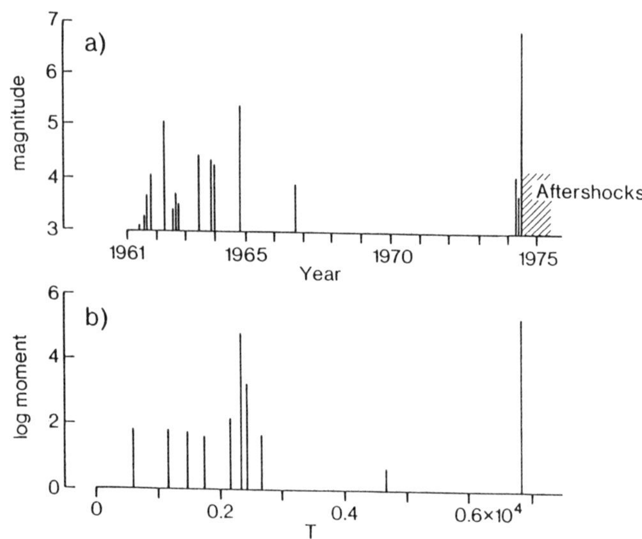

Figure 1 a) Seismicity in the Izu Peninsula, Japan, 1961-1975 (after Ohtake[11]) b) Sample seismicity sequence from simulation of crack fusion dynamics. The smallest events in the swarm episode are not plotted.

L is some length, so $K \propto \sigma_0 L^{1/2}$. The stress intensity factor at each crack tip is calculated by a pairwise approximation, and the cracks are allowed to grow. Because of the antiplane stress orientation, the cracks can only grow in their own planes; they cannot bend. The problem of crack growth and fusion under plane strain or under tensile stresses for a non–planar crack population is beyond our computational power at this time.

As the cracks grow larger, the gaps between them decrease in size, the stresses in the gaps increase, and in turn the rates of disappearance of the gaps accelerate. As long as the growth rates are slow compared to elastic wave velocities, the growth can be calculated quasistatically. At some point the velocity of growth approaches the sonic (elastic) velocity and the crack begins to radiate; this instant corresponds to the onset of an earthquake. We assume that the earthquake is instantaneous; *i.e.* we avoid solving the dynamical radiation problem and assume that the transition to a lower energy state by fusion of two adjoining cracks takes place on a time scale that is very short compared with the time scale of crack growth due to fusion or creep. The stresses in the redistributed crack system are now recalculated and the process continues until all cracks have fused into an array of master cracks that extend across the entire space. Frequently sonic (instantaneous) fusion extends across several adjoining cracks before resumption of the quasistatic phase of crack growth.

If we have only one plane of cracks, we find that the smallest gaps disappear first with the release of small amounts of deformational energy stored in the gaps and their neighborhood. Fusion with larger cracks or through larger gaps releases larger amounts of energy, and these events occur later in the history; seismologists refer to this shift in the energy spectrum as a lowering of the b–value prior to the mainshock. In fact most of the fusions come just before the formation of the master event and a pattern of number of emission events as a function of time appears that is not unlike a foreshock sequence.[13]

If we have several planes of cracks, then events of intermediate size occur over the entire array, each with its own sequence of foreshocks. Although most of these events do not involve creation of cracks that break completely across the array, sooner or later one of the planes of cracks must fracture by successive fusions completely across its own plane before the others. At this instant the cracks on the planes that are broadside to the through–going crack are cast in a reduced–stress shadow. The reduced stress lowers the rate of growth and fusion is effectively halted on the other crack planes. But inevitably, they must resume their growth, even under the reduced stress, and ultimately the velocities again approach sonic values at some later time. The period of quiescence terminates and finally a through–going fracture occurs on another plane. This event — also preceded by its own foreshocks — is stronger than any of the preceding events, because the second through–going fracture occurs on a plane that is stronger than the first. In the example, the similarity of the output from the simulation of our model, Fig.1b), to the episode of swarm–followed–by–quiescence that preceded the large Izu earthquake is remarkable; this is only one example out of many computer runs that the simulations have produced. The durations of the intervals of quiescence and the magnitude of the large shock are highly variable and depend strongly on the details of the choices of the initial locations of the cracks.

In the model the stresses are redistributed following a rupture event and these stresses remain localized essentially without change unless healing rates are rapid. The model requires that they are not. However healing rates are not well-determined experimentally.

This model has the features of a diffusion–limited aggregation model. We conclude that it is the largest cracks that influence the stress distribution of the other nearby crack arrays. The occurrence of a major earthquake effectively sets the stress fields of the nearby smaller ones to zero. In addition we note that the precursory episodes leading to the main shock involve stress fields that are extremely inhomogeneous. Models that do not have the feature of inhomogeneity are likely to be unsuccessful, and in my opinion, the inhomogeneity should be three–dimensional. We do not, as yet, know how to draw statistical conclusions from the simulations of fluctuations in these many d.o.f. problems.

5. Locked Faults

I give a second example of a many d.o.f. problem that is designed to simulate the problem of a locked fault. To elaborate on the relevant phenomenology, it has been noted that often — if not invariably — seismic activity is absent for many years on a fault segment that has ruptured in a great earthquake. This observation would seem to run contrary to the usual assumption that the broad distribution of earthquake sizes ranging over many orders of magnitude over a large range of time and space is an almost stationary process, should also apply to the faults that support the largest earthquakes. In other words, on the usual model, one would expect to see many small earthquakes on a fault that supports a very large earthquake, as anywhere else in a seismically active region. But the observations contradict this assumption. For example, from 1933, the date of the beginning of the instrumental era in Southern California, until recently there have been no earthquakes with magnitudes greater than 3 anywhere on the segment of the San Andreas Fault, California that broke in the 1857 Fort Tejon earthquake, considered to be the largest earthquake in history in Southern California. Whether or not earthquakes with magnitudes smaller than 3 occur at the present time on the fault segment that broke in 1857 is arguable. Yet there are many small earthquakes nearby. It is easier to assume that the nearby small earthquakes occur on their own nearby set of faults, and the largest earthquake and its repetitions take place on a separate fault, and that the two are exclusive sets.

Consider a two–dimensional array of lattice points with variable breaking strength at each site. Under the stress–corrosion model the variable breaking strength corresponds to a variable critical stress intensity factor. The variation in strength in 2-D can be considered to be a projection in some sense of the geomet-rical inhomogeneities in the fault system. Let the mean distribution of strengths be the same over the entire region but let the ratio of the rms dispersion of healing strengths to mean strength be high in one region and low in another. Let the applied stress at each lattice point be a scalar quantity that is inhomogeneous due to redistribution from neighboring ruptures, and further let the stress increase at a uniform rate at all sites, during the time intervals between rupture/fusion events. A fracture is initiated at a lattice site after a suitable time delay after the critical state is reached. The fracture grows by the usual crack growth process under the influence of stress corrosion, and stops when a site is encountered at which the stress difference between strength and applied stress is greater than the redistributed stress due to the fracture. The redistributed stresses are short–range terms. In the simulation, we find that the overwhelming majority of events occur in the high dispersion regions and that these have an energy (or moment) distribution that is a power law if the site strengths are fractally distributed. Because of the high statistical dispersion of strengths in this region, a growing crack has a good chance of stopping within a short distance of the point of initiation; in other words, the magnitudes in the high dispersion region are influenced by statistical effects and we can adjust these — via the dispersion — so that the crack almost never ruptures all of the way across the high dispersion array.

On the other hand, once rupture is initiated in the low dispersion region, the crack grows dramatically. Since the dispersion is small, the dynamical stress at the crack tip is sufficient to overcome the small fluctuations in strength in the path of the crack and the crack grows in a single event completely across the low dispersion array.

Thus, in the high dispersion region, we have many smaller events with a wide range of energies, while in the low dispersion region, we have no small events and only — in this case — periodic catastrophic events. The success of this model depends on the assumption that healing, that is the restoration of bond strength, takes place very rapidly after fracture; this is in contrast to the model of quiescence in which we have assumed that restoration or healing occurs slowly relative to the time–scale of the sequence.

In this model, the extended low–dispersion region supports only large earthquakes and no smaller ones. The high–dispersion region supports a number of small and intermediate sized earthquakes only, but no large ones. This is what we set out to reproduce. But the unsatisfactory aspects of the model are evident. In the two–dimensional system I have described, the extended low–dispersion region interacts almost insignificantly with the region(s) with the abundant smaller earthquakes because of the unfavorable two–dimensional geometry. Thus there are no premonitors or precursory clustering before the large events. As in the example of quiescence that was used in the preceding section, premonitory clustering can be expected to be a result of three–dimensional models of this genre or more complex two–dimensional models, but we have not yet done these numerical experiments.

The model is much oversimplified, especially because of its two–dimensionality as well as the scalar nature of the interactions. But I doubt that the broad conclusions will be much altered if these latter considerations will be taken into account.

6. Self-Organized Criticality (SOC)

The model of SOC is a many d.o.f. model that has received much attention in recent months.[14−17] In this model it is supposed that the stress distribution after many fracture events approaches some stable spatial distribution. In this case, the zeroth order moment of the distribution, namely the stable distribution of crack sizes can be determined, under the assumption that the stress distribution upon fracture is a critical and sensitive relation that has the form of a conservation law. The particular form of this relation in the model of Bak et al., is that the stress drop at a fractured lattice point is 4 units in 2-D and the increase in stress at the edge of the crack is 1 unit, *i.e.* at all bonds connecting broken and unbroken lattice sites. The barrier strength is also 1 unit everywhere. Fluctuations are introduced by a having a spatially nonuniform loading rate, which is taken to be random at each lattice site, but with spatially uniform average. Fracture at one lattice site may trigger fractures at adjacent lattice sites. The solution to the problem displays most of the features of critical point phenomena, and yields a distribution that is a power law extending over a "logarithmically" infinite range of sizes.

The SOC model has been proposed as a paradigm for avalanches as well as earthquakes in the geophysical arena. In these cases, at least, the probability that a crack grow beyond a barrier, or that an avalanche move past a protruding rock, depends on the relative sizes of the avalanche and the obstacle. In the case of growing fractures, the stress at the edge of the crack is a function of the size of the crack as we have already seen. Thus we can consider a modification of the above model in which the stress transferred to the edge is now $1 + J_0 N^\nu$, where N is the number of lattice sites already ruptured in the current episode, and ν is a positive constant, which we take to be $1/4$ for the case of 2-D cracks imbedded in a 3-D elastic medium. In our simulations,[18] we have taken as the fracture criterion the more familiar condition that the probability that an unbroken site at the edge of a crack fracture, varies as $\tanh(1 + J_0 N^\nu)$; we can thus insert a spectrum of sizes of barrier breaking strengths. In this case we find a finite cutoff in the spectrum of sizes; for magnitudes very much less than the cutoff magnitude, the usual power law relationship holds. Beyond the cutoff there is a spike in the distribution that corresponds to runaway; in other words, there is a finite probability that some cracks will runaway and never arrest within the lattice. Between the cutoff and the runaway spike there is a gap in the size distribution. Optimistically, these results represent even more realistic simulations of the real target of seismicity as we have discussed. In these models, the issues of fractality of the lattice or time delays in the fracture process have not yet been addressed.

7. Concluding Remarks

There are many unsolved problems, partly because of the large amount of phenomenology that we need to model, and partly because we do not yet know how to model the physics well. For example, we do not yet know how to incorporate the tensor character of the stresses satisfactorily into the models; once this is done we will have a better picture of the healing process, which remains a difficult issue. We are not mounting an attack on the problem of constructing genuine three–dimensional models — mainly because of the same computational limitations that afflict other areas of physics. If we try to stick with 2-D models, we run into the problem of learning how to project a complex three–dimensional structure onto a plane studded with asperity and barrier loci of variable strength. With regard to the physics, we do not have a good picture of how fluids migrate in a region whose connectivity or tortuosity may be a variable, and possibly an abruptly changing variable, according to the degree of temporal closeness to the earthquake catastrophe. Indeed we do not know whether the dominant process for migration of water is due to H^+ diffusion or molecular diffusion. Finally, I remark that these efforts that I have been describing, attack only isolated aspects of the full problem of patterns of seismicity on a piecemeal basis; we do not, as yet, have a satisfactory synthesis of most of successful modeling efforts. There is the annoying doubt whether the full problem will turn out to be one of low–ordered dimensionality. If not, we may be up against a task that takes us into the realm of reporting a widely variable computational phenomenology that generates as wide a suite of clustering phenomena as the natural targets themselves, but a modeling phenomenology one whose consequences may be strongly dependent on initial conditions and strongly model dependent as well. I take the optimistic point of view that the influence of the hierarchical structure of a fault system will serve to transform the many d.o.f. problem into one with a low order. But this remains to be tested.

References

[1] L. Knopoff, *A stochastic model for the occurrence of main-sequence earthquakes*, Rev. Geophys. Space Phys., **9**, 175, (1971).

[2] J. Lomnitz-Adler, *The statistical dynamics of the earthquake process*, Bull. Seismol. Soc Amer., **75** , 441, (1985).

[3] D.T. Griggs, *Experimental flow of rocks under conditions favoring recrystallization*, Bull. Geol. Soc. Amer., **51** , 1001, (1940).

[4] B.K. Atkinson, *Subcritical crack growth in geological materials*, J. Geophys. Res., **89**, 4077, (1984).

[5] D.T. Griggs and J.D. Blacic, *Quartz: anomalous weakness of synthetic crystals*, Science, **147** , 292, (1965).

[6] Y.Y. Kagan and L. Knopoff, *Spatial distribution of earthquakes: the two-point correlation function*, Geophys. J. Roy. Astron. Soc., **62** , 303, (1980).

[7] Y.Y. Kagan, *Spatial distribution of earthquakes: the four-point moment function*, Geophys. J. Roy. Astron. Soc., **67** , 719, (1981).

[8] W.I. Newman and L. Knopoff, *Crack fusion dynamics: A model for large earthquakes*, Geophys. Res. Letts., **9** , 735, (1982).

[9] W.I. Newman and L. Knopoff, *A model for repetitive cycles of large earthquakes*, Geophys. Research Letts., **10** , 305, (1983).

[10] L. Knopoff and W.I. Newman, *Crack fusion as a model for repetitive seismicity*, pageoph , **121** , 495, (1983).

[11] M. Ohtake, *Search for precursors of the 1974 Izu-Hanto-oki earthquake, Japan*, pageoph , **114** , 1083, (1976).

[12] T. Yamashita and L. Knopoff, *A model for seismic gaps and premonitory quiescence*, Science, submitted, (1989).

[13] T. Yamashita and L. Knopoff, *A model of foreshock occurrence*, Geophys. J. Roy. Astron. Soc., **96** , (1989).

[14] P. Bak, C. Tang and K. Wiesenfeld, *Self-organized criticality: An explanation of 1/f noise*, Phys. Rev. Lett., **59**, 381, (1987).
P. Bak, C. Tang and K. Wiesenfeld, *Self-organized criticality*, Phys. Rev., **A38**, 364, (1988).

[15] C. Tang and P. Bak, *Critical exponents and scaling relations for self-organized critical phenomena*, Phys. Rev. Lett., **60** , 2347, (1988).
C. Tang and P. Bak, *Mean field theory of self-organized critical phenomena*, J. Stat. Phys., **51** , 797, (1988).

[16] A. Sornette and D. Sornette, *Self-organized criticality and earthquakes*, Europhys. Lett., **9** , 197, (1989).

[17] Carlson and Langer, *Properties of earthquakes generated by fault dynamics*, Phys. Rev. Lett., **62**, 2632, (1989).

[18] J. Lomnitz-Adler and L. Knopoff, *Crack propagation model with size-dependent coupling and implications for self-organized criticality*, Phys. Rev. Lett., submitted, (1989).

Index

[†] Numbers in boldface type refer to chapters